MW00837022

The History of Geographic Information Systems

ISBN 0-13-862145-4
90000
9 780138 621452

Prentice Hall Series in Geographic Information Science

KEITH C. CLARKE,
Series Editor

Avery/Berlin, *Fundamentals of Remote Sensing and Air Photo Interpolation*

Clarke, *Analytical and Computer Cartography*

Clarke, *Getting Started with Geographic Information Systems*

Foresman, *The History of Geographic Information Systems: Perspectives from the Pioneers*

Jensen, *Introductory Digital Image Processing: A Remote Sensing Perspective, 2nd Edition*

Peterson, *Interactive and Animated Cartography*

Star/Estes, *Geographic Information Systems*

Tomlin, *Geographic Information Systems and Cartographic Modeling*

The History of Geographic Information Systems: Perspectives from the Pioneers

Timothy W. Foresman, editor
University of Maryland, Baltimore County

To join a Prentice Hall PTR Internet mailing list, point to
http://www.prenhall.com/mail_lists/

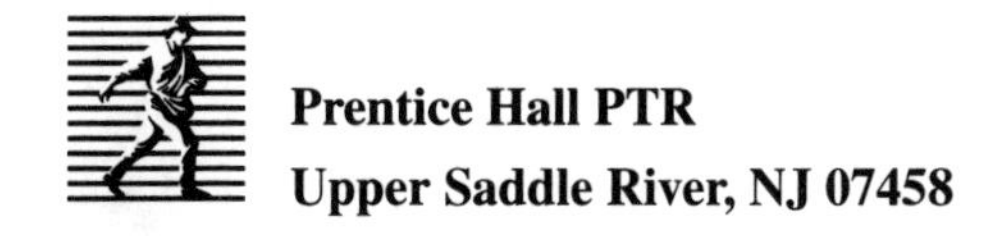

Prentice Hall PTR
Upper Saddle River, NJ 07458

Editorial/Production Supervision: *Precision Graphics*
Acquisitions Editor: *Bernard M. Goodwin*
Marketing Manager: *Betsy Carey*
Cover Design: *Joe School and Tim Foresman*
Cover Design Direction: *Jerry Votta*
Art Director: *Gail Cocker-Bogusz*
Manufacturing Manager: *Julia Meehan*

Prentice-Hall, Inc.
A Simon & Schuster Company
Upper Saddle River, NJ 07458

Prentice Hall books are widely used by corporations and government agencies for training, marketing, and resale.

The publisher offers discounts on this book when ordered in bulk quantities.
For more information, contact Corporate Sales Department, phone: 800-382-3419;
fax: 201-236-7141; e-mail: corpsales@prenhall.com
Or write: Prentice Hall PTR
Corporate Sales Department
One Lake Street
Upper Saddle River, NJ 07458

Printed in the United States of America
10 9 8 7 6 5 4 3 2 1

ISBN 0-13-862145-4

Prentice-Hall International (UK) Limited, London
Prentice-Hall of Australia Pty. Limited, Sydney
Prentice-Hall Canada Inc., Toronto
Prentice-Hall Hispanoamericana, S.A., Mexico
Prentice-Hall of India Private Limited, New Delhi
Prentice-Hall of Japan, Inc., Tokyo
Simon & Schuster Asia Pte. Ltd., Singapore
Editora Prentice-Hall do Brasil, Ltda., Rio de Janeiro

Contents

Foreword

Ian L. McHarg

As I consider writing the foreword to this collection of GIS histories, I recognize that throughout my professional career I have engaged in identifying and exercising the theoretical and philosophical underpinnings of what has become today's Geographic Information Systems (GIS). Of course, much of my early work predates the discovery of the modern computerized versions of GIS or even the term GIS which was later adopted by Roger Tomlinson and his Canadian colleagues. As a landscape architect, I devoted my career to integrating the environment into the urban and rural landscape design and planning processes. Placing man properly into his natural setting has been a lifetime passion requiring numerous innovations in academic organizations, interdisciplinary curricula, popular public forums, and landscape architectural planning methods. As the "father of ecological planning," I have tried to provide inspiration to a remarkable bunch of students and an international audience working towards improving living conditions for the people of this planet. My methods have popularized and advanced the application of multiple "cake layers" for incorporating the environment and social constraints into the myopic developer's dreams for paving the planet. While at the University of Pennsylvania, I was able to recruit a competent faculty of multidisciplinary talent who learned to work together under the new landscape architecture department. In addition, I worked with my colleagues at Wallace, McHarg, Roberts, and Tod to apply to ecological planning concepts using the multilayered approach. These experiences provided the scientific basis for my well-received book, *Design with Nature,* in 1969. Much of modern GIS can be directly linked to the ecological planning concepts developed therein.

GIS can be described as the systematic introduction of numerous different disciplinary data, connected by their shared location on the planet, which can be used to record an inventory of the environment, document observations of change and constituent processes, and permit predictions based on current practices and management plans. The details for these methods, based on a comprehensive, chronologically layered approach starting with bedrock geology and then surficial geology and so on, can be found in many of my publications. Prior to the origin of the

Canadian GIS, I assembled a team at Penn that was the origin of what would become computerized ecological planning, or GIS. Nohan Toulon digitized the natural resource data with punch cards for our Metropolitan Open Space Study that operated from 1957–1959. His computations provided area calculations for each environmental parameter used in the transportation and housing marketing analyses. I look back at the "primitive" tools we employed and marvel at our prescience for what has become a billion dollar industry as well as a significant field of study. The rapid transformation of this socially and ecologically significant field relies on the foundations of landscape architecture and ecology for much of its theoretical and practical underpinnings.

When I was asked to offer the foreword to this history of GIS, I admit slight amusement and pause as I reflected the existence of a "history" embedded within my own lifespan. My accomplishments in landscape architecture and the highlights of my career with *Design with Nature* and the 1990 National Medal of Art from President George Bush are more than a man without a high school diploma or Ph.D. can expect in a lifetime. But I recognize that my hopes for nature will be implemented through the efforts of my former students and many groups throughout the planet using GIS technology methods. The capability of the computer to contribute to integration of very complex problems is unequaled. It is therefore appropriate that we stop and recognize through the writings of this book the many contributing individuals, groups, institutions, and programs that represent our present understanding of GIS applications. Many of these authors have shared their vision of GIS with me at national conventions and workshops addressing the many ills, products from the industrial revolution, and increasing world population pressures. We find agreement that an interdisciplinary, or ecological, approach is most important to our capability of dealing with tremendously complex problems and managing our resources. This book will benefit scholars and students with an epistemological appreciation for the many contributors to our GIS history and the delightful variations for the paths followed to get where we are today. I only hope that ideas popularized in my book, *Design with Nature,* can be used to build upon the GIS historical foundation and that our best and brightest minds will begin to properly apply what GIS technology has to offer as they grapple with restoring nature to a sustaining and nurturing home for human existence.

While problems increase in complexity, it is gratifying to observe that our ability to understand and manage has been enormously expanded by the new prostheses—environmental science, sensors, satellites, and not least, computers. The one inadvertent benison of computation is its capability of integrating data and perceptions from the full range of environmental sciences. This may well be its most significant contribution. We are buried in data. We desperately need integration. Computers can assist triumphantly in this quest.

Preface

It has been quite a few years since the inception of my ideas that became congealed into a project and finally became this book. My recollections of the seeds for this effort begin with my academic career at University of Maryland, Baltimore County where I have been entrusted with teaching Geographic Information Systems (GIS) to some 30 or 40 students a semester since 1992. This humbling experience first required that I reengineer my technical skills in the laboratory and lay out a semester long curriculum of the finest quality I might muster. The semester's first lecture was always easy, what with the introductions of students, instructor, and syllabi. The second lecture was always the stumbler—the history of geographic information systems. What history should I attend to? Could I rely on the textbook version (I wrestled with choices and selected Star and Estes)? Or, should I share with the students my experiences which certainly did not follow the "textbook" story. As with all such human events, I hybridized my experiences with the textbook and gave proper caveats to the students not to take my word, or the word of others for that matter, as the gospel in GIS history. My students have been reminded and remonstrated to use their own critical thought processes and analyze on their own the information they are getting and define for themselves the likely influences past events forged on GIS. Indeed, what might the future hold if these trends continue? And so I continued for the first year at UMBC.

Upon receipt of my own copy of Maguire, Goodchild, and Rhind (1991) in spring of 1993, I was delighted that such a great reference book had been developed and pleased that Coppock and Rhind had written a chapter on "the history of GIS." As I voraciously read the chapter, I was again dismayed that Coppock and Rhind did not share my vision of GIS history. While I applaud their work (it's very scholarly), I became convinced that the reason others did not share my vision was that 1) either they had not stumbled along the same paths as I, or 2) there was not a sufficient body of literature to assist scholars and others interested in GIS as to perspectives of "my world." For example, nowhere was the raster side of GIS presented nor was the utility/engineering side. With this conviction in mind, I tracked down Dr. Roger Tomlinson at the 1994

ESRI Annual Convention (AKA the Palm Springs GIS Party) and discussed with him some of my concerns regarding the lack of good information on GIS's history. He was most generous with me and offered to support a book project by contributing his foundation chapter. After that conversation, I did not require additional encouragement for the project, just input and assistance from the many authors and collaborators involved. I will defer the many notes of thankful assistance to the Acknowledgments section.

So what was unique or different about my vision of GIS history? First, my start in GIS was not the vector GIS foundation that is ubiquitously cited in the literature. Second, my experience was intertwined with many developed and not-so-developed examples of geographic information studies from the early years. Allow me to recap briefly the mix of vantage points that led to my experiences. The reader is warned to look elsewhere if semi-autobiographies chafe.

My GIS education began while I was a graduate student in the ecology program at San Diego State University. My vegetation study of Camp Pendleton United States Marine Corps Base literally went up in smoke (a major conflagration swept through my coastal study site) prompting me to search for tools that would help me study and model large acreages of the chaparral dominated landscape. Professor Bill Finch of the geography department taught a course in 1975 on remote sensing that I thought might be useful for mapping vegetation (as discerned from the catalog course description). Caesar said it first, but indeed I came, I saw, and I conquered. I ate, drank, and slept remote sensing. There were no software programs in geography to use digital remote sensing and that's wherein serendipity is so wonderful. Dave Mauriello, fresh from his doctorate at Rutgers, was a visiting lecturer teaching ecological systems modeling. In his grab bag, again literally, was a deck of cards labeled GRID from Harvard. We put our heads together and figured out that we ought to be able to use GRID to analyze, classify, and generate maps of Landsat data. Using only the *Landsat Users Guide,* my fellow grad students, Ralph Brown and Greg Rhoades, figured how to write the interface programs to get Landsat data into the IBM 360 and generate land cover maps for my thesis. We did so well (our heads did swell just a little) that we started a consulting company called Ecographics of La Jolla, CA. My thesis documented the fire ecology of Camp Pendleton and off we went as young environmental entrepreneurs.

During this early stage many contacts were made, as we visited most every county in California to drum up business. As I was pursuing business possibilities in my home state of Florida, I met Wayne Mooneyhan, whose charge it was to transfer NASA technology to the southern states. He graciously invited my company—me and three graduate student colleagues—to visit the Earth Resources Center in Slidell, LA (the facility later moved to Bay St. Louis, MS). That two-week visit in the fall of 1997 effectively launched my career, as I learned for the first time that there was a whole world out there of scientists and engineers trying their best to harness digital satellite data into environmental information systems for the entire planet (of course we didn't call them GIS back then) and that I was competent enough to contribute to this very exciting and dynamic field. I had passed my initiation into the raster world of GIS. The excitement is hard to communicate in retrospect; after all, NASA still had its luster of the Apollo era and a lot of terrific people worked at the Slidell complex.

In 1977, Bill Finch recommended that I look up one of his former students at the University of California Santa Barbara, Jack Estes. Jack was a true gentleman the first time I met him and has been a fast friend ever since. I finished my thesis research on Camp Pendleton's vegetation dynamics in 1978 and published my first paper on the economics of remote sensing. In August of that same year, I accepted a job offer to become the Navy's first research ecologist from one of my Ecographic clients, the Naval Civil Engineering Laboratory (NCEL), Port Hueneme, CA. I was thrilled with the concept of a regular paycheck and the potential attached with being the first research ecologist tasked to investigate remote sensing and GIS (I knew what to call these systems by now) for land use management and environmental protection on 1.3 million acres of military land. My penchant for night school while working full time garnered another degree in engineering from the University of Southern California, immediately followed by entering the brand new UCSB doctoral program in geography through the cajoling of Jack Estes. In 1992, I was fortunate to be awarded a one-year fellowship by the Navy to complete my course requirements.

A wonderful opportunity arose as the Marine Corps Headquarters in Washington, D.C., requested assistance for a computerized system for the land use management needs of its base facilities. Environmental regulations were placing extreme pressure on the USMC's ability to administer the land management activities for training and support. As principal investigator of the Land Use Management System (LUMS) project, I put together a team of scientists that included Jack Estes, Jeff Star, Todd Streich, Ralph Brown, and others to investigate the best automated approaches to the Marine Corps requirements. This project led to the thorough evaluation/bench testing of the leading computer aided drafting (CAD) and GIS vendor products. ESRI's products were determined the best at the time and thereby developed a deep and lasting relationship with the vector GIS world and Jack Dangermond's young team at Redlands, CA. The Camp Lejeune LUMS remains one of the best engineered GIS around the country to this day.

While at NCEL I was involved with many triservice meetings and workshops, many visits to various military installations, and intra-federal communications. I served on the Federal Interagency Coordinating Committee on Digital Cartography, the forerunner of today's Federal Geographic Data Committee. I compared notes with colleagues in the U.S. Geological Survey (USGS), the U.S. Fish and Wildlife Service (USFWS), the Tennessee Valley Authority (TVA), the Department of Energy (DOE), and the Environmental Protection Agency (EPA). It seemed quite natural to friends and colleagues that I transferred jobs, and I accepted the challenge of leading the new GIS research program, which had been added to the remote sensing program, at the EPA's Environmental Monitoring Systems Laboratory in Las Vegas in 1984. This tour of government duty led to the first large-scale GIS research on an active Superfund site at San Gabriel, CA. ESRI was subcontracted for this research that yielded many firsts. ESRI's GRID module was used to link a GIS database to a groundwater model running a temporally reversed trajectory. The Superfund remedial contractor, CH2MHILL, monitored our progress and decided to make GIS the operational site investigation tool before we completed the research. This was indeed a success. During my tenure with the agency, I was able to visit with all ten of the EPA regions and discuss the role of GIS and remote sensing with many EPA scientists and contractors in addressing the nation's environmental problems.

It is one thing to design GIS research projects; it is quite another to build and design an operational system. When the opportunity arose to lead the Clark County, NV, GIS program in the late 1980s I jumped at the opportunity to "do it right." Located in Las Vegas, the Clark County GIS program has become a national example based on carefully engineered steps with dedicated cooperation of all consortium members and sufficient financial backing. With the luxury of designing and implementing a multimillion dollar GIS with a team of highly motivated and creative people, the Clark County GIS has become a recognized success. The enthusiasm and professionalism of Clark County soon combined with that of the Washoe County GIS teams, and Nevada's annual GIS conferences were born. As the state's GIS chairman, I was also involved in USGS A-16 planning efforts with my colleagues, such as Nancy Tosta, from California, and others from Montana, Idaho, and Arizona. During this time I was also elected to the AM/FM International Board of Directors, thereby balancing my environmental experiences with those of the utilities and engineering world. All my Ph.D. work paled in comparison to the knowledge I gained as the operational leader of the Nevada and Clark County programs.

As projects there became operational, I began to search for new challenges, leaving behind the important task of maintaining a major municipal GIS to a team of talented and capable friends and professionals. Through the wisdom of my wife, I landed at UMBC where a new Laboratory for Spatial Analysis had just been assembled by Tom Millette. Tom's legacy provided me with the materials necessary to build a regional GIS and remote sensing research center that currently provides multilevel support for local nongovernmental organizations (NGOs), county, state, and federal agencies. Our laboratory has created the Baltimore-Washington Regional Collaboratory, sponsored by NASA's Mission to Planet Earth, that operates via the Internet to link up and foster spatial data usage by a rapidly growing constituency of local, regional, and global users. It has been, in part, the experiences with this growing constituency in combination with the continued student education process that kept me focused on this history book project.

As this book neared completion, I was contacted by professor David Mark of SUNY Buffalo who had initiated the GIS History Project from the National Center for Geographic Information and Analysis' (NCGIA's) Research Initiative 19, with his colleagues Nick Chrisman, Andrew Frank, and John Pickles. Research Initiative 19, "GIS and Society: The Social Implications of How People, Space, and Environment Are Represented in GIS," recognized the importance of establishing an authoritative history of GIS to develop and maintain related materials. This present book has provided important impetus and foundation materials for the NCGIA project. The Library of Congress is also coordinating with this author and others to begin collecting GIS gray literature from throughout the industry and academia. Readers are encouraged to contact professor Mark or this author to help contribute resource materials to fill in the knowledge gaps on the history of GIS.

The path that I traveled over the past couple of decades has provided me with a wonderful education and the delightful experience of meeting and working with many fine individuals involved in the spatial sciences and technology. It is my hope that through the collective writings in this book, the reader will become acquainted with the rich and interesting combination of people and projects that helped to bring GIS technology to the forefront of many of our disciplines.

As an addendum to this book, I should like to point out that more than one form of English is used by the authors. Readers will see British English—used in the United Kingdom, Canada, and Australia—as well as the language (especially spelling) of the United States. We have adhered to the usage of the authors.

Tim Foresman

Acknowledgments

Many thanks are due to many people for completion of this book. Some notable thanks go to Roger Tomlinson for graciously agreeing to participate in this project. Dr. Tomlinson's patience and cooperation provided a steadying influence throughout many months of activity. Much credit is also due to professors Jack Estes and Mike Goodchild who provided both moral and technical support to me. These fine scientists/scholars not only helped with the design phase of the project but also contributed many hours of technical review for the chapters, in addition to their own chapter contributions. While I accept responsibility for all errors or omissions in this book, the attributes of quality must be shared with these colleagues from Santa Barbara. Jack Estes was especially consistent in checking on progress of projects from various airports around the globe.

I am sincerely grateful for the time and energy contributed to this project by the excellent team of contributing authors. It has been an honor and a privilege to work with these delightful individuals in crafting and constructing this history book. Each of these authors has a very busy schedule. From the start, I could sense that the team of authors was as motivated as I in capturing and recapping the events of the past few decades, in a sense as a celebration of what we have accomplished and where we have set the path for others to follow. Basically, we all had a real good time writing these chapters, even when it was a struggle to balance our schedules to accomplish this feat. So to the world of readers, I would like the authors to stand and take a bow for their excellent performance.

There are many people from the past, some no longer living, who were instrumental in my early GIS experiences, and I would like to recognize some of them for their impact on my history. From my Ecographic days, appreciation goes to Greg Rhoades, Ralph Brown, and Steve Rhoades as well as to Wayne Mooneyhan for discovering us. At U.C. Santa Barbara, Jeff Star, Mike Cosentino, Doug Stow, Charlene Sailer, Larry Tinney, and Joe Scepan were strong colleagues in tough times. A special remembrance is for Jeff Star for a warm friendship with my family and a wonderful collegial relationship teaching more quantification than I really wanted to know. Another remembrance goes to David Simonett who demonstrated levels of scholarship worthy of emulating. Others from Santa Barbara who helped me along the way include Donna Peuquet, Julie-Allen Jones, Dan Botkin, and Waldo Tobler.

Key individuals who contributed to the Las Vegas GIS adventures include Gary Shelton, Tom Mace, Lynn Fenstermaker, Mark Olsen, Dave Edwards, Clint Woods, Bob Kelley, and many others.

Many thanks to the ESRI team, beginning with Jack Dangermond who closely monitored this project, and Clint Brown, S. J. Camarato, Mike Broten, and so many others.

Constructive input and leads for tidbits of esoteric historic information were kindly provided by Mike Kevany and Pete Croswell of PlanGraphics and Carl Steinitz of the Harvard Graduate School of Design. Denny Parker, the founder of *GIS World*, provided a strong endorsement for this project during the initial stages. Keith Clarke also provided support for the project in his role as GIS series editor for Prentice Hall Publishers.

Much gratitude also goes to the many generous individuals who contributed entertaining vignettes of their first GIS experiences, many of which we were not able to include in this printing.

The team that helped with the numerous critical clerical support activities at the University of Maryland Baltimore County includes Darryl Byrd, Chris Steele, Mitch Jones, and Tracey Serpi. Significant assistance with the graphic arrangements was provided by my friend in the basement, Joe School of cartography and his student interns Amy Smith and Kevin Moore. Joe School was also responsible for the cover design.

Bernard Goodwin, the publisher for the Professional Technical Reference books, proved to be a terrific professional in getting this book published. His support of the project was critical. He was capably assisted by Diane Spina and the Prentice Hall team of professionals who helped with many of the necessary details for a professional book production.

I am deeply indebted to Margaret Foresman, my mother, for her expert editing assistance. It was quite evident that her 25 years as managing editor of *The Key West Citizen* had honed her skills as a professional. And finally, I would like to thank my dear wife, Joyce, for agreeing to assist with this project and remaining my spouse in spite of the journey. Her support and talents were instrumental in bringing this project to its completion.

PART I

Introduction

CHAPTER 1

GIS Early Years and the Threads of Evolution

Timothy W. Foresman

Introduction: A Confluence of Cartographic and Computerized Concepts

In recognition that three decades have passed since coinage of the term "geographic information systems," the question arises: how should the history of GIS be recorded? From an academic perspective, this question confronts instructors as they teach the next generation of GIS professionals. For the GIS professional, this question may stimulate acute awareness of GIS's rich heritage and its relationship to the information technology revolution. Is it appropriate, therefore, to trace GIS's lineage from the cognitive spatial awareness of Ptolemy and Immanuel Kant, or Louis Alexandre Berthier's hinged overlay maps at the 1781 Siege of Yorktown, or the *Atlas to Accompany the Second Report of the Irish Railway Commissioners* published in 1838 (Parent 1988; Star and Estes 1990)? Or should we follow the path of automation beginning with Herman Hollerith's census tabulation machine, the "Turing machine," and the ENIAC (Electronic Numerical Integrator and Computer) project of the University of Pennsylvania (Streich 1986)? How should we describe the roots or underpinnings of geographic theory and spatial analysis techniques considered responsible for modern GIS analytical processes? No simple or single explanation can be expected for a reality based on confluent influences from many ancient roots.

Dr. Tim Foresman is Assistant Professor of Geography and director of the Laboratory for Spatial Analysis at the University of Maryland Baltimore County. His eclectic career includes experiences as a remote sensing/GIS consultant, the U.S. Navy's first research ecologist, the U.S. EPA's first GIS research scientist, the GIS Manager for Clark County, NV, Nevada's first GIS Chairman, and an educator. The major theme in his career has been the integration of spatial technologies for improved land use management and environmental protection. *Author's Address:* Tim Foresman, Department of Geography, University of Maryland Baltimore County, 1000 Hilltop Circle, Baltimore, MD 21250. E-mail:foresman@umbc.edu.

Lack of a single evolutionary path should not prevent us, however, from examining some of the major influences that helped to forge the current setting for GIS professionals, applications, and supporting industry at the end of the millennium.

Techniques for geographic inquiry, hallmarks for the promise of GIS, have existed for several centuries in the form of mapping overlays and Boolean/cartographic analysis (Rice and Brown 1974; Parent 1988; Parent and Church 1988). Carl Steinitz and colleagues (1976) observed the historical use of hand-drawn overlay techniques, a fundamental GIS process, to be "a logical and obvious basis for analyzing relationships among different elements of the landscape." Evidence for antiquity of the logical overlay technique has been documented in a series of etched stones at Angkor Wat, a major temple of the eleventh-century Khmer Empire in northwest Cambodia.

Although various lineages can be debated, we can convey to students and GIS enthusiasts that modern GIS has been influenced, at a minimum, by landscape architects and planners since the early twentieth century on both the European and North American continents. Researchers have noted that by 1912 Manning had published thematic overlay maps of Billerica, MA, while city planners in Dusseldorf, Germany simultaneously produced time-series mapping overlays (Steinitz et al. 1976). Many more examples of the applications of mapping overlays can be highlighted along this trail of historical GIS foundations. Ian McHarg's (1969) discourse on map overlay techniques for landscape architecture is considered a seminal work that demonstrated to an environmentally awakened generation the legitimacy of the overlay technique for addressing environmental and social compatibility issues with land development planning (Foresman and Millette 1997). It should be noted, therefore, that performing spatial analysis with multiple map overlays was indeed promoted early and promulgated very effectively by the community of architects and planners. However, many examples outside this community can also be found, including early works by professor emeritus Waldo Tobler (1959), who as a graduate student published on the prowess of computer-based mapping.

Do these examples of GIS "layered cake" capabilities or the early recognitions that computers can generate cartographic products present the complete portrait of the historical influences on GIS? Certainly not. Many more pieces of the historic puzzle must be considered to provide a comprehensive view of GIS origins and the agents of influence that brought this impressive topic to the forefront of engineering, geography, business, and other disciplines.

Fundamentally, students should understand that modern GIS owes its phenomenal success to the advent of and dependence upon computer automation. Students should be aware that GIS is wholly dependent upon automated techniques embedded in silicon chips and metal oxides in conjunction with many computer science and engineering advances for the collection, processing, management, analysis, and production of geographically-referenced information.

In the computer-automation era of GIS history, a milestone of note occurred during the early 1960s. Under the leadership of Dr. Roger Tomlinson, the Canadian government sponsored the development of the first industry-scale computer-based GIS, known as the Canadian Geographic Information System or CGIS (Chapter by Tomlinson). For the first time, the term "geographic information system" became widespread. Many researchers have recognized the CGIS as representing the single most important program responsible for launching many of today's successful GIS enterprises (Peuquet 1977; Star and Estes 1990; Coppock and Rhind 1991). However, this represents only one thread, albeit a significant one, in the rich tapestry comprising the worldwide GIS community. For example, the use of automated geoprocessing techniques can be traced back to the 1890 census. Herman Hollerith's pioneering innovation entailed applying punched cards, a technique he adapted from the French loom industry, to code the U.S. Bureau of the Census 1890 population data (Huskey and Huskey 1976; Streich 1986). A convincing GIS lineage can be diagrammed from Hollerith's early efforts along the Potomac River to automate demographics to the current computer-based TIGER Census files. The TIGER (Topologically Integrated Geographic Encoding and Reference) system is arguably one of the most influential programs in the dissemination of digital spatial data and has encouraged the use of GIS throughout the business and marketing community as well as the social sciences (Chapter by Cooke).

In truth, the GIS phenomena is ineluctably linked to myriad influences. Therein resides the difficulty of passing along to future generations a comprehension of GIS's rich heritage. Landscape architects will dwell on one line of history highlighting Harvard and McHarg, computer scientists may dwell on Alan Turing's computational theory, and remote sensing enthusiasts will track a somewhat different history that begins with the Soviet Union's Sputnik satellite. Today's GIS professionals may be excused for not knowing the many pathways that comprise the current global mosaic of GIS systems, methods, people, and projects. Indeed, thousands of attendees at annual vendor-organized GIS user conferences may be patently ignorant of any specific vendor's place in this history. Recognition that most members of the GIS community are not aware of the many interesting GIS evolutionary trails provided motivation for creating this book. This book presents a collection of essays from some of the GIS community's foremost founders and experts who share past anecdotes and their perspective on the collective historical influences that shaped today's GIS industry.

The reader is reminded that these chapters represent the perspectives of the authors themselves. In some cases, they have solicited input from former colleagues and associates. In other cases, they are presenting their best recollections augmented by the literature (white and gray) available. As such, we hope that this book generates discussion in the community and that others will offer their perspectives on key developments for future editions as we attempt to improve literacy on the history of GIS.

Pathways of Progress

In researching the many pathways traversed during GIS's early evolution, a few notable diagrams were examined for historic interconnections. Antenucci and associates (1991) include a time-scale diagram that identifies some of GIS's major projects and products. Another interesting graphic representation provided by Don Cooke (Coppock and Rhind 1991) attempts to link key pioneering companies, agencies, universities, and individuals. Although each perspective warrants review, it is obvious that none of these charts can be viewed as comprehensive. David Sinton (1992) notes in his article on GIS reflections that while working on a detailed offering from his colleague, Ken Dueker of Portland State University, regarding a family tree of who influenced whom, he finally recognized the futility of his effort to categorize GIS historical relationships. Another notable and humorous offering is the unpublished "GIS Evolutionary Chart" presented by the Arizona Land Resources Information Systems (ALRIS) team at the 1996 Environmental Systems Research Institute (ESRI) Annual Convention in Palm Springs, CA. This chart cleverly introduces the reader to the GIS eras of "pre-CADrian, CADaceous, PRIMEozoic, UNIXene, and NETerary." The ALRIS evolutionary chart is in reality extremely informative to GIS veterans who would recognize the transitions from computer-aided-drafting (CAD) to GIS along with operating systems transitions from PRIME computers to UNIX and the Internet. But this chart too represents only a single vector of the evolutionary path.

Somewhat like a mnemonic collage, Figure 1.1 portrays many of the major milestones, groups, or trends that helped influence GIS. The figure is named the geomander after the convention of naming odd geographic and political features (the gerrymander). Topics covered by authors of this book are noted in bold italics. The figure is meant only to highlight some of the key elements or lines of influence, from Herman Hollerith's automation of the 1890 U.S. Census to the collaborative development of the National Spatial Data Infrastructure (NSDI). The geomander displays a linkage between Hollerith and the modern TIGER database intersected by professors Horwood and Garrison's influence at the University of Washington. The Washington team was influenced by Tomlinson's Canadian GIS (Chapter by Chrisman). Tomlinson's team included IBM, which was created in 1911 by consolidating two other companies and Hollerith's Tabulating Machine Company. On the European continent, Austrian GIS seeds were being planted by Otto Schaeffler who made improvements on Hollerith's punch card method for tabulating the census for Austria in 1890. The patent for a universal punch card machine was also given to an Austrian, Gustav Tauschek, in 1927 (Huskey and Huskey 1976). The reader should now appreciate the difficulty in diagramming these interconnections.

Horwood was instrumental in founding the Urban Rural Information Systems Association (URISA) in the mid 1960s. This organization was instrumental in transferring experiences and developing business contacts for the early state systems of Minnesota, New York, Wisconsin, and Maryland as well as the Australian GIS community. In 1973, Jack Dangermond kicked off his fledgling consulting firm, Environmental Systems Research Institute (ESRI) of Redlands, CA. Using a copy of GRID developed at the Harvard School of Design by then fellow graduate student Dave Sinton (now of Intergraph, Inc., Huntsville, AL), Dangermond contracted to develop Maryland's Automated Geographic Information (MAGI) system for the Office of Planning (under the leadership of John Antenucci, Jay Morgan, and others). Antenucci went on to found PlanGraphics, and Morgan established the GIS program at Towson University. Because of the limitations of a two-dimensional diagram, the geomander only hints at these many connections.

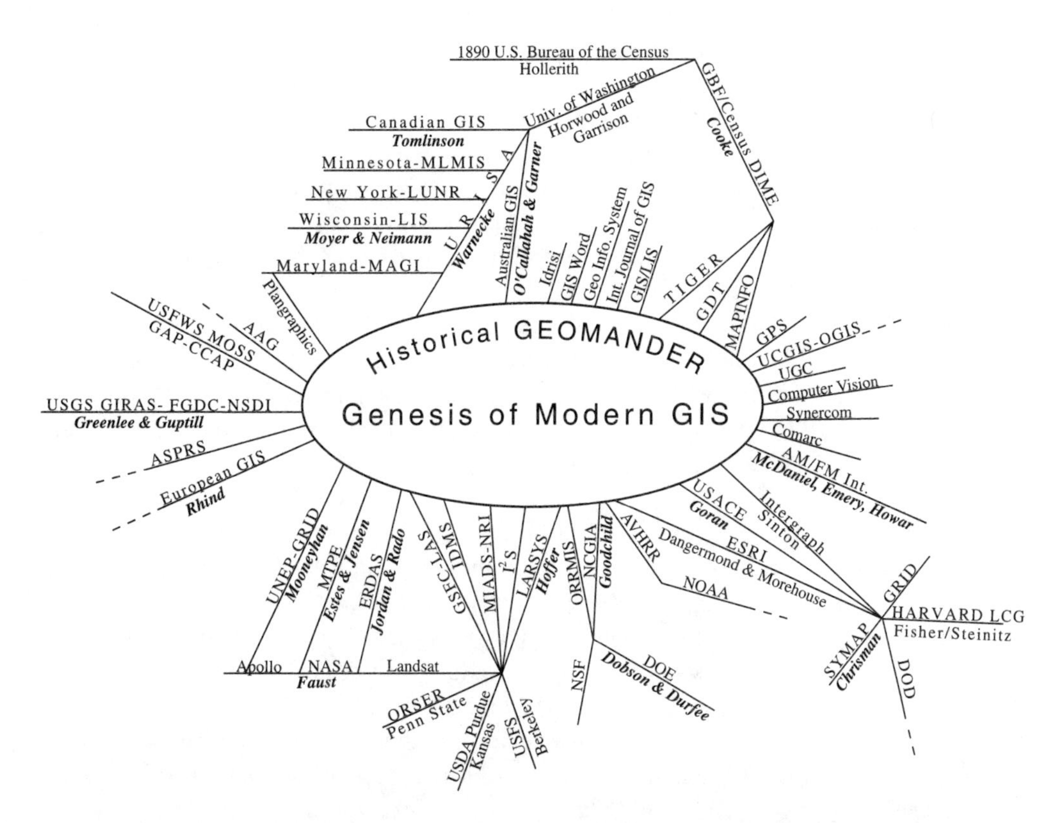

Figure 1.1 Diagram of historic pathways and connections for the genesis of modern GIS. Named the historical geomander after the convention of naming odd geographic/political features (gerrymander). Authors of history text are highlighted in bold italics.

Harvard's influence on GIS begins with Howard Fisher's creation of SYMAP (Synagraphic Mapping System, which actually started at Northwestern University where William Garrison later moved. It continues with the mentoring of professors Fisher and Carl Steinitz for early GIS applications and software (further developed by graduate students whose rebellion led to creation of IMGRID) that served the computational education needs of many key individuals who went on to form the nucleus of today's GIS commercial enterprises, such as ESRI, Intergraph, Synercom, and ERDAS (Earth Resources Data Analysis System, Inc.) (Chapters by Chrisman; by Moyer and Neimann; by Jordan and Rado). The Department of Defense financed many early GIS developments (and certainly supported early computer science developments, such as ENIAC at the University of Pennsylvania), helping to foster the engineering community's adoption of automated mapping (CAD and AM). These techniques were later combined with facilities management (FM) technology and provided initial financial support for the upstarts like Intergraph, Synercom, Comarc, ComputerVision, and others.

The National Aeronautics and Space Administration's (NASA) Apollo program and the National Oceanic and Atmospheric Administration's (NOAA) meteorological satellite program were instrumental seeds for the development of remote sensing and raster GIS technology. NASA's influence radiated out through people, programs, and software that traversed many paths of GIS evolution. Under federal and university cooperative applications in the late 1960s and early 1970s, agriculture and forestry provided a fertile environment for the creation of many GIS algorithms, some of which remain embedded in today's commercial software. Other federal agencies, such as the Department of Energy (DOE), Environmental Protection Agency (EPA), U.S. Fish and Wildlife Service, and U.S. Geographic Survey (USGS) provided additional impetus to GIS evolution through software development and enhanced applications development, database creation, and financing of fledgling commercial GIS companies. The federal scientific seal of approval was provided by the National Science Foundation's (NSF) establishment of the National Center for Geographic Information and Analysis (NCGIA) by the late 1980s.

Frequent national and international networking represented another critical component of early GIS evolution. Intellectual exchange was enhanced at professional gatherings such as the URISA annual meetings beginning in the mid 1960s, AUTOCARTO, EUROCARTO, and the International Geographic Union (IGU) meetings in 1970 and 1972 (which, under the influence of Roger Tomlinson, widely disseminated the term "GIS"). Professional associations such as the American Association of Geographers (AAG) and the American Society of Photogrammetry and Remote Sensing (ASPRS) have been augmented by vendor-specific annual conventions to hone expertise within the GIS community and proselytize the use of spatial analysis technology to an exponentially growing audience.

This geomander perspective is the author's version of influencing factors in the development of GIS and serves to broadcast the rich milieu of technical and social agents that forged our present technology-based, spatial-information community. The reader is invited to submit comments for revision of this historical geomander to improve our understanding of major lines of influence.

Pioneers and Champions

For posterity's sake, we should remind ourselves that there have been many pioneers and unsung "champions" who worked for federal, private, and university organizations that helped build the GIS phenomena byte by byte and line by line of code, first using the tools of FORTRAN, Basic, and COBOL, and then Pascal, C, and C++. There were many other pioneers who took chances in funding these early efforts, often risking their careers to see if the new GIS software could be applied to solve complex problems such as land management analyses, natural resources inventory and assessment, urban infrastructure mapping, Earth's surface analysis from satellites, population modeling, pollution monitoring, and community planning. More often than not, these early endeavors were based on the energy and promotion of a single individual or core group within an agency.

These intrepid pioneers were required to master elements of computer science, spatial data management systems, and their own agency's charter, plus the black arts of management, politics, and marketing. Faltering successes or outright failures created personnel employment dynamics that more conservative scientists and bureaucrats would never experience. It is tracking some of these career dynamics that provides for both an interesting and insightful look at how the technology and methods spread from region to region, company to agency, and so forth. For example, we find pioneers like Dr. Ray Boyle, now teaching at the University of Saskatchewan, leaving an established reputation in England during the mid 1960s and developing (in cooperation with David P. Bickmore) the first free-cursor table digitizer for the Canadian GIS program (Chapter by Tomlinson). Bickmore went on to establish the Experimental Cartography Unit (ECU) in the United Kingdom (Chapter by Rhind). Business start-ups were plentiful, breeding both failure and success. Today's major vendors can be traced back to a few key groups where individuals believed in the promise of the technology and whose fortunes were lucky in that they survived in a competitive and fickle market.

Filling in Historical Gaps

This book presents a select collection of writings from some of the GIS pioneers who have invested two or three decades of their energy and talents to the GIS field. While no claim is made that this represents a complete perspective or a definitive documentation of all historic events, the reader can be confident that a fairly broad knowledge base is provided, highlighting approximately three decades of automated GIS and identifying the major GIS influences and groups on multiple continents. An awareness of the historical roots for this dynamic community may prove useful to the many groups currently working together and using GIS technology to manage or solve local and global problems. As Francis Bacon reminds us, "Ask counsel . . . of the ancient time what is best, and of the latter time what is fittest."

Those individuals who seek an understanding of the depth and breadth of our GIS heritage will benefit from tracking the lines of progress that have been made in specific technical areas supporting GIS operations. These areas include the components of hardware, software, digital data, and spatial theory. To capitalize on these advances requires the professional to read broadly from a variety of disciplines and to appreciate the historic cross linkages of these disciplines. For example, many current solutions to integrating environmental modeling or remote sensing with GIS databases are closely linked to the early development of Harvard's GRID program, Purdue's LARSYS program, and other raster-based image processing systems (Chapters by Faust; by Estes and Jensen; by Jordan and Rado). This book brings to the reader a refreshing historical balance for vector and raster, urban and rural, engineering and natural resources, and the physical and human topics. These important gaps in our historical knowledge have heretofore not been filled in by the existing literature, for it has been the pieces and parts from these past endeavors that form the whole of what we are now witnessing in the GIS community.

Tracking the roots of geographic information systems has proven a challenging enterprise. Others' attempts to define a lineage of GIS have been helpful for this retrospective collection

(Streich 1986; Parent and Church 1988; Star and Estes 1990; Coppock and Rhind 1991). These efforts emphasize the nonlinear evolution of this field. So while we find ourselves asking, "Where did the major influences of GIS come from and what made GIS a success?", we must accept the obvious—that these influences represent rich human experiences, woven from a variety of international sources. An examination of these major GIS origins will set the stage for understanding the details of each historic path and the influences these events and people had on each other.

Genesis of Geographic Information Systems

How did GIS start? People, policies, and technology can all be linked to the evolution of this field. Figure 1.2 provides a chronology appropriate for examining the temporal relationships between social and technological agents in GIS evolution.

Computer technology represents the backbone for GIS evolution. As each technical component has advanced, computer systems have reached wider markets, motivating further applications into integrated solutions for handling spatial data. Thirty years ago, GIS pioneers, equipped with computing devices unrecognizable to today's students, faced the daunting task of automating maps and analytical processes. Their tools of three decades past included: punch cards for programs and data storage, limited core memory, magnetic drums giving way to multiple disk magnetic memory, central processing units with processing speeds capable of handling a few hundred thousand instructions per second, programming languages such as FORTRAN and COBOL, and output devices limited to character printed output sheets (Huskey and Huskey 1976; Gerola and Gomory 1984). An interesting dynamic confronted these early pioneers as it confronts us today. When selecting tools to handle large spatial-referenced data sets, the early GIS scientists and managers faced the remarkable phenomenon of exponential increases in computing hardware capabilities with comparable reductions in cost. Therefore, while at times these early pioneers seemed handicapped by technology, the more significant limitations were probably represented by advances in software, spatial theory, and data availability.

A study of policies and societal trends will illuminate the shifts in political thinking that began to influence government activities during the nascent stage of GIS. In the 1950s and early 1960s, Canada's government began to exert a stronger role in land use management than it had previously. This land use federalism set the stage for large-scale land inventory programs and hence the Canadian GIS (Chapter by Tomlinson). In the United States, the National Environmental Policy Act (NEPA) of 1970 continued the nationalistic trend toward increased land use management and environmental protection that led to a plethora of mandated programs and both fostered and funded GIS technological development. NEPA has been recognized as the most significant motivating factor for the use of GIS technology by many federal agencies. Land use management and environmental impact assessments were early applications of GIS and helped finance the start of many GIS software vendors (Chapters by Dobson and Durfee; by Goran). The Department of Defense (DOD) was especially keen to seek out methods of automation for environmental compliance. DOD requirements for missile-siting and air-defense strategies also

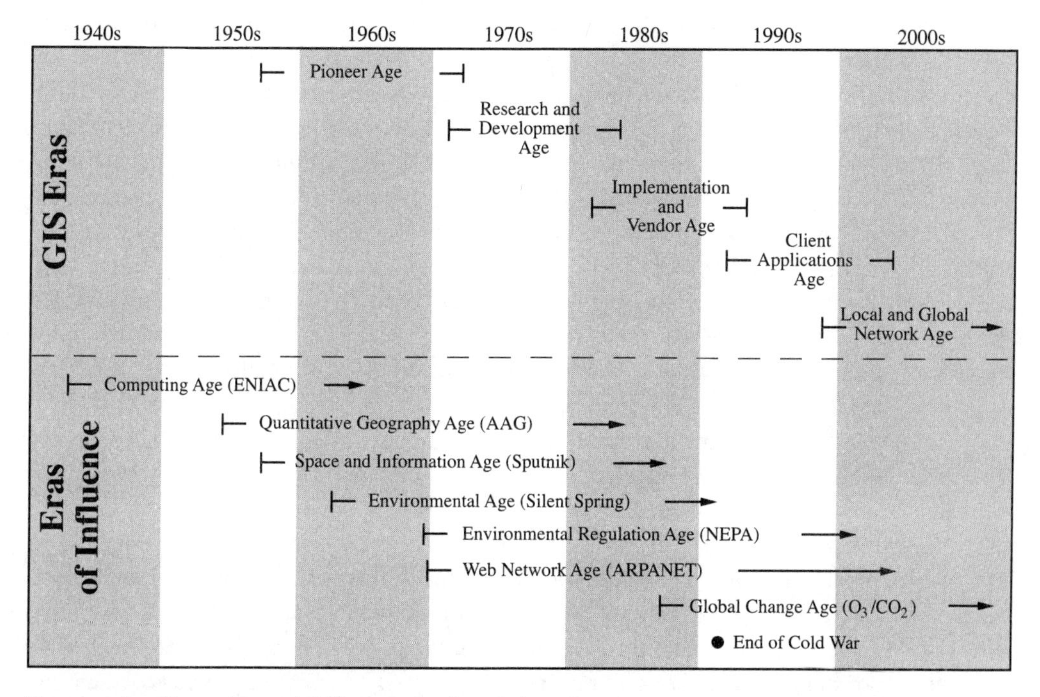

Figure 1.2 Chronology of GIS evolution in relation to major agents of change.

supported early GIS vendors and employed many technicians (Parent 1988). Along with NEPA, the formation of the U.S. Environmental Protection Agency served to motivate federal and state agencies to seek automated solutions for large-area resources inventory and assessments under the increasing litany of complex environmental laws. It should also be noted that major corporations represented important early consumers of GIS technology solutions. Many large corporations required economic approaches when applying resource mapping and inventory methods to deal with large-area forest lumber management and mineral exploration programs.

Major shifts in government policies toward land use management were not isolated to North America. International recognition of land use policy and land tenure implications for environmental protection initiated a variety of global programs that relied on GIS technology (Chapters by Rhind; by Garner and O'Callaghan; by Mooneyhan). Government policy initiatives accompanied major social transition in North America—from an industrial to an information society (Naisbitt 1982) and to an increasingly regulated public and private society.

Data remain the fuel for GIS evolution. Geographic data models have been seminal in promoting GIS advances. One of the most important advances for the automated spatial data model was the development and application of dual data systems that handled graphics and attributes separately. The advent of arc-node vector data structures arguably provided the next most important kernel for GIS development, that is, topology. These developments were first noted by the

Canadian GIS team (Chapter by Tomlinson) and then the DIME research team (Chapter by Cooke). Other data models and technical advances of the Canadian GIS team included raster to vector conversion; integration of scanning, digitizing, and keypunch data encoding; chain coding and compression; and automated edgematching, error detection, spatial coordinate systems, and command language for data overlay. Vector data models have a solid foundation in the commercial GIS evolution. While many unsolved problems remain in vector polygon overlay processing, most algorithms, including ARC/INFO, date back to the ODYSSEY code of the mid 1970s (Van Roessel 1997). Raster GIS processing advances, rooted in early remote sensing technology, followed a different evolutionary path. Raster based algorithms and large integrated array (or grid base) data models advanced along with the space-race programs. Computational efficiencies utilizing array processors enabled raster GISs to grow in parallel with the distinctly separate vector GIS community (Chapters by Faust; by Estes and Jensen; by Hoffer).

Key advances for information encoding and data model linkages were being made by the late 1960s for land information systems and utilities user communities (Chapters by Moyer and Neimann; by McDaniel, Howard, and Emery). Government agencies supported many GIS enterprises for regional spatial data handling and analysis. In the late 1960s for example, Systems Development Corporation, under contract to the Department of Housing and Urban Development, developed a transportation planning information system that closely resembles modern GIS in functional capabilities (Kevany 1968). Advances in digitizers (from Boyle's first free-cursor model), scanners, and raster datasets obtained from remotely sensed air and spacecraft continued to accelerate data conversion rates, thereby removing formidable data acquisition bottlenecks to GIS implementation. The expansion of data collection henceforth motivated advances in commercially available software to manage and manipulate these spatially referenced data resources (Chapters by Jordan and Rado; by McDaniel, Howard, and Emery). Advances in the man-machine interfaces augmented these software advances, thereby inducing more client-users to investigate the new cybernetic-based information trough (Wong 1982).

Contributors to GIS History

For this collection on GIS's historical roots, each author provides information on key people and organizations from their past experiences. These recollections, supported by a variety of published accounts and conferences, represent an important milestone for the pioneering development of GIS. After more than 30 years of development of what is currently known as GIS, a new phase of phenomenal expansion is occurring. The next milestone in GIS history may become more difficult to identify as the technology and discipline of GIS become subsumed into tomorrow's information highway on the World Wide Web (the network started as the ARPANET in 1969). Spatial maps and geographic analysis are becoming part and parcel of desk-top publishing, graphics, mapping, spread sheets, and decision-support systems. The GIS moniker will likely become diluted within the vernacular of future web-nurtured professionals whose cybernetic or office-management environs acclimate them to A/I "know-bots" and Internet-based personal assistants. Distinctive language for spatial sciences and processes, although by no means universal for the GIS community, may well be lost in the next millennium. Therefore, the words of these pioneers become paramount in documenting the first era of the GIS legacy.

Dr. Roger Tomlinson appropriately begins our historical review by sharing his knowledge of the key people and technical advances involved with Canada's effort to automate their natural resources records. His accomplishments go beyond the CGIS project where he initiated numerous conferences and proceedings to communicate worldwide the Canadians' advances. Dr. Tomlinson's impressive career continues today, as he consults with major organizations that desire to harness the power of spatial data. Canada's more recent GIS history is presented by another Canadian, David Forrest, who is well known for his monthly reports on Canada's GIS community. Forrest describes a progressive community of vendors, government agencies, and programs that marks Canada's continued leadership in GIS.

Professor Nick Chrisman offers an informed view of how university groups in the late 1950s through the 1970s contributed to fundamental spatial and cartographic theories that underlie many of today's GIS algorithms. Chrisman provides a glimpse of the intellectual gardens that existed during their pioneering days at the University of Washington, Northwestern University, the University of Chicago, and Harvard University. Many of today's GIS leaders were nurtured in these academic environments.

The activities of the U.S. Bureau of the Census team in the "Census Use Study in New Haven, Connecticut" were the prototype for today's automated urban planning (and marketing) systems. Don Cooke, who was an original member of the bureau's team and actively works with census and demographic information systems, describes the characters and developments that created today's TIGER files.

It is important to note that most but by no means all of the early developments of GIS originated in North America (Coppock and Rhind 1991). An important international perspective is provided with David Rhind's contribution in documenting the many pioneering efforts of the European community and noting the transatlantic cross-fertilization of ideas and people. An impressive list of well-known GIS researchers received their formal education in Europe. Unfortunately, Europe's significant contributions often have been lost during translation to the American continent. Continuing with the international perspective are Barry Garner and John O'Callaghan who document the GIS efforts and contributions from the continent and government of Australia. Many GIS leaders whose contributions are recognized around the globe were trained in their Australian homeland.

The commercial vendors that cater to today's billion dollar GIS industry were influenced by the developments of academia, computing science, and social needs. Lawrie Jordan and Bruce Rado, leaders in the raster-based GIS vendor community, continue with a delightful tale of their commercial venture beginning with its intriguing origins at the Harvard Graduate School of Design. Balancing the natural resources and planning perspectives is the historical perspective for utilities. Keith McDaniel and his colleagues, Chuck Howard and Hank Emery, who helped to pioneer the AM/FM side of GIS, document the early foundations of the engineering and utility industry that originated with CAD engineering software solutions.

Raster GIS heretofore has not been well documented compared with the vector history of this discipline. Nick Faust provides a view of the major components of raster development from a variety of government, university, and industry perspectives. Professor Roger Hoffer relates the first experiments in agriculture and forestry that fostered early remote sensing and raster GIS development. Professors John Estes and John Jensen further explore raster GIS, describing a variety of development programs fueled by NASA's Earth resources remote sensing initiatives.

Continuing to examine the U.S. government's early GIS contributions is the report by Jerry Dobson and Richard Durfee on the Oak Ridge National Research Laboratory's pioneering developments. William Goran also contributes to our knowledge of federal contributions with a chapter on the military's efforts to create GIS programs for land use management. The federal government's GIS contributions leading up to the present Federal Geographic Data Committee are presented by the USGS team of Dave Greenlee and Steve Guptill.

How the states and land-records communities worked with the early GIS developments is well defined by David Moyer and Ben Neimann. Going back to the early Egyptians, their chapter provides a critical review of the long history of land information systems (LIS) related to GIS. A perspective on the evolution of states' GIS and their contributions to the GIS community is presented by Lisa Warnecke, who helped foster many state-level GIS cooperative programs. Wayne Mooneyhan, one of NASA's early remote sensing and GIS leaders, extends this historical review to the global arena by documenting the United Nations' efforts to expand and transfer GIS technology internationally.

Finally, Mike Goodchild examines our place in history with an eye to the future. His key position in the GIS research community and his role as director of the National Center for Geographic Information and Analysis enables him to take an insightful look at what future challenges and advances may unfold for GIS programs in the next millennium.

The book's authors all have been significant contributors and participants during the early phases of GIS development, albeit there exist many other significant persons of notable status in GIS. By sharing their recollections of past GIS efforts, this book's writers hope to record some of the important contributions that they, and their colleagues and agencies, have made to this dynamic and evolving field. Many other important events and collaboration projects also helped foster early GIS development. However, the inclusion of all major contributions would soon overwhelm the resources of any single volume on the history of GIS. Some of these contributing activities would include the cross-fertilization of ideas and disciplines at many GIS and LIS conferences held in Canada, the United States, Europe, and Australia, and at the annual vendor User Group meetings by ESRI, ERDAS, Intergraph, and others. These frequently held meetings have brought together, in an unprecedented and unexpected fashion, civil engineers, urban planners, landscape architects, computer scientists, ecologists, rangeland managers, tax assessors, water quality chemists, geographers, and policy makers. The influences of numerous congregational affairs and projects make it difficult to highlight any single meeting from the past and define its significance on the present. In terms of sheer numbers of people involved and operational systems installed, corporate influences of ESRI or Intergraph on today's GIS community can be argued to be as significant as the influence of landscape architects, via Ian McHarg—although it can be demonstrated readily that founding members of both of these GIS market leaders built on the map analysis technique popularized by McHarg, one of the "spiritual" fathers of GIS. A key to GIS's apparent success has been the impressive exchange of ideas and rate of technology transfer generated by an unprecedented network of individuals, businesses, and agencies.

Why Does GIS Continue?

This book brings out many fundamental perceptions, unique for each author, regarding the current success of GIS. We can accept resolutely that the ability of spatial analysis to meet pressing demands is a common theme in the genesis of GIS. Overlaying maps (a fundamental step) is very old and very logical and can be linked from the Siege of Yorktown to the United States' National Biological Survey's biodiversity GAP program (Scott et al. 1993). Spatial data turns out to be everywhere, ubiquitously tied to just about every data record on the planet. Contemporary experts inform us that perhaps 80 to 90% of data collected has a spatial component. Verification of this theory simply requires one to identify data that have no spatial relationships—not an easy task. So what could be more natural than to stumble upon a wealth of untapped resources for scientific understanding and analytical processes? We can be humble and recognize the philosophical roots of this line of reasoning, as Immanuel Kant recognized almost two centuries ago the importance of the spatial component for science in the domain of human affairs (Prichard 1909; Bennett 1966).

Common needs often foster cooperation. Human impacts on land are now recognized as the primary agent for global change (Gore 1992; Vitousek 1994; National Research Council 1994). From the Stockholm conference to the Rio Summit (Agenda 21) and the Cairo World Population Conference, at local to global levels, international recognition has arrived regarding the issues of population growth and environmental degradation. These pressing human survival issues are threatening our capability to manage Earth's finite resources. When automated GIS began back in 1890 with Herman Hollerith, the world population was approximately 1.5 billion (Cohen 1995). Today, as we face the formidable task of managing the Earth's resources with more than 5.7 billion exponentially growing inhabitants, the tools of GIS are increasingly being challenged to address the complexities of human-land management interactions.

Cooperation works best when a common philosophy exists and common data models converge. Although variations do exist in how we should organize and exchange spatial data to support the decision-making process, these variations appear to be converging with respect to standards for policy makers (Foresman et al. 1996a), government (FGDC 1994), and global applications (NSF 1992; EarthMap 1995). Not to be underestimated are the economics of data sharing (Antenucci et al. 1991) that are forcing organizations to move in directions that promote data documentation standards and cooperative programs for developing global data exchange protocols (Estes et al. 1994; NSTC, 1995; Foresman et al. 1996). These unprecedented examples of peace-time cooperative activities will, it is hoped, breed their own success. Augmenting this is the trajectory of the business community embracing GIS technology and spatial data for economic gains. These inventions of humankind will require extreme focus and coordinated effectiveness to address global ailments as the mix of social/political, environmental, and economic pressures mounts. As we reflect on the contributions and development efforts of many along the historical GIS path, GIS technologies at local, regional, and global scales are currently being tested for their role in our future. Rachel Carson (1962) quotes Albert Schweitzer, who said, "Man has lost the capacity to foresee and to forestall. He will end by destroying the earth." We should hope that, based in part on our awareness of GIS's roots and impressive capabilities, we can be realistic about our potential to both foresee (model) and forestall (manage) human affairs on a global basis.

Bibliography

Antenucci, J. C., K. Brown, P. L. Croswell, M. J. Kevany, and H. Archer. 1991. *Geographic Information Systems: A Guide to the Technology*. New York: Van Nostrand Reinhold.

Bennett, J. 1966. *Kant's Analytic*. New York: Cambridge Press.

Carson, R. 1962. *Silent Spring*. Greenwich, CT: Fawcett Publications.

Cohen, J. E. 1995. *How Many People Can the Earth Support?* New York: Norton Publishers.

Coppock, J. T., and D. W. Rhind. 1991. "The History of GIS." In D. J. Maguire, M. F. Goodchild, and D. W. Rhind, eds. *Geographic Information Systems: Principles and Applications,* London: Longman Group UK Ltd.

EarthMap. 1995. *EarthMap Design Study and Implementation Plan: A Proposed Public-Private Consortium to Advance the Use of Geospatial Data and Tools for Decision-Makers.* Annandale, VA: Global Environment and Technology Foundation.

Estes, J., J. Lawless, D. Mooneyhan, et al. 1994. *Report of the International Symposium on Core Data Needs for Environmental Assessments and Sustainable Development Strategies*. 2 volumes. New York: United Nations Development Programs.

FGDC. 1994. *Content Standards for Digital Spatial Metadata.* Washington, D.C.: Federal Geographic Data Committee.

Foresman, T. W., and T. Millette. 1997. "Integration of Remote Sensing and GIS Technologies for Planning." In J. Star and J. Estes, eds., *Remote Sensing and Geographic Information Systems Integration*. Cambridge, U.K.: Cambridge University Press.

Foresman, T. W., H. Wiggins, and D. Porter. 1996a. "Metadata Myth: Misunderstanding the Implications of Federal Metadata Standards." First IEEE Metadata Conference, Washington, D.C.

Foresman, T. W., J. Estes, J. Garegnani, and D. Porter. 1996b. "Remote Sensing & Core Data Needed to Support Planning and Policy Decision Making," *International Geoscience and Remote Sensing Symposium,* 4: 2243–2245.

Gerola, H., and R. E. Gomory. 1984. "Computers in Science and Technology: Early Indications." *Science,* 225 (4657): 11–18.

Gore, A. 1992. *Earth in the Balance: Ecology and the Human Spirit*. New York: Houghton Mifflin Company.

Huskey, H. D., and V. R. Huskey. 1976. "Chronology of Computing Devices." *IEEE Transaction on Computers,* C-25 (2): 1190–1199.

Kevany, M. J. 1968. *An Information System for Urban Transportation Planning: The BATSC Approach.* Technical Memorandum 3920/000/01. Santa Monica, CA: Systems Development Corporation.

Maguire, D. J., M. F. Goodchild, and D. W. Rhind, eds. 1991. *Geographic Information Systems: Principles and Applications.* London: Longman Group UK Ltd.

McHarg, I. 1969. *Design with Nature.* Garden City, N.J.: Doubleday and Co.

Naisbitt, J. 1982. *Megatrends: Ten New Directions Transforming Our Lives.* New York: Warner Books.

National Research Council. 1994. *Science Priorities for the Human Dimensions of Global Change.* Washington, D.C.: National Academy Press.

National Science and Technology Council (NSTC). 1995. *Our Changing Planet: the FY 1995 U.S. Global Change Research Program.* Report by the Subcommittee on Global Change Research, Committee on Environment and Natural Resources Research of the National Science Technology Council. Washington, D.C.: NSTC.

National Science Foundation (NSF). 1992. *The U.S. Global Change Data and Information Management Program Plan.* Report by the Committee on Environmental and Earth Science. Washington, D.C.: NSF.

Noyce, R. N. 1976. "From Relays to MPU's." *Computer.* 26–29.

Parent, P. 1988. "Geographic Information Systems: Evolution, Academic Involvement and Issues Arising from the Proliferation of Information." Master's thesis, University of California, Santa Barbara.

Parent, P., and R. Church. 1988. "Evolution of Geographic Information Systems as Decision Making Tools." *San Francisco GIS '87.* Falls Church, VA: ASPES/ACSP, 63–71.

Peuquet, D. J. 1977. *Raster Data Handling in Geographic Information Systems.* Buffalo, N.Y.: Geographic Information Systems Laboratory, State University of New York.

Prichard, H. A. 1909. *Kant's Theory of Knowledge.* Oxford, UK: Clarendon Press.

Rice, H. C., and A. S. K. Brown, eds. 1972. *The American Campaigns of Rochambeau's Army.* Princeton, N.J.: Princeton University Press.

Scott, J. M., et al. 1993. "GAP Analysis: A Geographic Approach to Protection of Biological Diversity." *Wildlife Monographs,* 123: 1–41.

Star, J., and J. Estes. 1990. *Geographic Information Systems: An Introduction.* Englewood Cliffs, N.J.: Prentice Hall.

Steinitz, C., P. Parker, and L. Jordan. 1976. "Hand Drawn Overlays: Their History and Prospective Uses." *Landscape Architecture,* 66 (5): 444–455.

Sinton, D. F. 1992. *Reflections on 25 Years of GIS.* Fort Collins, CO: GIS World.

Streich, T. A. 1986. "Geographic Data Processing: A Contemporary Overview." Master's thesis, University of California, Santa Barbara.

Tobler, W. R. 1959. "Automation and Cartography." *Geographical Review,* 49: 526–534.

Van Roessel, J. W. 1997. "The Vector Overlay Puzzle: Where Do the Pieces Fit?" In M. Molenaar and S. De Hoop, eds. *Advanced Geographic Data Modelling.* Netherlands Geodetic Commission, no. 40. 1–18.

Vitousek, P. M. 1994. "Beyond Global Warming: Ecology Land Global Change." *Ecology,* 75 (7) 1861–1876.

Wong, P .C. S. 1982. "MMI: The Man-Machine Interface." *TRW Electronics & Defense/Quest,* Winter: 25–45.

PART II

Reflections on the Starting Points

Carl Steinitz, Harvard University, Graduate School of Design

My first contact with GIS activity came in 1965 at a lunch at the Harvard-MIT Joint Centre for Urban Studies during which, by chance, I was seated next to Howard Fisher, who was visiting Harvard while considering a move from the University of Chicago.

Fisher had recently invented SYMAP, a computer-mapping program which was based on line-printer technology. I immediately seized upon the relationship between the capabilities that he described and the needs of my doctoral thesis and convinced him to let me try some experiments with his basic program. With Fisher as my tutor, I gave SYMAP its first applied test in my study of the perceptual geography of central Boston. Partly because of this work, I obtained my first teaching appointment as Assistant Professor at the Harvard University Graduate School of Design and as an initial research appointee to the then new Laboratory for Computer Graphics.

Carl Steinitz was one of the earliest GIS principals at Harvard University and has continued to be an influential presence on both the American and European continents.

Peter S. Thacher, 54 Gold Street, Stonington, CT 06378-1229

In 1974, Roger Tomlinson passed through Geneva to brief me on his consultancy at FAO Rome. I was running the European office of UNEP, and Roger had been selected by FAO at our expense to see what could be done to integrate the various databases maintained there: soils, animal diseases, crop forecasting, forests, fisheries, etc., each of which had developed independently and could not be cross-compared or analyzed. Roger said there would always be difficult "turf" problems between sectoral departments, but technically he was confident we could aid FAO, and other UN agencies such as WHO, WMO, and UNESCO, to develop a common framework to help governments deal with the new set of environmental and development problems identified at the 1972 UN Stockholm Environment Conference. The key was something new, called GIS.

In 1975, with FAO, UNESCO, and the International Soil Science Council, a young American was selected to carry out a recommendation governments had approved at the Stockholm Conference: to create a methodology by which to measure the rate of loss of arable soil. In 1977, the results were published in the first such map ever produced, combining satellite imagery, soil characteristics, slope, aspect, climate data, erosion rates, et al. The man selected was Jack Dangermond. Jack's work was absolutely critical; the maps were splendid and showed not only the *rates* of loss but also the *risk* of loss for all of Africa north of the equator and through the Arabian peninsula into Pakistan.

There is much more that could be said about the use of GIS in the international community. With the experience of national meteorologic services, WMO was best placed to take advantage of GIS , but epidemiological work in national health departments at WHO and national and cultural parks in UNESCO's and IUCN's mandates were early beneficiaries. Scientific unions that had been involved in the IGY (International Geophysical Year, 1957–58) came together at ICSU, the International Council of Scientific Unions, and drew on this experience in the formation of IGBP (International Global Biosphere Program) many years later.

Peter S. Thacher served as the first deputy director of United Nations Environmental Program and was instrumental in the Stockholm Environmental Conference and the Rio Summit.

CHAPTER 2

The Canada Geographic Information System

Roger Tomlinson

Introduction

The advent of geographic information systems was the result not of academic inquiry but rather of the growing societal need for geographical information, of a change in the technology that made such systems possible, and of private sector vision and government foresight that initiated and sustained their development. This conjunction of conditions took place in Canada in the early 1960s.

Need for Geographic Data Handling

Canada in the late 1950s and early 1960s was concerned with the pressure on its natural resources. Although these resources had long been regarded as limitless, there was now competition among the potential uses of land in the commercially accessible parts of the country. Farmers on the many marginal farms had nonviable incomes, and rural depopulation was increasing. Government began to think the previously unthinkable: that it had a role to play in making decisions about land use, in planning the utilization of natural resources, and in monitoring change. Symptoms of this change of heart were evident. A special committee of the Canadian Senate was established in 1958 to examine the topic, "Land Use in Canada." The "Resources for Tomorrow" conference involving most of the country's senior resource scientists and attended by the prime minister was held in 1961.

Dr. Roger Tomlinson has a distinguished career hallmarked by leading the Canadian GIS in the mid 1960s. His career has continued as an international consultant with Tomlinson Associates in the design and implementation of large-scale GIS operations. *Author's Address:* Roger Tomlinson, Tomlinson Assoc. Ltd., 17 Kippewa Drive, Ottawa, Canada, K1S 3G3.

In order to undertake this new role, government required geographical data that would be of use for land management decisions. The types of data with a sufficient level of detail to support decisions of land adjustment and regional economic development at national (federal) and regional (provincial) levels are at map scales between 1:250,000 and 1:20,000. The number of maps at these scales to cover a country of Canada's size is extremely high. At that time of relative prosperity, Canada could afford to gather the data and make the maps. However, the people who would be making decisions based on the maps had to analyze and interpret the geographical data they contained in order to provide the information needed. The manual methods of map analysis then prevalent were highly labour intensive and time consuming, and Canada simply did not have enough trained people to carry out the analysis of such data using manual methods. There was thus a considerable economic incentive to develop geographic information systems.

With this new demand for geographical data pertaining to the capability and use of land and to subjects related to it, the time had come for a fundamental examination of hardcopy maps as a medium to hold and provide the amount of information needed. Although maps can be an efficient way to portray data concerning Earth's surface, they have two basic limitations.

Technological Considerations

The first limitation is that only a certain amount of descriptive data can be shown on a single map sheet. Choices have to be made about what is included on the map and, perhaps more importantly, what must be left off. One choice may be to limit the content of a single map to one particular set of features on Earth's surface (a choice that incidentally may influence the type of observations made to provide the data). To cover other features or provide more detail, more maps of the same area would have to be produced. Another choice might be to reclassify, simplify, and symbolize information from the original data. The risk here would be that in generalizing the map significant detail might be lost. In either case, the data content of hardcopy maps is limited by the size of sheet on which the information is recorded and the space required by each item of data so that it remains legible.

The second limitation is that data on a hardcopy map have to be read and analyzed by the human eye and brain. Although these work faster than even the best software on an equivalent amount of data, it would be an overwhelming task for humans to even read the map data in the volumes contemplated. Many people in the field today recognize how laborious, slow, and expensive are the manual techniques of subsequent measurement and qualitative comparison. Few people have actual knowledge of the cost and time involved, because, unlike those in the field 30 years ago, they have other choices. In summary, to store a large amount of data on maps, you have to produce many maps. To extract information visually and manually from a very large number of maps represents a formidable task of reading and measurement.

The critical change in technology that occurred at that time and provided the opportunity for development of geographic information systems was the advance from vacuum tubes to transistors. This change made computers faster, more reliable, and cheaper. More importantly, transistors gave them larger memories, so that they could be used for information storage as well as calculations. Thus, for the first time computers could store and handle comparatively large volumes of digital data economically. To put this advance in context, the first machine used for concept testing was the IBM 1401 with 8K of memory. The main development machine was the mainframe IBM 360/65 (worth $2 million in the mid 1960s). It had 512K of memory, which seemed enormous at the time. The technical challenge was to convert lines, shapes, and images into numbers, so that these elements of maps could be input, stored, and manipulated by digital computers.

The recording of shapes and lines as numbers was not new; in fact, descriptions of lines by numerical coordinates had existed for centuries. As early as the end of the 17th century, Francis Galton, the Secretary of the Royal Geographical Society, invented methods for recording line shapes using compact digital codes of line direction (later called Freeman codes), and he pointed out that they might be used to code maps. By the late 1950s, somewhat cumbersome instruments that used a moving arm for digitizing were in use to record curved lines such as missile trajectories, ships' lines, bubble chamber traces, and highway alignments. Also in the late 1950s and early 1960s, photogrammetric instruments started using digital encoders to record x, y, and z coordinates.

It is not a long step from using numbers to record lines and shapes to thinking how computers could be used to control the instruments for map-making. This indeed was the thrust of early systems of automatic cartography in the 1960s. The major change in such thinking that was needed was a set of concepts that laid the foundations for geographic information systems. The fundamental new idea was that computers could be used to ask questions of maps, to read them, and then to measure, compare, combine, and analyze the data that the maps contained or that could be related to them, in such a way that the computers would produce useful information from them. To be used in this way, maps had to be in digital form. That this was the desired product of map-making was certainly not the view of map-making agencies at the time. However, the concept opened the prospect that many maps in digital form could be linked together across Canada in a permanent database that could be widely available for analysis. In addition, the potential existed for these maps to be intelligently linked to digital databases of statistics. These ideas originated with Roger F. Tomlinson, they stemmed from Canadian requirements, and they led to the development of the first geographic information system.

Development of the Geographic Information System

The conditions that sparked the innovation were twofold, one arising in the private sector and the other in government. In 1960, Tomlinson was employed by a large aerial survey company, Spartan Air Services, in Ottawa, Canada. The company's strength was based on high-technology use of photogrammetric, geophysical, and map-making equipment. Some of its mapping projects had a requirement for subsequent manual map analysis. One project in particular was in East Africa. There, Spartan was asked to analyze (overlay) all available map series (including the ones produced by the company itself) in order to propose locations for new forest plantations and for a new pulp and paper mill. Careful estimates of the costs for this analysis were viewed as prohibitively expensive by the company and clients alike. Tomlinson could see that the prospects for success of this type of project would be rather uncertain unless he could find more efficient methods for map analysis. With the encouragement of the company management, particularly that of George Brown, chief of the Land Resources Division, he initiated experiments to see if computers could be used in this context. A small digital map, five inches by five inches, containing five polygons, was created. There were no digitizers available, so a photogrammetric plotter was used in reverse, and the values for each increment of head movement were read off the shaft encoders. Gradually, it was determined that two such digital maps could be overlaid graphically and that areas could be measured from the digital record. Spartan approached the major computer companies with offices in Ottawa at the time (Computing Devices of Canada, IBM, Sperry, and Univac) with the intention of starting a joint venture to develop the technology further, but with no success.

Later that year at a meeting of the American Society of Photogrammetry in Washington, D.C., John Sharp, a consultant to IBM, introduced Spartan to the digital photogrammetric research being done by IBM Poughkeepsie. This encounter led to further contacts with IBM Canada and the IBM Ottawa office and was the beginning of a fruitful relationship that was to develop significantly over the years. Tomlinson brought the geographical ideas to the work. IBM brought early experience of computers and programming.

The second event that led to the innovation was a chance meeting in 1962 between Roger Tomlinson and Lee Pratt, head of the newly-created Canada Land Inventory. To put this meeting in context, it should be noted that the federal government, in response to the competing pressures on natural resources mentioned earlier and in particular to the very low incomes then being earned by farmers in many parts of the country, had passed the Agricultural Rehabilitation and Development Act (ARDA) into law in June 1961.

Administration of this act, initially by the federal Department of Agriculture, required the creation of a uniform inventory of land use and land capability across Canada, with the cooperation of provincial governments. The Canada Land Inventory (CLI) was established in 1962 and planned to create about 1,500 maps with consistent classifications across the continent, covering the commercially productive parts of Canada at scales of 1:50,000 to 1:250,000. These maps would show the land's capability for agriculture, forestry, wildlife, and recreation, and additional maps would describe present land use and the boundaries of census divisions.

Lee Pratt not only had to set in motion this huge mapping effort, he also had to solve the problem of analyzing the maps when they had been created. The possibility of using computers for map analysis was very attractive. With Pratt's encouragement, Tomlinson laid out his ideas on the subject in a paper which was presented to the National Land Capability Inventory Seminar in Ottawa in November 1962. This was the first meeting under ARDA to bring together the federal and provincial agencies and scientists who would implement the Canada Land Inventory. The paper was well received, and in 1963 the Department of Agriculture gave Spartan Air Services a contract (Project 14007) to carry out a technical feasibility study for a "computer mapping system" for the CLI. The report, written by Tomlinson, was delivered to the department in August of that year.

That seminal report clearly established the functional requirements for a geographic information system. The purpose of such a system would be to analyze geographical data over any part of a continent-wide area. It would need to be capable of reporting the results of analyses in statistical (tabular) or graphic (map) form or both and accepting data from many maps of different kinds in a seamless, nationwide data structure. The data file organization was critical. The structure recommended would keep the descriptor data separate from the image (boundary) data. Thus the system was clearly not designed for automatic cartography. To handle the task of converting many maps to a digital form, the optimum input process was seen to be different for each fundamental data form: automated scanning for polygon boundaries, digitizing for selected points serving as identifiers inside polygons, and keypunching (typing) for input of descriptor and statistical data. These data types would be input separately and linked logically at a later stage. The paper examined requirements for the coordinate system for the main database and discussed the concepts of error and its effects on subsequent area calculation. It identified the requirement for compaction to reduce the volumes of image data that must be stored on magnetic tape. The document recognized that it would be essential to combine databases of socioeconomic data with mapped data. The capabilities required for data analysis and assessment would include area measurement and performance of multiple topological overlays. Many different assessments were prescribed over time, on all or any part of the database. All these required capabilities added up to a comprehensive geographic information system.

System Characteristics

Pratt and Al Davidson for the Department of Agriculture decided to act on this report and followed it up with an economic feasibility study and contracts with IBM and Spartan Air Services for development work. Tomlinson was invited to join the Agricultural Rehabilitation and Development Administration (ARDA) team and direct the development of the system. At that time, people and ideas moved rather easily between industry and government. More than 40 people worked on the development teams in the remainder of the decade.

Data Structure

A key role was played by Guy Morton, who designed the data structure now known as the Morton Matrix, which was the underlying strength of the system. It involved creating "frames" to hold data in the largest units that could be conveniently processed by the computers available at the time. This concept is the origin of the "tiling" employed in most modern GIS. Such units were larger than the sections created in the scanning process but usually smaller than a map, and they were the main unit of storage for the final databank. Their eventual size depended on the resolution of the data set to be stored and the CPU capacity. A near square (equal angular displacement in x and y) was found to be the best shape for the frame, both to maximize the length of line segments contained within it and to simplify subsequent calculations and grouping. The data structure was designed at the outset to be continent-wide. Each unit frame was assigned a unique number, starting at the point of origin (lat. 145°W and long. 40°N) in the Pacific Ocean, to the west and south of any part of Canada. From that point, frames were sequenced so that they fanned out from the origin as frame numbers increased.

The arrangement of the unit frames was crucial to operating efficiency of the system. Random access discs were not available in the early stages of system design, and searching a magnetic tape for data was a time-consuming sequential process. To overcome this constraint on the system's efficiency, it was desirable that frames close together in the sequence should refer to locations close together on Earth's surface and, similarly, that records from close locations on Earth should have the least possible separation possible in a sequential file. It was also important that the searching sequence is the most efficient method of adding area to the first selected frame. The Morton Matrix arrangement (Figure 2.1) had these characteristics.

An additional benefit of this file structure, which is not immediately apparent, is that the actual record addresses of a geographical location can be directly computed by manipulating the binary representation of the geographic x-y coordinates shown on the Morton Matrix diagram. The structure can be recognized as the quadtree used in later systems for different purposes.

Input Procedures

From the beginning, the system was designed to build large databases and hence to handle high volumes of input data. It was necessary to optimize the input processes, because tracing of lines by hand (manual digitizing) was seen to be error prone and laborious. A drum scanner based on the facsimile principle was proposed in the 1963 report for input of the line work (images). The scanner, manufactured for the project by IBM, was the first 48-inch by 48-inch cartographic digital drum scanner in the world. Statistical (descriptor) data related to polygons on the maps were input by keypunch (later keytape). This allowed data to be entered twice and be validated much more quickly than if descriptor data were entered conjointly with image data, as in a typical digitizing process of automatic cartography. The link between the descriptor data and the image data was established by digitizing a single coordinate point inside each polygon, with an arbitrary number assigned to the polygon and also to the descriptor statistics. These three separate operations (scanning, typing, and point digitizing) were logically linked later in the software. The principal of keeping image data separate from the descriptor database and combining them only as required led to key system efficiencies later adopted in other successful GIS. Many types of geographical analysis could be completed using only descriptor data.

Don Lever tackled the problems of creating the vectors from the scanned raster image. Because of the screw motion of the drum scanner head, the scanner actually produced pixels arranged in a helix. In the early years, the maps were primarily polygon coverages. Some coverages contained points (handled as infinitely small polygons). Line networks were handled in later versions of the system. The scanner produced a "cloud" of 0.004-inch bits for each line. Then the width of the cloud was reduced by the software, leaving line segments a single bit

Y																
	85	87	93	95	117	119	125	127	213	215	221	223	245	247	253	255
	84	86	92	94	116	118	124	126	212	214	220	222	244	246	252	254
	81	83	89	91	113	115	121	123	209	211	217	219	241	243	249	251
	80	82	88	90	112	114	120	122	208	210	216	218	240	242	248	250
	69	71	77	79	101	103	109	111	197	199	205	207	229	231	237	239
	68	70	76	78	100	102	108	110	196	198	204	206	228	230	236	238
	65	67	73	75	97	99	105	107	193	195	201	203	225	227	233	235
	64	66	72	74	96	98	104	106	192	194	200	202	224	226	232	234
111	21	23	29	31	53	55	61	63	149	151	157	159	181	183	189	191
↑ 110	20	22	28	30	52	54	60	62	148	150	156	158	180	182	188	190
101	17	19	25	27	49	51	57	59	145	147	153	155	177	179	185	187
Y 100	16	18	24	26	48	50	56	58	144	146	152	154	176	178	184	186
3 011	5 0101	7 0111	13 1101	15 1111	37	39	45	47	122	135	141	143	165	167	173	175
2 010	4 0100	6 0110	12 1100	14 1110	36	38	44	46	132	134	140	142	164	166	172	174
1 001	1 0001	3 0011	9 1001	11 1011	33	35	41	43	129	131	137	139	161	163	169	171
0 000	0 0000	2 0010	8 1000	10 1010	32	34	40	42	128	130	136	138	160	162	168	170
	000 0	001 1	010 2	011 3	100 X	101	110 →	111								

Figure 2.1 The Morton Matrix

wide. The lines were logically followed, vertices were identified, and thus arcs and nodes were created. The enclosed polygons were "coloured," and the related perimeter segments could then be identified. Each line segment was identified with the polygons on each side of it, designated as right and left according to the direction of the line first traced during cloud elimination. This was the first effective use of the arc-node concept, coupled with automatic creation of polygon topology, directly from the scanned image.

It must be remembered that this work was carried out in the main without the benefit of computer screens. Even as late as 1967, computer graphic screens cost approximately $90,000 and required a dedicated port on the mainframe. Very few were available. Batch processing was the order of the day, and the use of topology was originally necessary as an edit function in the creation of reliable polygon figures. Later systems of automatic cartography which work solely in the graphic domain overlook the benefit of the rigorous edit involved in creating topology.

Edgematching and Error Recognition

Bruce Sparks and Peter Bédard made major contributions to creating the capability for automatic edgematching of map sheets, a capability that topologically maps polygons seamlessly over very wide areas. The initial limit was 8,000 frames (over 1,000 maps), although this was improved in later versions of the system. The process started by matching frame edges, resolving polygon identities, assigning equivalences, and calculating new centroids for the resolved polygons. The process moved on to handle the map boundaries, occasionally returning to the frame level for final solution. Where necessary, line segments at edges were automatically split and joined within tolerances related to frame size. The process could handle complex situations such as the classic "snake" polygon which extended over numerous frame and map boundaries, although it could take several passes to complete the match.

Art Benjamin played a major part in designing the capability for automatic recognition of topological map errors which allowed links between image data and descriptor data to be established when errors had been resolved. This capability conceptually recreated the source map in computer storage. The links between image data and descriptor data were established by searches around the linking coordinate to confirm the line segments that specified the feature.

A wide variety of errors was recognized by the software, including validity errors, map border gaps, edgematch mismatches beyond limits, missing interior boundaries, double polygon values like adjacent polygons, unclassified polygons, map exterior classification, overshoots, occurrence of areas too small to be real, and so on. Error messages were output on the printer. The necessary fixes were submitted by punchcard to the error correction program designed by Bob Whittaker, which checked the new submissions for error and then made deletions and changes appropriate to the fix or update.

The system was able to carry out transformations on the input images to correct maps warped at an angle (requiring "rubber sheet stretch") or the simpler case of linear distortion (shrinkage in one direction) to produce automatically a straight-sided reoriented map. The inaccuracies incurred were not significant in system terms. The maps input to the system had been scanned at a resolution of 250 points per inch of map. The data bank resolution varied but would seldom be more than 50 points per linear inch of map. Line weeding was developed to lower the resolution. These operations were accompanied by routines to accomplish gap closing and line smoothing.

Limitations of storage and CPU size demanded further data compaction for effective usage. A simplified version of Galton coding with incremental (delta) measures was introduced. The technique employed sequences of two-bit codes denoting by context either a change in direction or a number of steps in an established direction. Four direction changes or three units of distance could be represented in two bits, and because a distance of zero has no meaning, zero was utilized to give a fifth direction: no direction change. A typical result of first compaction was to reduce a line recorded in 864 bits to compact notation occupying 76 bits, of which 32 bits were taken up with the start point x-y coordinates, that is, about a 10:1 reduction in data volume. Bob Kemeny developed the data compaction methodology. The combination of resolution reduction and data compaction yielded 50:1 or better reductions in data volume.

Coordinate System

Quite early in the development of system procedures, it was necessary to equate the coordinate system resulting from drum scanning with the table coordinate system used for the reference points of each piece of digitized descriptive data. It was found to be most efficient to convert both systems to degrees of latitude and longitude, and this natural Geographic Coordinate System (GCS) was used in the main storage for the databank. This process favoured no map projection, as transformation to and from the main databank for any projection involved only one step. There was no difficulty in going from one part of the continent to another.

The question arose whether Earth should be taken to be a sphere or an ellipsoid. The effects of errors were calculated, and it was found that if 0.25% errors in both area and length calculations for areas of up to 3,000 square miles were acceptable, then representing Earth as a sphere was adequate. These errors were virtually independent of the size of the quadrilaterals contemplated.

The remaining questions were the appropriate methods for computing areas and distances in the natural coordinate system. These were resolved, and, in area measurement for areas within a range of 500 miles from any meridian within Canada and 300 miles in the parallel of 60° latitude, the maximum error incurred was less than 25 parts in 100,000. Distance calculations in GCS caused no difficulty. If regions were sufficiently small, the Pythagorean theorem could be used, whereas for greater distances, a great-circle approximation could be employed for an equivalent sphere whose radius was the mean radius of curvature at a latitude midway between the two points. Frank Jankaluk was responsible for the investigation of these issues and for the programming underlying the use of GCS.

Also included in the system were the functions of dissolve of lines between like polygons and merge of the descriptor data of the contents, scale change, circle generation, and new polygon generation, all operating in the topological domain. Simple buffer zones could be created using the circle and new polygon generate routines.

Data Manipulation and Retrieval

One of the earliest design criteria for this geographic information system was that to the greatest extent possible, the manipulation and retrieval of information from the database should be initiated by commands that could be understood by a wide group of potential users. It should also be possible to use the system without extensive (and expensive) computer programming.

That object was achieved, to the extent that someone with a basic knowledge of one high-level computer language (PL/1) could assemble English-type key words from a command language into a program that allowed quite complex operations to be carried out without further intervention. Using PL/1 tied the system to IBM machines, but IBM had a dominant position in the marketplace at the time, particularly in government institutions, so this was not regarded as a significant limitation. PL/1 had several benefits. It had excellent list processing capabilities for the time, and GIS made heavy demands on list processing (then and now). This facility probably reduced the overall programming effort by 30%. PL/1 also was the first computer language to make macros easy to produce and employ. This capability made it possible to devise the command language and provide a user-easy interface with the system, which was a significant step toward the more user-friendly systems of today.

Retrieval commands included such terms as READ, SELECT, ASSESS, DISSOLVE, MERGE, CLASSIFY, EQUATE, ACRE, SQMI, CIRCLE (generate), POLYGON (generate), COMBINE, SORT, OVERLAY, PLOT, and another set of utility commands for addressing specific coverages, creating lists and reports, and so on. The commands acted under a "monitor system" that contained the command language interpreter, the librarian function, and the facility to create the programs needed by the system in response to the commands. The user could make repetitive use of a previously designed request, called up by very simple key words taken from a form. Alternatively, the user could write a special-purpose program using the command language or step outside the command language, program requests in plain PL/1 (or assembler language), give them a key word, place them in a command library and call them by key word thereafter. The command language was designed by P. Kingston, K. Ward, B. Ferrier, M. Doyle, J. Thacker, F. Jankaluk, H. Knight, and P. Hatfield.

One of the most powerful functions was the topological OVERLAY, which was vital for comparisons of land capabilities for agriculture, forestry, recreation, wildlife (ungulates and waterfowl), present land uses, and socio-economic statistics by census enumeration area. The overlay function enabled the user to lay one map coverage directly on top of another, producing a third coverage that was the merged image of the two outputs. Up to eight input map coverages could be overlaid in one operation. The descriptor data for the new image were concatenated and the geodetic data, area, and centroid were recalculated to reflect the new polygons. Any system query could then be addressed to the newly-generated coverage. Peter Kingston was responsible for the overall design of the data retrieval system and the efficient polygon-on-polygon overlay.

Several advances in cartographic instrumentation emerged from the project. D. R. Thompson at IBM Poughkeepsie was responsible for engineering the first automated scanner for cartographic use. It cost approximately $180,000 and was delivered to CGIS in 1967. The delay in delivery was because the U.S. machine tools necessary to produce the ultra-accurate lead screw for the reading head were committed to the war in Vietnam, and Canada's need had a lower priority. It continued to work well for 15 years until replaced by a newer model, and the original is preserved in the National Museum of Science and Technology in Ottawa. Also, Ray Boyle, then working for Dobie McKinnis in Scotland, designed and manufactured the first high-precision

(±0.001) 48-inch by 48-inch free cursor digitizing tables ever produced. They incorporated the project's requirement that feature codes be input from a moveable table-top set of buttons (the forerunner of today's cursor buttons) to improve operator efficiency.

In describing the project, the term "computer mapping" became clearly inappropriate by the end of 1963. Up to that time, the system was being described as the ARDA Data Coordination System. Various new terms were debated by the staff, among which "spatial data system" was rejected as far too general and "land information system" far too restrictive in view of the data types it was supposed to handle. Finally, the term "geographic information system" came into use. The working teams described their system unofficially as a "geo-information system" or "Geo-IS" and more formally as the "Canadian Geographical Information System" (CGIS). In 1966, the name was shortened to "Canada Geographic Information System," matching the popular parliamentary style of labeling agencies with the prefix of Canada as a synonym for the federal government. On reflection, the choice of "geographic information system" perfectly sums up the system's capabilities. The name has been widely adopted for such systems throughout the world since that time.

System Evolution

The Canada Geographic Information System continued to evolve through the 1960s. In 1967, the National Film Board of Canada traced its progress in a work entitled *Data for Decision.* Descriptions of the geographic information system were published by 1968. All the capabilities discussed in this chapter were demonstrated the following year, and the system became fully operational in 1971. The institutional context of GIS development is worth mentioning. Responsibility for the development and maintenance of CGIS moved through several institutions in the first two decades. It started in the Department of Agriculture, then moved with ARDA to the Department of Forestry and Rural Development where the emphasis changed from marginal farms to the problems of rural development as a whole. It next moved to the Department of Regional Economic Expansion, when rural problems and urban problems were addressed in a regional context. It then moved to the Department of the Environment as the government's emphasis changed to land management coupled with environmental protection. In the process, there were several federal elections, and the project came under the responsibility of numerous ministers, deputy ministers, and directors general, residing within various divisions of departments. Each year, the fledgling system had to have its objectives defined, its existence justified, and its budget and staff approved. It would not have been possible to sustain the development unless the original vision had been viable and the objectives well defined; these were specifically to provide the ability to carry out multiple types of geographic analysis on a growing database of many types of data. Further developments continued into the 1980s, with interactive graphics added in 1974 and a new scanner and related scanner software in 1983. As GIS spread in popularity and other systems grew in capability, links to commercial systems (SPANS) were developed in 1986. Although the system ceased receiving input of new data in 1989, in 1994 it had in excess of 10,000 sheets of more than 100 different types of geographical data in digital storage, making it the largest national archive of digital geographical data in Canada.

Bibliography

Kingston, P. P. C. H. 1968. "A Computer-Oriented Geographic Information Processing System." Paper presented at the 10th Annual Conference of the Canadian Operations Research Society. Toronto, Canada. May 1968.

Thompson, D. R. 1967. "A Cartographic Scanner." *Proceedings of the 27th Annual Meeting of the American Society of Photogrammetry.* Falls Church, VA: American Congress of Surveying and Mapping.

Tomlinson, R. F. 1962. "Computer Mapping: An Introduction to the Use of Electronic Computers in the Storage, Compilation and Assessment of Natural and Economic Data for the Evaluation of Marginal Land." Paper presented to the National Land Capability Inventory Seminar, Agricultural Rehabilitation and Development Administration, Canada Department of Agriculture. Ottawa, Canada.

Tomlinson, R. F. 1963. "Feasibility Report of Computer Mapping System." Contract Report from Spartan Air Services Ltd. Project 14007, Agricultural Rehabilitation and Development Administration. Ottawa, Canada: Canada Department of Agriculture.

Tomlinson, R. F. 1967. *An Introduction to the Geo-Information System of the Canada Land Inventory.* Ottawa, Canada: Canada Department of Forestry and Rural Development.

Tomlinson, R. F. 1968. "A Geographic Information System for Regional Planning." In G. A. Steard, ed. *Land Evaluation.* South Melbourne, Australia: MacMillan of Australia, pp. 200–210.

Tomlinson, R. F. 1988. "The Impact of the Transition from Analogue to Digital Cartographic Representation." *The American Cartographer,* 15:3. (Extracts used with permission of the American Congress on Surveying and Mapping.)

CHAPTER 3

Academic Origins of GIS

Nicholas R. Chrisman

Introduction

While some of the most important developments of early geographic information systems occurred through a collaboration of the commercial sector and a public sponsor (as described in the previous chapter by Roger Tomlinson), the academic sector played a critical role in developing the tools that we now call GIS. In part, the academic sector has the luxury to try new approaches of high risk, and also, in part, that sector combines many of the elements required to develop new technology, particularly when an innovative team makes new connections across disciplinary lines. This chapter will try to show that some of the development in the academic sector is the fruit of a conscious policy of public research funding.

Overall, the strategy of GIS development resembles two brothers who play all afternoon constructing a bridge over a stream. The wiser older brother sends the younger brother to test the bridge on the principle that he weighs less. The academic sector plays the role of the daring younger brother, taking on the various risks through the research grant competition. Sometimes the academics get wet in the stream, but they almost always get to the other side first.

This chapter will cover the role of the academic sector in the development of GIS from the early glimmerings in the 1960s until GIS became established in the early 1980s. This account will concentrate on the earlier period, showing how that period influenced later and more successful developments. Only a part of this account is based on personal experience. I started producing geographic information products for government clients when I was an undergraduate

Nicholas R. Chrisman is an Associate Professor of Geography at the University of Washington in Seattle, WA. He continues to explore all aspects of GIS, from the details of algorithms and data structures to the social construction of knowledge, that make it all meaningful. *Author's Address:* Nicholas R. Chrisman, Associate Professor, Department of Geography, Box 353550, University of Washington, Seattle, WA 98195. E-mail address: chrisman@u.washington.edu student in 1971. The prior decade I understand through secondhand accounts. Like any observer, my vantage point influences my account. I have worked at some of the institutions involved in the earliest period, so I can construct the larger sequence of events using these institutions as a base. However, there certainly were developments elsewhere, so this account must be taken as an example of a larger process.

The analytical problems of handling geographic information had been recognized long before the development of the modern computer. At that time, the dominant procedure for analysis of geographic distributions did not involve multiple maps. Large land inventories were organized in the 1930s, such as Dudley Stamp's Land Use Inventory of Great Britain and the multifactor inventory for the Tennessee Valley Authority (TVA). These surveys tried to incorporate all the factors of a study onto one map, a method continued in the "land evaluation" techniques of Australia through the 1950s and 1960s (Christian and Stewart 1968). The interesting lesson from these precursors of current methods is that a single map could incorporate an evaluation of many factors. The TVA's "fractional code" evaluated land cover, land use, soil fertility, and specific limitations all at once. The technology of map making clearly constrained this work, but the intention was to deal with the combination of various distributions.

Steinitz and his students (1976) had charted the hundred-year process in which the "map overlay" was developed for environmental planning. In the work of Warren Manning (1913) and the British town and country planners (Tyrwhitt 1950), the integration of various factors seems to involve visual inspection rather than any particularly analytical procedures. The work of John K. Wright, particularly in his "dasymetric" map of Cape Cod (Wright 1936), shows the clearest evidence of a precomputer demand for the numerical treatment of multiple spatial distributions. Wright had to publish a precalculated table with his article to assist others in applying his procedure, a form of what we would now call areal interpolation. The disciplines involved in handling geographic information were many: planners, landscape architects, surveyors, geologists, foresters, soil scientists, and geographers. Much of the development referenced above did not occur in university contexts, perhaps because there were not so many research universities.

The construction of modern computers is widely recognized as a precondition for the development of geographic information systems. The innovations were not just at the hardware level. World War II unleashed an unprecedented flood of research funds into universities. While the best-remembered projects built atomic bombs, Aitken built an externally sequenced calculator at Harvard to compute naval gunnery tables. The University of Pennsylvania built the ENIAC computer with army research contracts. Universities could not develop these machines into their full form, since these developments required capital, marketing, and time. These university-developed prototypes had to be taken up by the commercial sector to develop a workable computer industry, a process that happened later with GIS.

This era also produced a major shift in the outlook of universities. The G.I. Bill created a huge increase in student enrollment and a trend to democratize what had been an elitist sector of society. Postwar universities came to rely on government funding for research and for increased enrollment. In return, universities were more closely tied to the changing agenda of public policy.

These changes had impacts throughout the academic community. Increased enrollments increased the demand for faculty and permitted greater specialization. Cartography arose as a specialization, as did photogrammetry and photointerpretation. Sources of research funding also influenced many developments. In part to justify their funding in the competitive scientific arena, the social sciences went through the discovery of quantitative methods at various rates. Geography began this transition in the middle 1950s.

Quantitative Geography and Early Computer Mapping

To move from the generalized history of academic research to the specific history of GIS, a few institutions must be mentioned. John K. Wright at the American Geographical Society had contributed his observations on cartography and quantitative geography from the 1920s through the 1950s. Wright's work on cartography recognized many dimensions of geographic measurement that were simplified in A. H. Robinson's (1953) influential textbook, *Elements of Cartography,* as point, line, and area. William Warntz joined the American Geographical Society and began to connect geography to the "social physics" movement. This school of thought sought to explain many social phenomena using population as mass, attenuated by distance in a gravity model. Social physics provided a basis for early urban simulation models, transportation models, and theories of regional economics. The emphasis on space as an explanatory factor was a part of the background from which GIS developed.

The University of Washington was a key location for some of the earliest work in GIS, though it may not have been recognized as a coherent whole at the time. In the 1950s, the Department of Geography was run by Professor Donald Hudson, who had worked at the TVA on the multifactor land inventory and other kinds of applied research. Among other faculty, Edward Ullman studied transportation, economic geography, and cities. Ullman used numbers but little in the way of formal statistical methods. John Sherman, trained as a climatologist, had taken on teaching cartography. William Garrison was assigned by Hudson to teach a course in quantitative geography, based on a short course he had taken with Walter Isard (Morrill 1994, personal communication). The crop of new graduate students in 1955 included a number who took to the topic with great interest. Richard Morrill, Brian Berry, Duane Marble, Waldo Tobler, Michael Dacey, and John Nystuen all collaborated with Garrison and Marion Marts on their grant to assess the impact of highway bypasses on small town shopping districts. William Bunge arrived a bit later. In a few short years, this group of students developed a new approach to mathematics, mostly statistical methods, for geographic information. With assistance from others such as Warntz and some British academics, the movement now termed the "quantitative revolution" in geography arose. Their approach to quantitative methods was largely borrowed from sociology, psychology, and economics. Berry's Ph.D. dissertation, for example, applied factor analysis to the retail structure of Spokane, WA. The single IBM 601 computer for the University of Washington campus became a fixture in the lives of these geography graduate students. Berry's computer processing required a separate grant of ten hours on an IBM 650 at a Western

Regional Computing Center in Los Angeles. For the most part, the computer processed geographic attributes in the social science form that would later be called the "geographical matrix" (Berry 1964).

Tobler worked with Sherman, the cartographer. In 1957, they produced a short paper (Sherman and Tobler 1957) that illuminates Sherman's approach to photographic map production. Sherman viewed the map as a database as well as a communication medium. At a time when others treated color separation as a technical step, with one image for each color, this article called for a multipurpose cartography based on one negative for each attribute value for each variable on the map. In Sherman's map of the University of Washington, he used dozens of layers for a two-color map, so that he could produce variants for different purposes. With photographic techniques, these layers could be combined to create a map with whatever combination of features a user would need. The photographic technology limited the application of this method, but this paper provides a glimpse of the multilayer approach that did not arise until many years later. Tobler's Ph.D. study dealt with the mathematics of map projections, a topic he pursued throughout his academic career. By the time this initial crop of quantitative Ph.D.s finished, Garrison moved on to Northwestern University.

On the University of Washington campus at roughly the same time, Dr. Edgar Horwood joined the Departments of Civil Engineering and Urban Planning to work on transportation and urban problems. At this time, the tradition of operations research from World War II had created much interest in mathematical models of transportation and other urban problems. Horwood fits into this background, but he took an interest in detailed geographic distributions that makes his work particularly important for the history of GIS. Horwood used the computer to develop databases and to link records to geography using what he called geocoding. Before 1960, he was offering a course on geocoding and computer mapping techniques. I know of no earlier case of an academic offering on the computer processing of geographic information. In 1963, Horwood founded the Urban and Regional Information Systems Association (URISA). He and his group gave a week-long workshop on this new technology in Chicago around 1964. This presentation was a part of a series that covered regional science methods such as input-output analysis. Horwood's session was well-attended by regional scientists, quantitative geographers, and academic cartographers. The academic cartographers argued that the line printer output was too crude (they were right), but the promise of computer processing excited many. Horwood's software package was rather clunky, as most software of the time was.

Founding the Harvard Lab

One attendant at the Chicago event was Howard Fisher, a Chicago architect. Fisher saw a potential in computer cartography and thought he could write a better package. He found some programmers to cobble together a "synagraphic" mapping package (SYMAP). He persuaded the Ford Foundation to give him a fairly large sum of money to develop the concept. Fisher asked two universities in his native Chicago to allow him to develop SYMAP, but neither one wanted to take in this retired architect, even if he brought a few million dollars. Fisher's *alma mater,* the Graduate School of Design at Harvard University, was more accommodating. In 1965, Fisher founded the Laboratory for Computer Graphics, and the work on SYMAP began in earnest. SYMAP implemented the reductionist simplicity of Bunge's (1962) metacartography. The package took points, lines, and areas as input. After some cruder attempts, the program acquired remarkably sophisticated techniques for interpolating through the work of a Harvard undergraduate, Donald Shepard (1968). Version 5 of SYMAP, which first appeared in 1968, had a large range of possibilities inside the restrictive nature of its line printer output. SYMAP was originally distributed for free, but as the Ford grant dried up, it was sold for a few hundred dollars per copy. Eventually, over 500 institutions bought an official copy, and many more copies circulated unofficially.

In 1968, William Warntz became director and expanded the name to the Laboratory for Computer Graphics and Spatial Analysis. His appointment to Harvard University titled him Professor of Theoretical Geography, although the Department of Geography had been closed 20 years earlier. Warntz ran a number of large research projects on the fundamental nature of surfaces for the Office of Naval Research, but the mathematics did not connect directly with new software packages. Theoretical contributions, such as Ernesto Lindgren's (1969) treatment of cost surfaces, deserve more attention in current software packages.

Alongside Fisher and the SYMAP programmers together with Warntz and his theoretical geographers, there was a third group at the lab with a more applied planning focus. This group, including Carl Steinitz, a landscape architect, did projects on powerplant siting, transportation corridors, and suburban expansion. They became involved in designing grid-based inventories for New York State (LUNR), Bonneville Power Administration, and other clients. This group had a pressing need to integrate information from different sources. The GRID (Graphic Display of Rectangular Grid Information) package was developed from parts of an older version of SYMAP by David Sinton in 1969. GRID used the grid cell as the storage construct, not just the output. Overlay of multiple layers became the core procedure as the package developed. The Honey Hill project (Murray et al. 1971) is an early example of the computer-based analysis done by this group.

Steinitz, Sinton, and the other landscape architects at the Harvard Lab had different disciplinary roots from Warntz and his theoretical geographers. Landscape architecture had moved from designing gardens for the rich to considering vast regional landscapes. During the 1960s, landscape architects like Phillip Lewis (1963) at the University of Wisconsin–Madison took on massive projects covering whole states requiring the simultaneous analysis of a number of factors. The map was an important tool in their emerging analytical technique. The research process was driven by specific applications projects, so the documentation of the landscape architecture roots for GIS is much more difficult to track. However, a key event was the publication of Ian McHarg's (1969) *Design with Nature.* McHarg was a professor at the University of Pennsylvania and, like most landscape architects of the time, kept a healthy private practice alongside his academic duties. His book, written for a popular audience and promoted in the *Whole Earth Catalog* (Portola Institute 1969), set a clear agenda for the integration of map layers to create plans. The timing of this book coincided with the emergence of the environmental movement and the creation of various environmental regulations. More than any other event, this book created a demand for the kind of logic that was implemented by Sinton's GRID program at Harvard.

From 1968 to 1971, the Harvard Lab was a large enterprise with as many as 40 staff of various job descriptions. SYMAP was joined by SYMVU, a three-dimensional perspective viewing program, and CALFORM, a plotter-oriented choropleth mapping package. Students like Jack Dangermond (a landscape architecture master's degree student) were given an exposure to an exciting field of study as it was being discovered. Only a few became as prominent as Dangermond, but lab alumni became influential in a number of other organizations. The greatest legacy of the early lab came from the students at hundreds of universities where SYMAP was used to teach basic computer cartography.

GIS Developments at Other Universities before 1970

During the same period, Harvard was certainly not the only center of innovation. On many campuses, researchers experimented in writing software to display maps on the computer. A major focus for innovation in Great Britain was the Experimental Cartography Unit (ECU), housed in the Royal College of Art, though far from a traditional academic unit. The leadership of David Bickmore at ECU has been described by Rhind (1988). ECU did influence the direction of the Ordnance Survey, but their software did not reach a large audience. The package MAP/MODEL produced at University of Oregon (Arms 1970) was a prototype for the early (brute-force) vector approach to polygon overlay. This software served as a model for PIOS (Polygon Information Overlay System) at Environmental Systems Research Institute (ESRI). At the University of Kansas, a combination of geographer George Jenks and geologist John Davis sparked the creation of SURFACE II, a contouring and surface-plotting package distributed at least as widely as SYMAP.

Another major thread in the diverse origins of GIS starts in the disciplines of soil science and geochemistry. A number of research workers in France, Great Britain, and the United States developed methods of spatial analysis that are now called geostatistics. A key development was the "optimal" interpolation method called *kriging* (Matheron 1971). The work of P. A. Burrough and others (1986) extended this approach and applied it to soils and other aspects of land evaluation. During the same period, quantitative geographers had developed methods to treat spatial autocorrelation (Cliff and Ord 1981). Taken together, academic research on geostatistics created a further demand to rethink the methods used to collect land inventories.

A Second Chance at the Harvard Lab

In 1971, the Harvard Laboratory shrank dramatically. The Ford Foundation grant funds had been spent, and SYMAP had to survive on its revenues. Warntz found that his goals for theoretical geography did not fit the goals of the School of Design. He resigned and moved to the University of Western Ontario. The landscape architecture group pulled out to do their work through their department and their professional practice. The lab deflated to six employees, and it looked like it was finished.

Some of the most important research from Harvard came from the next period: 1972–1981. The landscape architecture group, still led by Steinitz, trained successive classes of graduate students in year-long studio classes that performed serious analytical projects. The databases were handcoded and keypunched far from modern scanners. The cells sizes were crude, but the principles applied were carefully considered. David Sinton led the software development as IMGRID (Interactive Manipulation GRID) emerged from GRID. This software still ran on mainframe batch computers, limited in memory and disk space. Dana Tomlin added some critical operations to IMGRID and developed the thinking he used to write the classical MAP package during his Ph.D. experience at Yale (Tomlin 1983). In the same class of Harvard landscape architecture students, a number of GIS pathways originated. Lawrie Jordan and Bruce Rado placed these software ideas at the kernel of their original product in a company called ERDAS (Earth Resources Data Analysis System). Ted Driscoll, perhaps the most probing programmer amongst them, contributed to the early software at I^2S. Sinton's concepts, linked to Tomlin's, form the basis for much of the raster-oriented software in current use, such as Idrisi, MAP II (Map Analysis Program), and ARC/GRID. Sinton's (1978) essay at the 1977 Endicott House meeting shows the clarity of the instructional message presented to these students in a period considered to be the dark ages from the current perspective. Sinton saw the distinction between raster and vector not as programming concerns but in terms of the approach to measurement.

Although the residual lab was just one floor above Sinton's office, the connection was not very close. The research agenda upstairs focused on vector databases and topological data structures (Peucker and Chrisman 1975). Yet the connection was close enough that polygon overlay was recognized as a key requirement for vector processing (Chrisman 1975). This research led to the design of the ODYSSEY system (Dutton 1978b), a prototype for the processing engines behind the current generation of commercial systems. ODYSSEY made the switch to interactive control through a flexible command language (Dougenik 1978). The interactive machines of the period were smaller in memory than the mainframes, but careful programming permitted quite large problems to pass through the memory available (Dougenik 1980). The history of this project has been related elsewhere (Chrisman 1988; Chrisman et al. 1992).

In this same period, the Harvard Lab hosted an important research conference at Endicott House in 1977. This event assembled many of the people who remain important in the current GIS world. It included Berry, Marble, and Tobler from the original quantitative geographers. Ken Dueker and Jerry Schneider represented Horwood's legacy. The U.S. Geological Survey and the Census Bureau were cosponsors, sending the teams that developed their in-house topological software. James Corbett and Marvin White discussed ARITHMICON, the predecessor to TIGER (Topologically Integrated Geographic Encoding and Referencing). Steven Guptill and Rogin Fegeas reported on the LUDA/GIRAS (Land Use Data Analysis/Geographic Information Retrieval and Analysis System) software that preceded the DLG developments.

Michael Goodchild (1978) published his first piece on the errors involved in polygon overlay. There was considerable attention to three-dimensional problems, with various triangular systems presented by Thomas Peucker, David Mark, Kurt Brassel, and Chris Gold. Since the focus was on topological structures, there were fewer from the raster community, but Donna Peuquet and Jack Estes did defend the raster approach. The proceedings of the Endicott event (Dutton 1978a) have long gone out of print, but they set a high standard for publication in the GIS arena.

The Harvard Lab also ran "Computer Graphics Weeks" from 1978 to 1982, events of great enthusiasm, considering the nature of the tools available. Most of the papers (Moore 1979-81) presented applications from a wide range of fields, mixing academics with commercial and government authors. The proceedings were printed in 18 volumes. It is interesting that all this activity preceded the availability of what we would now identify as a full-function GIS software package. These events helped create the energy for the next stage.

On the promise of polishing ODYSSEY for a commercial launch, the Harvard Lab built up to 40 staff by 1981. Scott Morehouse managed the ODYSSEY project in this final stage. The deal fell through (Chrisman 1988) and the team dispersed. Morehouse went off to ESRI, where he set to work designing ARC/INFO. As always, the transfer from the research sector to the commercial sector is far from simple. It is easy in retrospect to fault the Harvard Lab for trying to venture into the commercial arena. Research in GIS now seems quite distinct from software development. However, in that era, the SYMAP model of software distribution seemed perfectly viable. Academic research combined basic innovation and implementation. ODYSSEY, as a prototype geographic information processing package, was quite ready to ship in 1981 with many of the features that ARC/INFO 1.0 had three years later. It did not have the crucial database interface, but perhaps that would have developed somehow.

Other University Developments in the 1970s

A major portion of the university research efforts in the 1970s was devoted to remote sensing and imagery sources. These developments will be covered in other chapters (see Estes and Jensen in this volume). The remote sensing developments were not very closely articulated with the developments discussed in this chapter. Even the gridbased overlay programs such as GRID and MAP were not designed to handle remote sensing imagery; they were operated on hand-coded data entered by a legion of students at keypunch machines. The gap between the early GIS developments and the early remote sensing developments remains present in the current disciplinary structure.

One major development of the 1970s is the blossoming of computer science. This discipline barely existed as an identifiable academic pursuit in the 1960s, but by the 1980s it had taken its place as a much more prominent component of academic life than geography or landscape architecture. Computer scientists have contributed greatly to the development of GIS, though very few academic computer scientists would identify themselves as GIS specialists. The original work on computer graphics, such as the classical hidden surface problem (Sutherland et al. 1974), was a critical element in developing computer cartography as it moved off the line printer. The image processing group at the University of Maryland has been a major contributor to the basic algorithms used in remote sensing as well as GIS (Rosenfeld and Kak 1976, for example). Quadtrees originated here in many variants, along with various forms of spatial indexing (Samet 1990).

During the 1970s, a number of other universities became involved in applied pilot projects for a large number of government agencies. Many of these systems produced creditable results but failed to take hold because the technology to maintain the database in its real complexities did not exist. In the field of natural resources, 34 states started natural resource information systems during the 1970s, often beginning on campus (Mead 1981). A few projects, such as the Minnesota Land Management Information System, did transfer from a university into state government, but the Minnesota success is one of few. Most attempts to create integrated databases in this era failed for institutional as well as technical reasons. In the 1980s, the success rate has been higher. Perhaps the technology finally produces on its promise, and perhaps the university role has been better defined.

University research did develop software of various kinds. The Minnesota EPPL, a grid-based analysis package, survives, but the more innovative bitmapped GRASP developed for a similar project at the University of Wisconsin did not survive to transfer off the outdated Univac mainframe. In the vector realm, Dueker and Goodchild collaborated on PLUS, a polygon overlay package written at Iowa and Western Ontario in 1975, but this was not developed for distribution. Few university groups had the size of team and the focus on developing a package that characterized the Harvard Lab.

The main role of academics in the 1970s was in promoting the possibility of computer-handling of geographic information. Many ideas were explored, and the basic principles were discovered. Each academic center had to wrestle with all the issues from start to finish in building a database and developing software. Many fundamentals were discovered over and over again; that was part of the fun. After all, these academics chose to try to cross the rickety bridges in order to be first on the other side.

Summary

The field now known as geographic information systems developed from the confluence of different disciplinary origins. The academic sector provided an environment of exploration and innovation that was critical in the formative period. The history of technological development is full of other examples of simultaneous discovery at separate locations. Basically, the time was ripe, and the fundamental ingredients were there. Each university had a different mix of disciplines and individuals, so the results varied dramatically.

Bibliography

Arms, A. 1970. *Map/MODEL System: System Description and User's Guide.* Eugene, OR: Bureau of Governmental Research and Service, University of Oregon.

Berry, B. J. L. 1964. "Approaches to Regional Analysis: A Synthesis." *Annals of the Association of American Geographers,* 54: 2–11.
Bunge, W. 1962. *Theoretical Geography.* Lund, Sweden: Gleerup.
Burrough, P. A. 1986. *Principles of Geographical Information Systems for Land Resource Assessment.* Oxford, U.K.: Clarendon Press.
Chrisman, N. R. 1975. "Topological Data Structures for Geographic Representation." *Proceedings AUTO–CARTO II,* 1: 346–351.
Chrisman, N. R. 1988. "The Risks of Software Innovation, a Case Study of the Harvard Lab." *The American Cartographer,* 15: 291–300.
Chrisman, N. R., J. A. Dougenik, and D. White. 1992. "Lessons for the Design of Polygon Overlay Processing from the ODYSSEY WHIRLPOOL Algorithm. *Proceedings 5th International Symposium on Spatial Data Handling,* 2: 401–410.
Christian, C. S., and G. A. Stewart. 1968. "Methodology of Integrated Surveys." In *Aerial Surveys and Integrated Studies*, 233-280. Toulouse, France: UNESCO.
Cliff, A. D., and J. K. Ord. 1981. *Spatial Processes: Models and Applications.* London, U.K.: Pion.
Dougenik, J. A. 1978. "LINGUIST: A Processor to Generate Interactive Languages." In G. Dutton, ed. *Harvard Papers on Geographic Information Systems*, vol. 7. Reading, MA: Addison Wesley.
Dougenik, J. A. 1980. "WHIRLPOOL: A Geometric Processor for Polygon Coverage Data." *Proceedings, AUTO-CARTO IV,* 304–311.
Dutton, G., ed. 1978a. *Harvard Papers on Geographic Information Systems.* Reading, MA: Addison Wesley.
Dutton, G. H. 1978b. "Navigating ODYSSEY." In G. Dutton, ed. *Harvard Papers on Geographic Information Systems*, vol. 2. Reading, MA: Addison Wesley.
Goodchild, M. F. 1978. "Statistical Aspects of the Polygon Overlay Problem." In G. Dutton, ed. *Harvard Papers on Geographic Information Systems*, vol. 6. Reading, MA: Addison Wesley.
Lewis, P. 1963. *Recreation in Wisconsin.* Madison, WI: State of Wisconsin, Department of Resource Development.
Lindgren, C. E. S. 1969. "A Study of the Movement of a Point on a Plane and in Space." *Harvard Papers in Theoretical Geography, Geography and the Properties of Surfaces Series 36.* Cambridge, MA: Laboratory for Computer Graphics and Spatial Analysis, Harvard University.
Manning, W. 1913. "The Billerica Town Plan." *Landscape Architecture,* 3: 108–118.
Matheron, G. 1971. "The Theory of Regionalized Variables and Its Applications." *Cahiers du Centre de Morphologie Mathematique de Fontainebleau.* Paris, France: Ecole National Superieure des Mines de Paris.
McHarg, I. L. 1969. *Design with Nature*. Garden City, NY: Natural History Press.
Mead, D. A. 1981. "Statewide Natural Resource Information Systems—A Status Report." *Journal of Forestry,* 79: 369–372.
Moore, P., ed. 1979–81. *Harvard Library of Computer Graphics.* Cambridge, MA: Harvard Laboratory for Computer Graphics.
Morrill, R. 1994. Personal communication.
Murray, T., P. Rogers, D. Sinton, C. Steinitz, R. Toth, and D. Way. 1971. "Honey Hill: A Systems Analysis for Planning the Multiple Use of Controlled Water Areas." *Institute of Water Resources 71-9; NTIS AD 736 343 & 344.* Army Corps of Engineers.
Peucker, T. K., and N. R. Chrisman. 1975. "Cartographic Data Structures." *The American Cartographer,* 2: 55–69.
Portola Institute. 1969. *Whole Earth Catalog: Access to Tools*. Menlo Park, CA: Portola Institute.

Rhind, D. W. 1988. "Personality as a Factor in the Development of a New Discipline: The Case of Computer-Assisted Cartography." *The American Cartographer,* 15: 277–289.

Robinson, A. H. 1953. *Elements of Cartography*. New York: John Wiley & Sons.

Rosenfeld, A., and A. Kak. 1976. *Digital Picture Processing*. New York: Academic Press.

Samet, Hanan. 1990. *The Design and Analysis of Spatial Data Structures.* Reading, MA: Addison Wesley.

Shepard, D. 1968. "A Two-Dimensional Interpolation Function for Irregularly Spaced Data." *Proceedings, Twenty-Third National Conference, Association for Computing Machinery,* 517–524.

Sherman, J., and Tobler, W. 1957. "The Multiple Use Concept in Cartography." *Professional Geographer,* 9(5): 5–7.

Sinton, D. F. 1978. "The Inherent Structure of Information as a Constraint to Analysis: Mapped Thematic Data as a Case Study." In G. Dutton, ed. *Harvard Papers on Geographic Information Systems*, vol. 7. Reading, MA: Addison Wesley.

Steinitz, C., P. Parker, and L. Jordan. 1976. "Hand-Drawn Overlays: Their History and Prospective Uses." *Landscape Architecture,* 66: 444–455.

Sutherland, I. E., R. F. Sproull, and R. A. Schumacker. 1974. "A Characterization of Ten Hidden Surface Algorithms." *Computing Surveys,* 6: 1–55.

Tomlin, C. D. 1983. "Digital Cartographic Modeling Techniques in Environmental Planning. Unpublished Ph.D. dissertation, Yale University.

Tyrwhitt, J. 1950. "Surveys for Planning." In Association for Planning and Regional Reconstruction, ed., *Town and Country Planning Textbook.* London, U.K.: Architectural Press.

Wright, J. K. 1936. "A Method of Mapping Densities of Population with Cape Cod as an Example." *Geographical Review,* 26: 103–110.

PART III

Advances in Data Structures and Computing Environments

Scott Morehouse, 813 Clifton Ave., Redlands, CA 92373

My first exposure to the excitement of GIS was through the Harvard Lab for Computer Graphics & Spatial Analysis *Lab Log* catalog of mapping research papers and software in 1976. I thought it was cool to combine computers, visual communication, modeling, the landscape, and tool building into a single endeavor. I knew it would be a fun and challenging environment to work in and would attract good people to work with.

Scott Morehouse is the principal programmer for Environmental Systems Research Institute and the creator of ARC/INFO.

Michael J. Kevany, Senior Vice President, PlanGraphics, Inc.

My first recollection of the term GIS was as a candidate name for a subject. I was working in 1968 with the Urban Systems Group at System Development Corporation (SDC), a government think tank spin-off of the Rand Corporation. We were writing a book for the Department of Housing and Urban Development on use of information system technology in regional and local governments, and we needed a term. Geographic Information Systems was a candidate, as were Geocoding and Geoprocessing, but we wound up with the term Urban and Regional Information Systems in the title of the book. It went on to be adopted as the name for the professional association URISA, but in the meantime GIS has eclipsed it as a much more prominent term!

We were doing a lot of interesting work at SDC in those early years—interactive graphics, linking graphics with attribute data, developing a DBMS independent of application programs, the first GBF DIME based network analysis, and others. Technology was very limited then and the products, though rudimentary by today's standards, were in many cases the ancestors in the genealogical tree of current technology.

Michael J. Kevany has made a career of designing and implementing spatial data systems for municipal and state agencies.

CHAPTER 4

Topology and TIGER: The Census Bureau's Contribution

Donald F. Cooke

Introduction

The Census Bureau is conspicuously absent from the roster of chartered mapping agencies of the U.S. government. Nevertheless, it was the bureau and not the U.S. Geological Survey or Defense Mapping Agency which led in implementing topological data structures, developing street address geocoding, and building the first truly useful nationwide general-purpose spatial data set.

The Census Bureau never set out explicitly to accomplish this. How it did is a story of an agency performing its mission with exceptional concern for its constituency, in an ever-changing technological environment and affected at crucial junctures by individuals with extraordinary talents and temperaments. The story that follows describes how luck, timing, and temperament determined the evolution of key GIS technology much more than the orderly, step-by-step progress historians would like to relate. As Mark Twain said, "Of course, truth is stranger than fiction. Fiction, after all, has to make sense."

Background

Moore's law says that computer processing power doubles every 18 months: a 100-fold increase each decade. The U.S. Census Bureau, required by Article I of the Constitution to enumerate the

---

Donald Cooke is founder and president of Geographic Data Technology, Inc., a supplier of digital maps to commercial markets. Cooke was a member of the Census Bureau team that developed the Dual Independent Map Encoding (DIME) system in 1967. Besides his family, his interests include windsurfing, ice hockey, vintage sports cars, astronomy, and the learning process. *Author's Address:* Donald Cooke, Geographic Data Technology, Inc., 11 Lafayette St., Lebanon, NH 03766. E-mail: don_cooke@mail.gdt1.com.

country every ten years, is periodically torn between the need to keep census statistics comparable from decade to decade and the need to adapt to the alien data processing environment engendered each decade by two orders of magnitude growth of computer technology.

The bureau has always played an important role in the evolution of data processing. The most often-cited instance is the invention of a punch card tabulation system by the Census Bureau, which enabled the bureau to complete processing the 1890 Census before it had finished with the last non-automated (1880) one.

Though the bureau had bought the first non-military digital computer in 1950, the 1960 Census was still—to the outside world—primarily a paper operation. Census made extensive use of computers for internal processing in 1960, including designing and building FOSDIC (Film Optical Sensing Device for Input to Computers) to scan the 1960 questionnaires. But the bureau published 1960 statistics solely in printed tables. Only a handful of intrepid analysts requested 1960 Census results in machine-readable form; these requests were filled on punch-cards.

Preparations for the 1970 Census

For the past three decades, the Census Bureau has shown exemplary sensitivity and care for the needs of users of its data. Much of the intelligence about data user requirements was supplied by a panel of outsiders called the Census Small Area Data Advisory Committee.

This committee recommended formation of two groups in 1966: the Data Access and Use Laboratories (DAULabs), led by Jack Beresford at the bureau's Suitland, MD, headquarters, and the New Haven, CT, Census Use Study, located at the site of the April 1, 1967, dress-rehearsal test of new census enumeration procedures.

The first time that the country was enumerated primarily by mail was in 1970. The bureau planned to buy commercial mailing lists on computer tape, print mailing labels, and mail questionnaires to each household. People were to fill out the forms "in the comfort and privacy of their homes," then drop them in the mail to a regional processing center. Census had mailed out forms in 1960, but enumerators visited each home to pick them up, recording the location—Census Tract and Block—of each household on a map. The 1970 mail-out/mail-back plan promised to be more efficient, but its success hinged on being able to "geocode" each questionnaire to the appropriate census block *without* sending people into the field with maps.

Increasing computer power, geographical requirements of a new enumeration technique, and a sincere desire on the part of the bureau to serve data users better resulted in a permanent change in accessibility of census data on one hand and a vital contribution to GIS technology on the other.

Beresford's DAULabs group facilitated a smooth transition from the paper era of the 1960 Census to the machine-readable era of 1970 and beyond, by setting standards for tape files and documentation, establishing satellite Summary Tape Processing Centers to assist data users, and conducting hundreds of workshops and training seminars. DAULabs' efforts created today's demographic analysis industry. Companies like Claritas, National Planning Data, and Urban Decision Systems were among the first of Beresford's Summary Tape Processing Centers.

In contrast to Beresford's measured success, New Haven was to become the first battleground in a technical and management revolution that kept the census geographic operations in turmoil into the 1980s but ultimately led to creating a nationwide spatial data resource which anchors the nation's spatial data infrastructure and has spawned Business Geographics, currently the fastest-growing segment of GIS.

The New Haven Census Use Study in 1967

The Small Area Data Advisory Committee posed five challenges to the Census Use Study:

1. Could a useful transportation survey be run using a sample of census households, keyed so that the transportation data could be augmented by individual data from decennial census questionnaires?
2. Could the same be done with a health questionnaire?
3. What would it take to be able to generate *any* cross tabulation of census data, taking into account the need to preserve privacy of respondents? How useful would it be to local data users to have this capability?
4. What about matching two data files where the common element in each is a street address? Can this be done efficiently even with varying addressing conventions? Would the ability to geocode data sets by address matching be a useful function?
5. What about computer mapping? Census and local data are inherently spatial; could new computer graphics technologies be used to map the data?

The Census Bureau appointed Caby Smith as study director. Smith was a Mississippian who had joined the bureau as an entry-level typist. He had risen in the ranks and developed a reputation as a "can-do" manager, if a bit of a maverick. Perhaps this assignment would fit his temperament and keep him out of the way of the more sober statisticians and administrators who were busy gearing up for the 1970 enumeration.

Smith's deputy on the New Haven site was George Leyland, a recent Harvard graduate whose wife Mary was a key manager in an IBM-funded urban information system project at the City of New Haven. Smith and Leyland set about to staff the study, Smith bringing on Joyce Annecillo, a shrewd and experienced administrative assistant with skills to meet bureaucratic complexities to come. Leyland hired Bill Maxfield and the author, both of whom were finishing studies at Yale, to work on mapping and special tabulations and engaged Jack Sweeney of Cambridge Computer Associates to tackle the address matching initiative. All told, the New Haven staff stabilized at about a dozen people, with several more commuting occasionally from census headquarters in Suitland.

Using computers at Yale and the City of New Haven, the staff started on the five tasks outlined by the Small Area Data Committee. Although Sweeney had considerable programming experience, neither the author nor Maxfield had formal training in geography or cartography. They both had nominal classroom exposure to FORTRAN but no background in computer graphics, data analysis, or demographics. In addition, both preferred to plunge in and start programming, eschewing a literature search which might have uncovered, for example, Robert Dial's 1964 Ph.D. thesis on Street Address Conversion System (SACS). Dial's contribution would come a dozen years later when, as an Urban Mass Transit Administration (UMTA) administrator, he provided funding for a critical prototype GIS at the Census Bureau.

The New Haven staff started working with the New Haven Address Coding Guide (ACG), produced by the Census Geography Division with local assistance, as a prototype of the geographic base file needed to geocode mailing addresses in 144 metropolitan areas to be enumerated by mail. The ACG contained block-face records, each of which supplied:

Street Name

ZIP Code

A low-to-high address range

Census tract and block number corresponding to the range of addresses

Conceptually a block face was one side of a city block, a usable definition in regular downtown street patterns but one that fell apart in curvilinear suburban developments. William Fay, chief of the Census Geography Division, had in 1965 described the ACG as an "ideal foundation for a computer mapping file." The Census Bureau's in-house engineering department had built a digitizer (from scratch; digitizers were not yet a marketplace product), and operators had digitized a coordinate measurement at the middle of each block face.

Though Sweeney's fledgling address matcher could use the digitized ACG to assign census codes and coordinates to addresses, Cooke and Maxfield could do little with the ACG to map census data that were summarized to tract or block. The use study requested that the geography division redigitize the New Haven ACG, this time taking two coordinate measurements, one at each end of the block face. This would allow the budding mapping programmers to draw lines from one end of the block face to the other, recreating the street network and displaying the census blocks.

The result was disappointing for two reasons. First, the block face digitizing technique meant that the coordinates of each downtown street intersection were measured eight separate times—and, because of operator variability and drift in the newly-built digitizers, usually there were eight different coordinate readings. Urban maps, while recognizable, looked unacceptably crude. The second problem was more serious. Operators were instructed to digitize both ends of a block face, which became an impossible task in the curvilinear suburbs. One computer plot of a suburban New Haven tract was christened the "ruptured eagle" by George Farnsworth, a use study's Washington staffer—scant reward for the geography division workers who had struggled with the use study's digitizing requirements.

Maxfield, ever optimistic, set about trying to program around the ACG's flaws, searching for and averaging nearby coordinates. The results were better, but everyone shuddered at the prospect of digitizing each of the roughly four million intersections in ACG areas eight times each, using prototype digitizing boards, then averaging the coordinates. There had to be a better way.

Digitizing efficiency became a focal point. How could one digitize each point just once? Perhaps one could analyze the ACG and come up with lists of intersections to present to the operator. But what about a street that crossed another one, then circled back and intersected it again? What about streets that turned and curved without intersecting?

In early June of 1967, James Corbett of the bureau's Statistical Research Division (SRD) presented the use study staff and Technical Steering Group with a terse and opaque overview of the topology of maps, describing how zero-, one-, and two-cells could be related through incidence matrices. The New Haven staffers did not understand this, but Corbett insisted that it was important. Finally one of the staffers admitted his bewilderment to Corbett and said, "I just want to know if we have to number the nodes." Corbett replied in the affirmative and the logjam was broken.

The procedure is relatively simple. Take a census map which has streets labeled and tracts and blocks numbered. Start anywhere and assign unique numbers to street intersections (nodes) in any order. Number the nodes at the end of dead-end streets. Put nodes anywhere that a street makes an appreciable bend. You do this because these are the points that you will eventually digitize.

Now notice that the nodes define "objects," to use today's terminology, that are straight lines between nodes. Each line segment can have only two nodes and can be between only two census blocks. Even better, all of the information about each line fits on one punch card:

Street Name
From Node
To Node
Left Tract/block
Right Tract/block
Left Address Range
Right Address Range
ZIP Left
ZIP Right

Conventions quickly evolve: If the line segment is part of a street and has addresses, then the "From" node is the one at the low address end of the segment. Otherwise, it doesn't matter. "Left" and "Right" orientation is determined by standing on the "From" node and looking toward the "To" node.

Maxfield and the author numbered the nodes on a map of Tract 1 in New Haven, manually encoded each line segment, keypunched the segments, digitized node coordinates from graph paper, merged in node coordinates for the proper node numbers, and plotted the resulting file. A couple of zingers appeared due to miscoded node numbers, but a corrected file plotted perfectly, demonstrating the usefulness of check-plotting to detect errors.

Then Corbett's presentation started to make sense: The "zero-cells" were nodes; the "one-cells" were the line segments; the "two-cells" were the blocks. Recording the "From" and "To" nodes was really building the zero/one-cell incidence matrix. Recording the "Left-Right" block numbers built the one/two-cell incidence matrix. Didn't Corbett say you could check the fidelity of the coding by multiplying the incidence matrices?

The researchers couldn't figure out how to multiply incidence matrices. Instead, Cooke wrote a 30-line Michigan Algorithm Decoder (MAD) program which read the Tract 1 database and attempted to chain the line segments together around each census block by linking the nodes together. The program insisted that there were errors—that it could not chain each block, even though the plot check had led to correcting all node errors. But there were still errors in block numbers, which did not appear on the plot. For example, block 101 might be keyed 110 in the "Left" block field on one of the line segments. What the MAD program would see is a missing segment for block 101 and a superfluous segment for block 110. One correction fixed both problems.

Topological editing was born. The existence of a program that could systematically detect and flag clerical coding errors ignited a small but bright hope that it would be possible to create huge mapping databases at the block level in large cities in such a way that you could assure that *all* boundaries of *all* polygons would close without error, a requirement for automated mapping.

In short order, the New Haven staff made mapping files for the entire cities of West Haven and New Haven. Many agencies supported this effort: Don Luria of the IBM/New Haven project supplied clerical staff and Bob Barraclough at Tri-State Transportation Commission (New York) had his staff digitize West Haven (on digitizers that Tri-State had paid ITEK corporation $150,000 to design and build). The State of Connecticut Highway Department ran check-plots. The "ruptured eagle" development in western New Haven now looked like a map, not a joke, giving the Census Use Study success in cartography that one would have expected from the geography division. This contributed to a growing schism between the use study and Census Geography Division.

In August 1967, Farnsworth christened the new process DIME (Dual Incidence Matrix Encoding, later Dual Independent Map Encoding). "Dual" reflected the two incidence matrices; "Independent" was taken by the New Haven Staff to mean without the help of the geography division. On short notice, the author and Maxfield wrote up the DIME process (Cooke and Maxfield 1967), and Barraclough squeezed their DIME presentation into his computer graphics session at the September 1967 Urban and Regional Information System Association (URISA) conference.

By fall of 1967, Sweeney's Admatch program was running well; the health initiative was supplying numerous databases for geocoding and mapping tests; the Census Bureau had processed the April dress-rehearsal census and delivered prototype summary tapes for the use study to map. A Yale administrator declared the Census Bureau a threat to individual privacy and cut off the use study's account at the computer center. The result was that the mapping effort turned north to Harvard.

The Harvard connection coincided perfectly with the SYMAP boom at the Harvard Lab for Computer Graphics (Chapter by Chrisman). During the fall of 1967, the New Haven staff produced reams of maps with SYMAP and any plotting equipment that was available at Harvard and MIT. Caby Smith sensed an opportunity to capitalize on DIME and took the New Haven staff on a tour of federal agencies, promoting DIME and computer mapping and promising to perform groundbreaking research projects that could be funded through interagency transfer of funds unspent at yearend. This blitz (six New Haven-to-Washington round trips in November 1967 alone for one staffer) assured the finances—and independence—of the Census Use Study.

DIME in the 1970s

The Census Use Study's influence was immediate: Samuel Arms, author of the exquisite and little-known "Map Models" system, who was in the 1967 URISA audience, promised to implement DIME concepts upon his return to Oregon to eliminate sliver problem that plagued his polygon system. Jack Dangermond, founder of ESRI, went so far as to say that the New Haven people "invented topology," the sort of hyperbole one might expect from a generous Californian. Ken Deuker and Ed Horwood attended the URISA DIME presentation, so the innovation was immediately disseminated to the University of Washington as well as the Harvard Lab, which at the time were the two major academic GIS centers.

The Census Use Study had been charged to get maps out of computers; it discovered that the real problem was how to get the maps into the computers in the first place. The primitive data processing environment of the time channeled innovation to useful ends. The 80-column punch-card and focus on geocoding demanded that the line segment be the fundamental object of DIME, in contrast with the polygon focus of virtually every other GIS project (CGIS, the Harvard Lab, PIOS, Map-Models, etc.). The batch processing environment necessitated DIME's topological consistency edits to trap clerical errors, and the topological purity attainable through DIME's edits allowed algorithmic generation of error-free polygon files from DIME files.

The study's chauvinistic competition with the geography division led to recording left and right address ranges in the New Haven DIME file, breaking reliance on the error-prone ACG. The undisciplined "not-invented-here" temperament of census researchers saved them from a possible technical sidetrack which Dial's SACS system might have afforded. Jack Sweeney's mentoring of the New Haven apprentices grounded them in good data processing practice, far more useful to the development of DIME than formal training in geography or analytic geometry. DIME, after all, turned out to be an exercise in data management and processing, not a computer graphics or cartographic problem.

The Census Bureau's response to innovation at its out-of-control research outpost was predictable. Managers all the way up to Associate Director Morris Hanson, who had pioneered use of computers at the bureau, were drawn into a turf battle against the use study. Caby Smith hired Booz-Allen Hamilton to document the New Haven findings, then put a continent between bureau headquarters and his operation by moving his staff—funded by interagency transfers—to Los Angeles, reconstituting the use study as SCRIS (Southern California Regional Information Study). Smith continued to recruit excellent staff, notably Matt Jaro, who honed Sweeney's Admatch work and expanded it into the more general UNIMATCH. (Sweeney had left in 1968 to start Urban Data Processing, Inc., with the author and Maxfield; Jaro now heads Matchware Technologies Inc.) Another early SCRIS hire was physicist Marvin White, who would make a crucial contribution to DIME a decade later.

Back in Washington, the weight of outside funding on the Census Geography Division forced them to "add DIME features" to the existing ACGs and create DIME files in 90 new areas. Bill Fay, who had championed ACG and resisted DIME, was replaced as geography division chief in 1971. The use study initiated a series of week-long DIME training workshops which competed for mindshare with ACG/DIME meetings sponsored by geography division.

Friction between Smith and the census establishment boiled over in 1974 into a demoralizing, scorched-earth bureaucratic battle complete with FBI investigations of all SCRIS and Census Use Study personnel. Smith sidestepped the fray by founding the National Computer Graphics Association (NCGA), which drew 1,800 attendees to its first conference, 8,000 to its second, and the international Segment-Oriented Referencing System Association (SORSA), while serving out his federal career as a chief scientist at the National Parks Service. He still heads the World Computer Graphics Association, an offshoot of NCGA.

SCRIS returned to Suitland and completed its remaining contracts as the "Center for Census Use Studies," headed by Don Luria from the IBM/New Haven study. Luria, who had also run both the Charlotte, NC, and Wichita Falls, TX, USAC urban information system projects, later moved to New Mexico where he founded the largest catering organization in the state.

DIME becomes TIGER

As the 1980 Census approached, the geography division updated the 1970 ACG/DIME files to make the 1980 GBF/DIME (Geographic Base File) files. The Correction, Update, and Extension (CUE) process was a labor-intensive, batch-oriented process involving thousands of workers at hundreds of local agencies. Turnaround times for updates and edits were measured in weeks, and the bureau was faced with redigitizing all the expanded GBF coverage.

Though batch-mode update and editing were barely feasible, off-line digitizing with no graphical feedback to operators was a nightmare. Fred Broome, the geography division manager in charge of digitizing, faced immense obstacles to progress as the Census Systems Division insisted that all computer processing be done on UNIVAC mainframes. Broome broke into interactive minicomputer technology only by acquiring a DEC PDP-11 through Intergraph Corporation, under the guise of purchasing a digitizing system.

Marvin White, now in the Census Statistical Research Division, faced the same obstacle. He had inherited a prototype on-line DIME file editor called ARITHMICON from Corbett (now retired) and had run a comprehensive test on the economics of managing a citywide DIME file as an on-line database with an interactive graphics interface. But the systems division's mainframe mandate prevailed, and White's commercial time-sharing account was terminated.

White persevered by calling a friend from his California days, Frank Lockfeld, who ran the Center for Urban Analysis in San Jose. Lockfeld was tired of struggling to maintain the Santa Clara County DIME file as a sequential database and had purchased a Z-8000 Onyx supermicrocomputer as an alternative. He needed software; White needed interactive computer time. The continentwide gap was spanned by the Federal Telephone System. White obtained funding from Bob Dial at UMTA for a pilot on-line DIME demonstration project and had his prototype 2D system running before federal accountants caught up with the outrageous surge in long-distance phone usage.

Though it was used for years by Lockfeld and a Baltimore DIME pioneer, Fred Westerfield, White's 2D system really proved its worth as a design document. It demonstrated in detail an on-line, topologically structured paradigm for managing large map databases—salvation from the purgatory of batch processing of huge spatial databases. How could White get this innovation adopted by the bureau?

The geography division was reeling from problems which had forced cancellation of local quality-control procedures. Another division chief was replaced, this time by a monthly rotation of middle managers. Redistricting battles were unearthing geographic discrepancies between the 1980 Census statistics, paper maps, and the GBF/DIME files. Broome's digitizing project fell hopelessly behind schedule.

Joe Knott, a middle manager at geography division, recognized the value of the 2D model and helped White turn 2D over to Broome at the geography bureau (White left the bureau in 1984 for a "temporary" assignment at ETAK, a car navigation firm; he's still there). Broome immediately implemented enough of 2D to convert the digitizing process to the Direct-Dig online paradigm and brought his project to completion on schedule.

The success of 2D technology had a profound effect on a new generation of savvy, computer-literate geography division managers. Acting as the "Coffin Twelve" (a reference to their windowless meeting room), they produced in 1982 a technical manifesto committing the geography division to integrate all of the bureau's spatial knowledge into a single, nationwide, online database called TIGER (Topologically Integrated Geographic Encoding and Referencing), organized along the lines of White's 2D prototype. Following successful demonstration of a "Tigger" prototype, Bob Marx, a Minnesota geographer, emerged to lead the geography division to the fulfillment of the DIME vision in the 1990 TIGER files.

Epilogue

The story of TIGER is well documented (Marx 1986). Corbett's persistence in stressing the importance of applied mathematics gave administrators at all levels confidence that the new TIGER technology was a sound investment. Knowledgeable, hands-on geography division managers involved the U.S. Geological Survey in providing rural coverage to extend TIGER to representing the whole country. Even Marvin White and the author were called back in to contribute as their companies (ETAK and Geographic Data Technology) performed TIGER digitizing and editing contracts between 1986 and 1988.

Marx claims he never wrote a contingency plan in case TIGER proved intractable. He insists that the geography division had no choice but to succeed in implementing TIGER; no other course was technically or administratively feasible. He may be right, but that does not diminish the magnitude of the risk that the geography division undertook in the early 1980s. Its success has put the world's most useful general-purpose spatial database into the hands of more users than any other GIS data resource. The current boom in business geographics is only possible because of the groundwork laid by the Census Geography Division in building TIGER.

Summary

Today's nationwide TIGER file is the backbone of the adoption of GIS in business geographics applications. TIGER frees business users from the drudgery of map digitizing and allows them to concentrate on applying GIS technology to business problems.

TIGER is the serendipitous by-product of a modern computerized census process. The U.S. Geological Survey put its energies into computerizing topographic maps designed by John Wesley Powell a century earlier, but the result (Digital Line Graphs) has not had anywhere near the impact that the Census Bureau's accidental by-product has.

TIGER's precursor, DIME, turned the traditional cartographic paradigm on its head. The classical cartographic process started with photogrammetry and careful scribing of linework, upon which annotation is later applied. In contrast, DIME started by recording all of the annotation, cycling through topological edits and corrections, and finally—almost as an afterthought—inserting digitized node coordinates. The coordinates, formerly the framework of the entire cartographic construct, are simply attributes of the nodes and shape points of TIGER's topological structure.

But the story of DIME is more than one of technology and introduction of the mathematics of topology into managing spatial databases. The DIME idea and its topological data structure were inevitable. DIME emerged by accident of history in New Haven in 1967 scarcely influenced by prior developments.

What is extraordinary in this story is not so much the technical innovation but the importance of key personalities in getting the innovation adopted: Caby Smith forcing the Census Bureau to abandon the ACG for DIME, Marvin White's timely and modest delivery of the technological key to making nationwide TIGER feasible, and Marx's leadership and confidence each made a crucial and effective contribution appropriate to the time.

Bibliography

Broome, Marx, Tomasi, et al. 1990. *Cartography and Geographic Information Systems,* 17 (1), entire issue.

Cooke, D. F., and W. H. Maxfield. 1967. "The Development of a Geographic Base File and its Uses for Mapping." *Proceedings of the Urban and Regional Information System Association (URISA)*. Washington, D.C.: URISA.

Marx, R. W. 1986. "The TIGER System: Automating the Geographic Structure of the United States Census." *Government Publications Review,* 13: 181–201.

U.S. Bureau of the Census. 1968–69. *Census Use Study Reports 1–12.*

U.S. Bureau of the Census. 1974. *DIME Comix.*

CHAPTER 5

Raster Based GIS

Nickolas Faust

Introduction

Raster based GIS is a major component of commercial and public domain GIS on the market today. Raster processing of spatially referenced data has many advantages for handling spatial information; however, significant disadvantages can also be identified when compared with vector data processing. This chapter outlines the historical development of raster based geographic analysis and investigates how the integration of imagery into GIS has influenced the general acceptance of the raster system.

Raster and vector formats are inherently different data structures for the storage of spatial information. Each structure tries to preserve as much information as possible related to spatial location available in the source data. The source data for GIS may be existing cartographic hard-copy output such as the 1:250,000, 1:100,000, or 1:24,000 scale map products that are available from the U.S. Geological Survey, existing digital vector files such as the Digital Line Graph series (DLGs), or raster images representing aerial photography or scan digitized map products. Traditional GIS applications involved a major amount of work initially in capturing spatial information from maps using a line digitizer system. Only recently has previously digitized map information become available to the GIS user.

Nickolas Faust is Associate Director of the Center for GIS and Spatial Analysis Technologies and is the head of the Image Analysis and Visualization Branch within the Georgia Tech Research Institute (GTRI). He is a principal research scientist at GTRI with more than 30 years experience in remote sensing and GIS analysis. He is cofounder of ERDAS, Inc. high tech spinoff company from Georgia Tech. He heads research in visualization and integration of spatial and temporal data and the development of new techniques for collaborative analysis of spatial data. He was previously an Aerospace Engineer/Physicist at NASA/JSC. *Author's Address:* Nick Faust, Georgia Tech Research Inst., Electro-Optics, Environment, and Materials Laboratory, 925 Dalney St., Baker Building, Atlanta, GA 30332.

A raster data structure for spatial information may be thought of as a computer photograph where each image picture element (pixel) has a value that is associated with a specific geographic location. Each pixel is associated with a discrete area on the surface of the Earth, and the area of one pixel is the resolution of the raster. A horizontal string of pixels in the image is an image row, and a vertical string of pixels represents an image column. The spatial location of any pixel in the image may be determined by finding the row and column within the image and relating that location to the geographic position of the top left corner of the image. There may be a number of layers of raster data that relay different information about the same spatial area. For example, multispectral image data is a number of layers that show the same ground area as seen through specific spectral filters. The size of a raster data set is found by multiplying the number of columns, the number of rows, and the number of layers by the computer storage associated with one data value. To represent accurately the information contained in a scanned map or a photographic image, very small pixel sizes are often needed, leading to very large data storage requirements. For raster data sets that were digitized from hardcopy products, the final pixel resolution in terms of ground units is a function of the original scale of the hardcopy product and the resolution of the scanner.

Vector data, on the other hand, includes spatial location information explicitly for every point digitized. Vector GIS data are usually represented as points, vectors, or polygons. For example, a road may be digitized from a U.S. Geological Survey (USGS) quadrangle map using a manual digitizing cursor by recording an x and y map location for each point. Sequential points represent a data vector that may be associated with certain attributes such as the road name, the state road number, and the width of the road. A polygon represents a closed area on the map that may be associated with other attributes such as area, perimeter, and the like. Other data structures such as arc and nodes were introduced to represent more effectively the topology of vectors enclosing areas. In recent times, vector data sets have been created using raster data as an intermediate product. A hardcopy map is scan digitized, and semiautomated tools, such as line following, are used to develop vector and polygon data sets. A significant amount of human interaction is currently necessary to clean, edit, and attribute the vector data.

Ideally, vector data more closely represents the smooth curvature of lines such as elevation contours of the existing hardcopy products, and therefore it is often preferred as a digital method of data storage. Raster data has a drawback—that at some point the user must choose a resolution of the image pixel. If the selected resolution is not very fine, a curved line on a map will have a jagged appearance in a raster data file. If users wish to change the resolution to capture more of the curvature, they must rescan the hardcopy product at a smaller pixel size. As the pixel size is decreased, the size of the resulting raster data file grows dramatically.

The use of satellite and aircraft imagery in GIS is becoming prevalent in commercial and public systems. Almost all GIS uses imagery, at least as an underlay for the plotting of vector data. Users relate well to imagery as a method of understanding the spatial relationships of GIS data. High resolution imagery is the method that allows update of GIS data to represent an area as it is today, rather than as it was when a map was compiled. Digital imagery is inherently a raster product, and, for it to be used effectively in vector GIS, methodologies must be developed to extract vector information rapidly from the raster gray scale or color images.

The battle between the proponents of raster and of vector has been raging since the earliest days of GIS development. Advocates and detractors of both raster and vector GIS have exhibited an almost evangelical fervor concerning the correct form for the analysis of spatial data. In 1979, for example, buttons were being passed out at the International Symposium on Computer Assisted Cartography (Auto-Carto) Conference that read "Raster is Faster." In reality, similar functions generally exist in both systems, making the choice between them complex and confusing for potential users. Only recently have systems been developed that facilitate the processing of data in both raster and vector forms. The GIS research community continues to address the issue of raster/vector integration.

Early GIS Development

Some of the earliest computer based GIS development occurred in Canada. This development was associated with the Canadian Geographic Information System (CGIS). CGIS began operations in 1964 and involved both vector and raster data structures (Chapter by Tomlinson). R. F. Tomlinson, of the Canadian Department of Agriculture, had performed a feasibility study for the development of a GIS for Canada that subsequently led to the operational development of CGIS. CGIS stored data in polygon form, but much of the initial data came from scanned maps (raster data) that were then subjected to raster to vector conversion routines.

In the United States, early GIS work included the development of the Minnesota Land Management Information System (MLMIS) from 1967 to 1971 (Tomlinson et al. 1975). MLMIS was a large-area Geographic Information System based on a gridded data structure (i.e., raster) with 40-acre cells. This structure is especially appropriate under the township and range U.S. Public Land Survey system. Interpreted aerial photography was used to provide land use information for Minnesota. MLMIS was implemented on a CDC mainframe. A second major system was developed by the state of New York and called the Land Use and Natural Resources Inventory of New York State (LUNR). LUNR was a raster system implemented on an IBM mainframe that employed a 1km x 1km cell size. LUNR was developed around a computer package called SYMAP (Synagraphic Mapping System) developed by the Harvard Computer Graphics Laboratory to manage grid cell based data sets. SYMAP was principally a data display and query system with limited analysis capability (Chapter by Chrisman).

A third pioneer GIS was implemented starting in 1970 at the Oak Ridge (TN) National Laboratory (ORNL) (Chapter by Dobson and Durfee) as a part of a project with the National Science Foundation-Regional Environmental Systems Analysis (NSF-RESA) program. The Oak Ridge Regional Modeling Information System (ORRMIS) was intended to incorporate modeling for planning and environmental purposes within a spatial information system context implemented on an IBM mainframe. ORRMIS was based on a fine resolution raster structure of 2.68

acres, or 3.75 arc seconds of latitude and longitude. Data were gathered at various resolutions and stored in a nested hierarchy (Tomlinson et al. 1975).

Harvard helped set the early phase of raster-based GIS origins (Chapter by Chrisman). A complex geographic modeling system known as GRID (Graphic Display of Rectangular Grid Information) was developed in 1968 at the Harvard School of Design in the Department of Landscape Architecture and the Laboratory for Computer Graphics and Spatial Analysis. GRID was a simulation model for determining the distribution of human population given geographical, physical, and political constraints. Based on an increase in population for a candidate area, the model would use a gravity function to determine the distribution of jobs and housing. Harvard had received a research grant from the National Science Foundation to apply GRID modeling for assessment of the practicality and effectiveness of zoning regulations in influencing the population distribution in an area including the Boston suburbs.

GRID was a sophisticated FORTRAN based modeling system that ran only on IBM mainframes at that time (Sinton and Steinitz 1971). In addition to GRID's use in the NSF research project, it was also utilized in teaching within the Department of Landscape Architecture. A student might have to develop his own computer routines to interface to the large, cumbersome GRID to be able to perform required class assignments. Students had to invest a considerable amount of time discovering the intricacies of the GRID simulation model, learning how to program in FORTRAN, and keep up with their classwork. In the early 1970s, interactive processing did not exist on mainframes. Normally, a user would study results from an analysis computer run from the previous night, determine what changes were necessary in the computer code or the data sets, personally keypunch change cards, insert those changes into his or her card deck, and submit another run for that night. Computer printouts were the only output mechanism available for the runs, so detailed print statements were used to track program flow and to isolate errors. Initial attempts were made to use computer printers to produce maps of spatial information in terms of character maps and gray level maps created by retarding line feeds on the printer and overstriking characters. These were often hand-colored for visual appeal.

During this early 1970s time period, computer memory was very expensive, limiting even large systems for storage addressable by user programs. This impediment had a large impact on the size and resolution of geographic data sets that could be analyzed at any given time. This was the case because the programs employed required that the entire data set to be analyzed reside in memory at one time. Project areas for research and teaching were, therefore, limited in area or grid size. If researchers wanted to analyze a large area, they would have to make choices such as the reduction in the spatial resolution of grid cells, or they would need to undertake the complex process of dividing a geographic area into multiple study areas. The painful logistics associated with large-area analysis limited the routine development of comprehensive geographic data sets. For spatial functions such as proximity analysis, the separated subsets of a geographic region would have to contain overlap on all sides so that these functions would be meaningful. This limitation made a complex situation even more complex, and it was difficult to determine how the interaction should occur between subregions of the larger database.

As a result of these and other factors, graduate students in the School of Design rebelled in 1973 and refused to use the GRID program (Chapter by Jordan and Rado). They contended that they were in graduate school to learn about Landscape Architecture and Regional Planning, not there to learn FORTRAN programming and complex simulation models. The rebellion led to the development of Conversational GRID (CONGRID) (U.S. Forest Service–Southern Region 1976) and subsequently the first interactive geographic analysis tool, Interactive Manipulation GRID (IMGRID). CONGRID was paving the way for interactive analysis by using a conversational language for the setup and execution of models, but time-sharing computing was just becoming available, and most functions were still required to be preselected and nested in machine specific job control languages. IMGRID was still written in FORTRAN, but functions were developed for the most used GRID analysis techniques. IMGRID was designed with a menu/keyword structure that allowed students to select the desired technique by a simple and easy to understand name. Initially, IMGRID was still run using punched card decks, but in fact it was well suited to work in an interactive computing environment.

IMGRID was the forerunner of all raster GIS, although similar functions were available in early image processing systems (Chapter by Estes and Jensen). IMGRID was initially developed by David Sinton working under Carl Steinitz, so that the students in Landscape Architecture in the School of Design could use a simple, key word interface tool to test out concepts of spatial analysis on actual data sets. The IMGRID program was made available to researchers and teachers of landscape architecture. Some of the graduate students with a strong background in planning and an ability for software development finished at Harvard and used IMGRID as a starting point in their development of academic, public domain, and commercial software systems. A number of these students (Jack Dangermond, Lawrie Jordan, Bruce Rado, Dana Tomlin, and Joe Berry) hold high positions today in academic and commercial enterprises. IMGRID embodied functions that are still the core of modern raster GIS, and the menu/keyword interface was ahead of its time in user interfaces. However, the original IMGRID, implemented on an IBM mainframe, had severe limitations as to the size and/or resolution of geographic databases it could accommodate. With IMGRID, databases covering area of one USGS quadrangle were traditionally analyzed with grid cell sizes from four to ten acres. Each database employed in an analysis could have almost any number of layers or themes such as land use, elevation, transportation, and the like. For larger area coverage, the border cell replication strategy was often used.

Data sets for IMGRID were initially developed by creation of a grid overlay in plastic that would exactly cover a USGS 7 1/2 minute quad. Grid cells were sized in equal arc second representation. A student would transfer base information to a quad and subsequently geocode the information manually using the plastic overlay. Later, the student would keypunch the cell values and read the data into IMGRID, after which base data and analysis maps could be printed out on a computer printer using the overstrike capability of most high-end computer printers. The normal computer printer of the time had a rectangular character set that was longer in the vertical than in the horizontal direction, with a proportion of ten to six. For the latitude of the Boston area, this was good because the cell proportion approximated the proportional difference between lengths of equal latitude and longitude cells. However, the same technique used at lower or higher latitudes would result in spatially or geometrically distorted maps.

IMGRID follows the strategy outlined by Ian McHarg in his book *Design with Nature* (McHarg 1969). In this book, a process is described for creating layered overlay sets and using a photographic technique for combining the layers into "suitability/capability analyses"(Steinitz et al. 1976). Hand-drawn overlays for suitability analysis date back at least to 1912 when they were used in the town of Billerica, MA (Manning 1913). Concepts for variable weighting of interpreted map layers to produce composites, as well as the key concept of the spatial proximity value, were subsequently explored (Abercrombie and Johnson 1922). Suitability maps were created using both proximity information and value weightings.

Initial implementation of IMGRID was in the traditional method of a single computer program with all functions residing within one executable layout. The size of the programs as well as data memory required by them was an impediment to the early transfer and conversion of IMGRID to other computer systems. In the mid 1970s, Georgia Tech restructured a version of IMGRID to be modular, with each keyword being associated with a separate subroutine program. A menu driver system allowed the user to select options or keywords, and subsequently the selected function module was executed. Each module contained IMGRID input/output functions as well as the algorithms necessary for analysis. This modularization of functions paved the way for implementation of IMGRID functionality on computers other than mainframes.

Extensions to the functionality of IMGRID were made based on contributions of Dana Tomlin, Joe Berry, and other motivated graduates of the design school in developing the Map Analysis Package (MAP). MAP includes many features for easy raster manipulation and can still be found at many universities as an inexpensive way to teach GIS.

Other raster and vector GIS development occurred in the early 1970s with work by Roger Tomlinson in Canada and Marble at State University of New York at Buffalo (SUNY-Buffalo). Duane Marble and his students, including Carl Reed, developed a GIS independently. The Map Overlay and Statistical System (MOSS) was developed along with the Wetlands Analytical Mapping System (WAMS) for the U.S. Fish and Wildlife Service in the middle 1970s by Autometrics, Inc. MOSS was primarily a vector based GIS with some grid capability. Whole polygons were represented as vectors, giving rise to the "sliver" problem in polygon description. WAMS included a relatively sophisticated digitizing system for data entry of GIS information. Autometrics, Inc. subsequently developed the Analytical Mapping System (AMS) which was turned into a commercial product.

During this period of the early 1970s, the competition between vector and raster GIS grew in intensity. Harvard Computer Graphics Lab developed ODYSSEY, a vector based GIS capability. Jack Dangermond and Scott Morehouse of Environmental Systems Research Institute (ESRI), early users of a gridded (raster) GIS representation, developed the Polygon Information Overlay System (PIOS), a vector based system used to manage GIS information for the Comprehensive Planning Organization, county of San Diego, CA. PIOS was replaced in the early 1980s with ARC/INFO, a topologically based GIS that remains the market leader for systems today. Under the direction of David Sinton, who had left Harvard, M & S Computing (now Integraph) moved from an electronic drafting (vector) company to a company with a powerful vector based GIS. For forestry applications, a specialized system was developed by Comarc, Inc. which allowed interactive management of forest stands in a vector GIS.

Early Remote Sensing and Image Processing

Another powerful technology was also being developed during the 1970s. Image Processing and Analysis were being used to analyze photographs of Earth's surface. In 1972, the United States launched the first Earth Resources Technology Satellite (ERTS) (Chapter by Estes and Jensen on history of development in this area) that dramatically changed the way we perceive the natural environment and human impact on that environment. Earlier photographs from the Mercury, Gemini, and Apollo space missions and the TIROS and NIMBUS meteorological satellites had begun to awaken the science community to the potential of satellite photography for monitoring Earth's environment. ERTS was a major commitment to provide relatively high resolution, repetitive, digital multispectral imagery on an ongoing basis. ERTS became a source of information that could provide data on a regular basis to scientists studying Earth's fragile environment and the anthropogenic effects on that environment. Applications of ERTS data ranged widely from oceanographic studies to land cover determination with disciplines in many fields, including geology, geography, biology, forestry, and the like. In 1976, the name ERTS was dropped in favor of Landsat. This satellite series continues today with Landsat 7 scheduled for launch in 1998.

Digital images of Earth's surface (from air or satellite) represent geographic areas in a gridded manner very similar to raster GIS. An image may be digital or analog, but analog images (photographs) can be easily scanned into a digital file. An image may consist of a number of spatially coregistered layers of information that correspond to information recorded by a photographic emulsion as converted by a detector into electric signals and recorded on film or in digital form. Sunlight consists of many wavelengths of energy that can be selectively recorded by the use of filters or specific detectors sensitive to only small portions of the electromagnetic spectrum. Other image layers may be obtained by recording the self-emitted Earth information in the ultraviolet and mid- and long-wave infrared portion of the electromagnetic spectrum. All of these techniques are regarded as passive remote sensing since the recording device only records reflected light from a source such as the sun or energy emitted from Earth. Imagery data also may be recorded using active remote sensing, whereas energy in longer wavelength portions of the spectrum can be provided by a manmade source such as a radar. In this case, the remote sensing system provides its own source of illumination and records reflected information from Earth's surface.

Inherently, remote sensing data does not necessarily have an absolute geographic position recorded at the time of data acquisition. Each point within an image (a pixel, or picture element) is located with respect to other pixels within a single image, but it must be geocorrected before it can be accurately inserted into a GIS. A major advantage of remote sensing information is the ability to provide a basis for update of GIS information on a repetitive and inexpensive basis. While this chapter does not go into great detail on the sources, analysis tools, and application areas of remote sensing data in relation to GIS, the integration of remote sensing and GIS is a major research area that is being studied by the National Center for Geographic Information and Analysis (NCGIA) in Research Initiative 12 (Star, Estes, and McGwire 1997).

The early history of the automatic processing of digital images is of interest because of the similarities in the spatial nature of raster GIS and remote sensing data. Image processing of multispectral remote sensing data, in particular, will be examined and later will be related to raster GIS functionality.

Digital image processing techniques were developed in the mid 1960s by a number of universities, research institutes, and private companies primarily under sponsorship of the National Aeronautics and Space Administration (NASA) (Chapter by Estes and Jensen). Research into digital recording of portions of the electromagnetic spectrum had been conducted in support of the Department of Defense since the development and use of radar systems during World War II. During the Korean conflict and the first stages of the Vietnam War, technology was being developed to sense information in various regions of the spectrum. Digital scanning systems which became available over that time period allowed the accurate rapid scanning of the surface of Earth and the recording of that information in a number of spectral regions into digital images. NASA pioneered the use and transfer of this technology for the assessment of environmental and natural resources applications.

During the late 1960s, the Willow Run Laboratories of the University of Michigan, which during the Vietnam era became the independent, not for profit, Environmental Research Institute of Michigan (ERIM), developed an optical-mechanical multispectral scanner system that was successfully used for early detection of corn blight in the corn belt states of the United States. At Purdue University, David Landgrebe and Phil Swain established the Laboratory for Agricultural Remote Sensing (LARS) and developed the image processing software system LARSYS which was used in the analysis of the multispectral image data for the program (Chapter by Hoffer). Indiana, Ohio, and Illinois were suffering under a blight that had significant implications for the agricultural stability of the region and the ability of the country's breadbasket to provide adequate food to the nation. The LARSYS system was developed and run on an IBM mainframe computer sponsored by NASA at Purdue. Many of the techniques in use today in the analysis of remote sensing data were developed as a part of the LARSYS package. Multispectral image processing techniques developed as a part of this system generally regarded each image pixel as having an N channel vector associated with it that was characteristic of a type of vegetation or material on Earth's surface. Statistical parameters were derived for groups of image pixels that could be delineated by a human on a computer printout or a sophisticated color display system. These pixels were used to determine a mean value and a variance in terms of digital gray scale values in each of the spectral regions for particular categories (such as tree type) that could be visually identified by a user both in the digital image and on the ground by field work. Many early image processing techniques fundamentally regarded each pixel as being independent of any other image pixel when a decision process was used to categorize automatically each image pixel for a given image. This concept of pixel independence is important in the development of low level raster image and GIS analysis.

One characteristic of multispectral remote sensing data is the large volume that must be accepted, analyzed, and produced to provide enhanced imagery and categorical products which could be used to assess, for example, vegetation health, existence of valuable minerals, or maps of land cover for discrete areas of the Earth (Jensen 1996). Memory constraints on the largest of mainframes required a concept for processing that would not require the computer to have all data from all spectral channels and all lines and columns of imagery resident at the same time. The concept of the independence of each image pixel, described above, allowed an organized approach to the analysis of very large image data sets. Images may be considered as a large matrix of data with N columns and M rows. Each row of image data may be considered as a "line or raster" from a continuous strip of an image. This representation of image data is satisfying since the methodology used in acquisition of such data follows the same pattern. An optical-mechanical, multispectral scanning system acquires data across a single scan line from left to right or right to left across its field of view. As the satellite or aircraft platform moves forward along its prescribed path, the scanning system creates a continuous strip image in the direction of flight. Practical implementation of scanning systems actually has a scanner check multiple lines at a time to avoid any possibility of any data being missed on the ground through scanning and flight parameters.

During the late 1960s and early 1970s, several special purpose image processing systems for the analysis of remote sensing data were developed by private enterprises. Bendix Corporation created a multispectral scanning system with 24 channels and a computing and display system for the analysis of the multispectral data. The software system developed was known as the Multispectral Data Analysis System (MDAS). MDAS was based on a minicomputer system, not a mainframe, and was one of the first uses of newly-developed minicomputer technology for civilian image processing applications. MDAS also had a high quality color display system for viewing and analysis of the data. Display was based on cathode ray tube (CRT) technology that allowed the presentation of multiple bands of multispectral image data on either the red, green, or blue color guns of a color monitor. The raster nature of image processing allowed the development of sophisticated analysis tools on a relatively small computing system.

During the same period, Richard Economy and General Electric created their special purpose image processing system called Image 100. This system was also based on minicomputer technology using a specially designed computing system and unique, high speed image processing hardware to allow rapid analysis of imagery. Image 100 was made commercially available for approximately $1,000,000. Even though it was ahead of its time in comparison to other multispectral image processing systems, the cost of Image 100 was prohibitive for most potential users.

The Jet Propulsion Laboratory (JPL) with Fred Billingsly, Al Zobrist, and Nevin Bryant, under NASA funding, was in process of developing an image processing system which had been developed for the analysis of medical images (i.e., for chromosome counts) and was being used for the analysis of image data sent back by unmanned probes of the solar system. The Video Information Communication and Retrieval/Image Based Information System (VICAR/IBIS) was a sophisticated analysis and display system that was implemented on a mainframe. It allowed for the enhancement, geometric correction, and mosaicking of large numbers of digital images. Functions within VICAR/IBIS included raster processing tools as well as a relational database system. JPL's raster system became the first major research and instructional GIS and image processing system.

Image processing systems led the way in minicomputer implementation of analysis tools for raster data. Early image processing systems operated on imagery with little emphasis on the accurate geographic positioning of the images. Image registration was more important than geometric rectification. Tools were developed to locate image pixels between images but not necessarily to site the geographic coordinates of the pixels in the image. The image itself was important, not its geographic positioning. VICAR/IBIS was used by JPL scientists to enhance and mosaic large numbers of digital photographs taken during planetary missions, and sophisticated mosaicking and feathering logic were designed to facilitate these operations. For many years, VICAR/IBIS had some of the most sophisticated rectification and registration algorithms of any image processing system. As VICAR/IBIS was used more and more for Earth resources, GIS data became important for interpretation and was converted initially from map-based coordinates to an image-based coordinate scheme.

Jack Estes and his colleagues, including Earl Hajic, Dave Simonette, Jeff Dozier, and Alan Strahler, at the University of California at Santa Barbara (UCSB) developed course curricula in remote sensing, working closely with NASA in the further research applications of remote sensing data. The University of Kansas, with significant expertise in radar mapping by Dave Simonette before he moved to Santa Barbara, Stan Morain, Tony Lewis, Roger McCoy, Farouz Ulaby, and John Rouse, developed software for the analysis of microwave and radar data. The Kansas system was called IDECSS for Image Discrimination Enhancement, Combination, and Sampling System. This system was originally developed as a part of a doctoral project by George Dalke for the Army Engineer Topographic Laboratories (AETL). Development was assisted by researchers including Jack Estes, who was getting his Ph.D. in geography at UCLA while working in the School of Forestry at the University of California at Berkeley. Estes assessed the applications of the system.

By the mid 1970s, the premier system for the analysis of remote sensing data was the Interactive Digital Image Manipulation System (IDIMS) developed by Jim Nichols and others at ESL, Inc. This raster system was based on minicomputer technology combined with special purpose array processors to provide sophisticated image processing tools for the handling of multispectral image data. Most of the major oil, gas, and mineral exploration companies purchased IDIMS as a basic tool for their work. NASA Goddard Space Flight Center used IDIMS extensively, and it was the precursor to the Landsat Assessment System (LAS). The principal drawback to IDIMS, its price tag of more than $1.5 million, limited its use outside the mineral exploration community. Another drawback to broader use of the system at the time was the proprietary nature of its software.

The common thread to the special purpose image processing systems was a methodology based on the independent analysis of raster data—one line of individual pixels or cells at a time. This methodology allowed the implementation of complex image analysis functions on minicomputer systems with a limited memory size.

In 1980, Earth Resources Data Analysis Systems (ERDAS) Inc., a high tech spinoff company from Georgia Tech, developed the first integrated image processing system and raster GIS on a microcomputer (Chapter by Jordan and Rado). ERDAS was formed in 1978 by scientists and natural resource professionals from the state of Georgia and Georgia Tech. ERDAS initially developed a raster implementation of the functionality in the IMGRID package on a minicomputer and combined it with image processing functionality modeled after the LARSYS system and the NASA developed Algorithm Simulation and Test Evaluation Program (ASTEP). The late 1970s saw the development of microcomputer systems such as the Zilog Z80 and the Motorola 6800, which were the basis of the hobby computer trend. Software developers formed clubs that allowed the sharing of software for games and useful utilities. Microcomputers at that time were characterized by an 8-bit word length and a small capacity for direct memory access compared to the minicomputer systems and mainframes of the day. Programs for the microcomputers were generally written in BASIC or other interpretive languages, but a trend had started toward compiled BASIC and FORTRAN languages. The use of FORTRAN on microcomputers allowed the development of the type of complex programs necessary for image processing and GIS analysis. The individual IMGRID like functions were coded in FORTRAN as separate executable programs that were chained together with a menu/help file interface. The system initially included two 8-inch floppy disks, 64 kilobytes of memory, and a specially modified color television display driven by a color graphics board with 256 by 256 elements and 24 bits of color depth. A light pen was used as an interactive pointing device, and a color printer was used to output scaled map products.

Recent Trends

In the past few years almost all image processing systems have included raster GIS and at least some vector GIS capabilities. Similarly, almost all traditionally vector GIS has included the capability to bring in imagery and other raster GIS data. The ability to have totally integrated vector and raster functionality, however, is still in its infancy. A fully integrated system would have the capability of including vector or raster GIS information directly in the performance of image processing tasks such as multispectral image classification as well as the ability to use image analysis functions directly within a raster or vector GIS analysis. Initial attempts are being made within modeling functions to transform automatically between data types in an analysis without overt action required by the user. In the multispectral classification process, vector data such as city boundaries or historical land use polygons may be used to direct a pattern recognition process to consider only certain classes that are expected to occur within an urban environment. Grass areas within a city boundary, for example, would be categorized as lawns or parks as opposed to pasture that might be expected within a rural area. If accurate historical land use information were available, a smart classifier would be able to use that information to assess the likelihood that a geographic area has changed from one land use category to another. For example, it is unlikely that an urban concrete category will change to a forest category. Expert classifiers are being developed that use vector and raster GIS data as part of a knowledge base that will promote more accurate classifications of multispectral imagery.

GIS should directly be able to take imagery, raster GIS, and vector GIS information and transparently perform suitability analyses that include image processing functions such as the computation of the Normatized Difference Vegetation Index (NDVI) without needing to have those functions precomputed by an image processing system. Overlays and intersections of raster and vector data sets should also be able to be performed without restriction. These types of functions are being considered whereas vector to raster and raster to vector conversions are being hidden within a model.

High resolution imagery that is now becoming available commercially will provide a major data source for image and GIS analysis. Numerous commercial companies are poised to provide one to three meter multispectral satellite imagery on demand. This high resolution image data will provide a rich source of imagery for the update and creation of GIS data sets. These will be inexpensive and far more accessible than previous satellite data. The high resolution of these data will go a long way toward the accurate representation of manmade and natural boundaries that has always been a point of contention between raster and vector proponents. The volume associated with the imagery data, however, will be large. Luckily the speed of computing and the price of memory in computers has radically changed over the years to be able to provide significant analysis capability at the desktop level.

One of the most important changes that has occurred to raster and vector GIS recently is the move toward the desktop. Image processing systems historically have been focused on providing extensive analysis capability to a small segment of university, government, and military users. Traditional vector and raster GIS enjoyed a larger market, but this was still limited to professionals and educators. The power of computing and the advanced operating systems now available on Personal Computers (PCs) have allowed a rapid expansion of the capabilities that are available to a much wider potential user base. Major initiatives are being developed by GIS and image processing vendors that attempt to exploit this much larger market by providing tailored and streamlined products which provide a somewhat more limited capability than their scientific counterpart products but also provide the most widely used GIS and IP functions with an easy to use interface. These products are also priced significantly lower than the unix based scientific counterparts. Integration of raster and vector GIS with relational database tools such as ACCESS and spreadsheets such as EXCEL are occurring in a Microsoft office environment.

Another major change is the explosion of the use of the Internet as a medium for access and potential analysis of spatial data. Major GIS and image databases are being made available over the Internet, providing a rich source of shareable data for the general public. Several states have created GIS data clearinghouses that allow interactive browsing of spatial data sets developed for the states as well as direct download capability of those data sets in traditional raster or vector formats. An example is the Georgia GIS Data Clearinghouse at http://gis.state.ga.us. A large amount of spatial data over Georgia is available for view and download. Currently, the data may be downloaded as ARC/INFO E00 files. Digital Orthophotos of the state at approximately one-meter resolution are available in raster form for state university and government users.

Several GIS vendors are investigating the provision of GIS analysis capability over the Internet with some provision for charges for the service.

Conclusions

Raster based GIS has experienced a resurgence in interest over the past several years. Major GIS companies have developed competitive products that have extensive functionality for analysis of spatial data sets. Vector GIS, once the cornerstone of any comprehensive GIS package, is still strong, but a new acceptance as to the role of imagery and raster data in analysis systems has made the raster processing of spatial data especially important. Since the majority of the costs in developing a GIS lies in the data acquisition, and since map data is inherently out of date by the time it is entered into a GIS, image data is becoming acknowledged as the only method capable of keeping GIS information "up to date" and "accurate." Image data is currently used as an underlay to vector GIS information, but integrated analysis using both data structures is necessary to achieve the full functionality of GIS. The computer handling of raster data structures is efficient and powerful, and newly developed parallel processing techniques will continually enhance the performance advantages that raster GIS currently has over vector based systems. Raster GIS functionality is similar to that of vector GIS, but the acceptance of analysis products with jagged features is yet to supplant the smoother vector representation of a final output product. Raster to vector and vector to raster conversions are now commonplace, but a true integrated GIS would allow functions to operate completely independent of the data structure of the input GIS data.

Bibliography

Abercrombie, P., and T. Johnson. 1992. *The Doncaster Regional Planning Scheme.* London: The University Press of Liverpool, Ltd., Hodder & Stoughton Ltd.

Jensen, J. 1996. *Introductory Digital Image Processing.* Englewood Cliffs, N.J.: Prentice-Hall.

Kauth, R. J., and G. S. Thomas. 1976. "The Tasseled Cap—A Graphic Description of Spectral-Temporal Development of Agricultural Crops as Seen by Landsat." *Proceedings, Symposium on Machine Processing of Remotely Sensed Data.* West Lafayette, IN: Laboratory for the Applications of Remote Sensing.

Manning, W. 1913. "The Billerica Town Plan." *Landscape Architecture,* 3 (3): 108–118.

McHarg, I. 1969. *Design with Nature.* Garden City, N.Y.: The Natural History Press.

Sinton, D., and C. Steinitz. 1971. *Grid Manual, Version 3.* Laboratory for Computer Graphics and Spatial Analysis. Cambridge, MA: Harvard University.

Star, J., J. Estes, and K. McGwire, eds. 1977. *Remote Sensing and Geographic Information Systems Integration.* Cambridge University Press.

Steinitz, C., P. Parker, and L. Jordan. 1976. "Hand-Drawn Overlays: Their History and Prospective Uses." *Landscape Architecture*, 66 (5): 444–455.

Tomlinson, R. F. 1963. *Feasibility Report of Computer Mapping System.* Contract Report from Spartan Air Services Ltd. Project 14007, Agricultural Rehabilitation and Development Administration. Ottawa, Canada: Canada Department of Agriculture.

Tomlinson, R. F., H. W. Calkins, and D. F. Marble. 1975. *Computer Handling of Geographic Data.* Unesco Press.

U.S. Forest Service–Southern Region. 1976. *CONGRID and DBMANG Users Manual.* Atlanta, GA.

CHAPTER 6

Investments in Personal Computing

Lawrie E. Jordan, III, and Bruce Q. Rado

Introduction

Some readers may fondly remember the 1960s TV series, "The Rocky & Bullwinkle Show," featuring among other things a short vignette known as "Peabody's Improbable History." Mr. Peabody (a canine inventor) and his young assistant, Sherman (and their dog), had an ingenious time-travel device known as "The Way-Back Machine." They'd dial in a particular year in history, push a button, and presto! they were there! Events of long ago were instantly relived, with clarity and context, not to mention a healthy dose of political satire. When the authors of this chapter were contacted about contributing to a book on the history of GIS, we found ourselves delighted with the opportunity to participate, yet wishful that we had our own "Way-Back Machine" to recall memories of some of the early key events that led us into our current profession.

Background

Like many others who have been involved with this GIS profession for 20 years or more, our history takes us "way back" to the Harvard University Graduate School of Design (GSD) during the early 1970s. The pioneering efforts of people like Howard Fisher at the Harvard Laboratory for Computer Graphics and Spatial Analysis; Carl Steinitz, a true visionary and cornerstone for the GIS Industry's existence today; and Jack Dangermond, founder of Environmental Systems Research Institute (ESRI) and former GSD student, are well covered elsewhere (Chapters by

---

Lawrie E. Jordan is president and founder of ERDAS, a leading remote sensing and GIS software development company. He was educated at the Harvard University Graduate School of Design where he inspired the development of user interfaces for GIS.

Bruce Q. Rado is vice president and cofounder of ERDAS. He received his graduate degree at the Harvard University Graduate School of Design.

Authors' Address: ERDAS Inc., 2801 Buford Highway, Suite 300, Atlanta, GA 30329.

Chrisman and by Neiman and Moyer). However, a little-known event in the form of a "student revolt" at Harvard occurred during that time which later proved to be an instrumental milestone in the adaptation of GIS technology to personal computing.

The setting was the school year 1973-74 in Cambridge, Massachusetts, and the primary computing environment consisted of IBM 360/370 Mainframes located at the university's Computer Center running the Geographical Interpolation (GRID) and Synagraphic Mapping System (SYMAP) software. Harvard Landscape Architecture students participating in a large National Science Foundation-sponsored project were actively engaged in building digital databases by hand, using the grid-cell encoding method. Vegetation and other data layers were created in part using air photo interpretation techniques and guidance from another pioneer in the field, faculty member Doug Way (now at Ohio State University). These hand-built databases would consume endless numbers of student all-nighters and untold volumes of coffee to construct, in preparation for the basic GIS inventory and mapping phase. Students would jokingly accuse zealous faculty members overseeing the data collection phase as having gridded contact lenses. One industrious student who was always searching for ways to spend less time in the studio and more time partying stumbled across an entirely different way of collecting vegetation data that did not require manual encoding—it was already digital, and it came from something new called an imaging satellite that had just been launched by the National Aeronautics and Space Administration (NASA). The spacecraft was known as the Earth Resources Technology Satellite (ERTS, also known as Landsat 1). But that's another story for a different chapter, told in part in Chapter by Jensen and Estes.

Once the GIS inventory phase was complete, problems arose when attempts were made by students in the class to execute analytical runs on the mainframe. This was because the GIS models themselves and the user-interface consisted of a deck of IBM cards key-punched in FORTRAN and its arcane job control language. Often a batch run was submitted in card-deck form at the Harvard Computing Center counter followed by an overnight computer run. Those students who were fortunate received a black and white chain printer map many hours later (Figure 6.1).

However, the more frequent occurrence was the discovery of a user key-punch error or a card out of order that required resubmission and more waiting. One particular student, who shall remain nameless, was terror-stricken once when he overheard a highly distraught senior professor say that some student's GIS analysis model (his) sent the university's mainframe into an endless "do-loop tailspin" and consumed several hundred dollars of the school's limited computing budget before finally crashing the machine in the middle of the night along with all the other students' runs. To add insult to injury, a small number of vocal students who chose to ignore the GIS curriculum and instead pursue the department's traditional courses in Landscape Architecture (planting plans, horticulture, residential design, etc.) never seemed to miss an opportunity to tell the rest of us that there was no future job market for GIS. Instead, their advice was to "forget computers and get back to the basics" of memorizing the botanical names of shrubbery. Each crash of the mainframe always seemed to trigger their "See, I told you so . . ." chorus. Such was

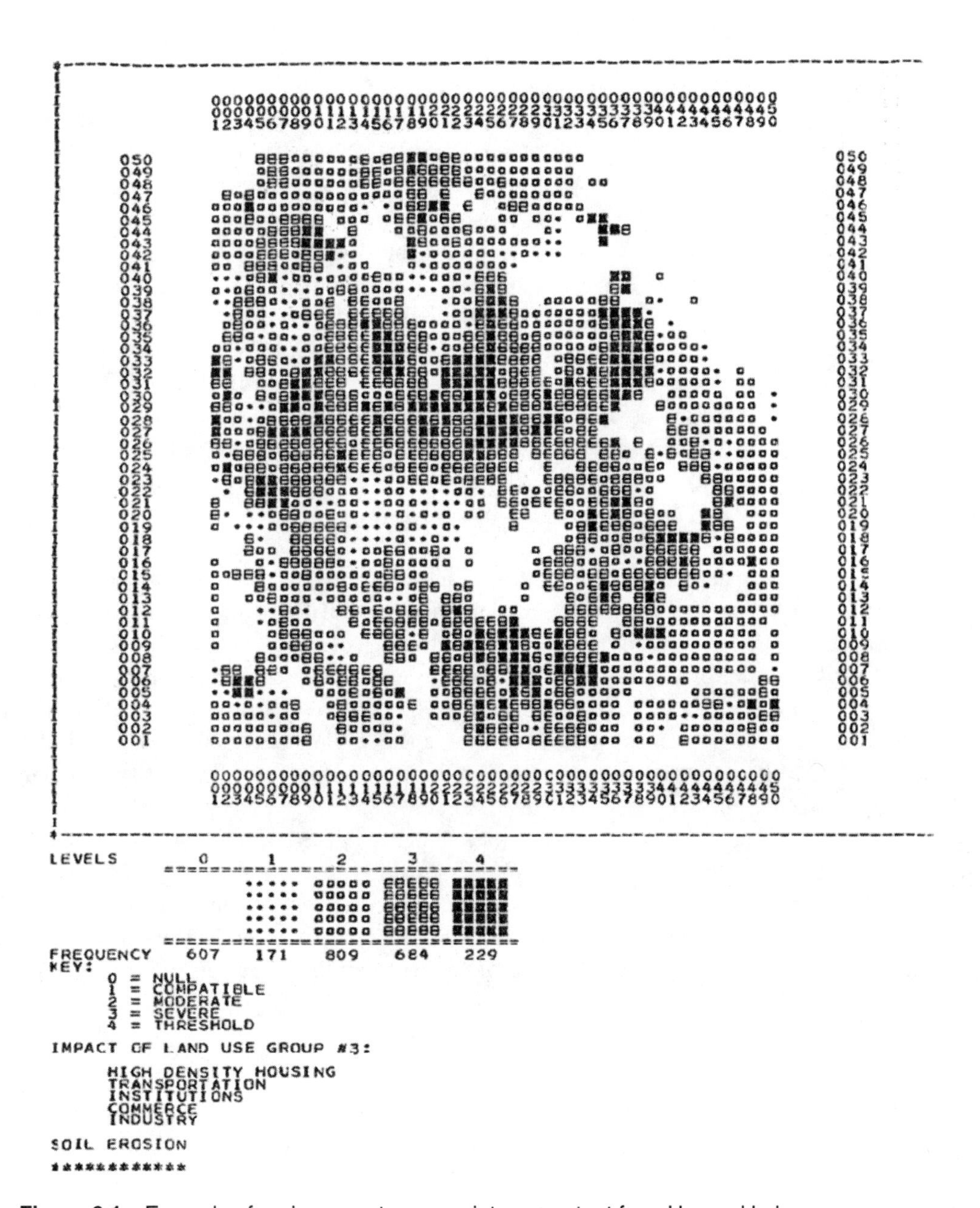

Figure 6.1 Example of early computer overprint map output from Harvard Lab.

the state of interactive computing way back then. Occasional partying, by comparison, certainly seemed to have some major advantages over spending the night with a 370.

Faced with these frustrations and the added burden of having to learn FORTRAN programming, the students in the Landscape Architecture GIS Studio soon revolted against the technology of data handling, boycotting classes and demanding that the faculty investigate alternative approaches for making computer-based geographic analysis easier and more understandable. The faculty listened and responded promptly and decisively under the leadership of Department Chairman Charles Harris and Professor Carl Steinitz. They tapped the technical genius of another early pioneer and earlier student of Steinitz, Professor David Sinton (now of Intergraph Corporation), along with talented student assistants, such as Dana Tomlin and Chuck Killpack, and before long a new program authored by Sinton called IMGRID (Information Manipulation on a Grid) was born. IMGRID operated on English keywords that were generally descriptive of their operation, such as OVERLAY, SEARCH, RECODE, etc. (An actual example is shown in Figure 6.2.) Students could use IMGRID to quickly construct an understandable GIS model, run it, modify it, and understand the consequences of parameter changes on the results.

The student revolt, with the faculty's help, had succeeded in accomplishing its objective, and soon the classrooms were filled with students generating IMGRID-based GIS analysis maps and debating the pros and cons of one team's resource analysis plan versus another. Dedicated classmates like Bruce Rowland, now head of GIS systems at the Tennessee Valley Authority, spent days at a time in the school's photo lab to create beautiful color analysis maps for the NSF final review, which was a resounding success. It was a very exciting and productive time, and it also seemed to be a much more appropriate use of those big tuition payments we were sending in. We also noted that the traditional "I told you so" crowd had become conspicuously silent and instead began hovering around the GIS studio late at night. Although key-punch cards were still required, a major step forward had occurred for non-programmers to develop early GIS resource analysis techniques using keywords without getting bogged down with complex programming languages. This simplified software interface would later become one of the key concepts that set the stage for the future commercial linking of GIS with personal computers. It is also worthy to note that the GSD students' admiration for the inspiration and vision provided by Carl Steinitz in those early classes continues today, some 25 years later.

Commercial Hardware Developments

Software notwithstanding, a major barrier to personal computing within GIS still existed in the hardware arena. A quick check with Mr. Peabody's "Way Back Machine" would reveal that in the mid to late 1970s several of today's most familiar signposts in the GIS landscape did not even exist. Sun Microsystems, ARC/INFO, standards such as X-windows, MOTIF, Postscript, the notion of networked UNIX workstations, "open systems," and even the IBM-PC itself were still only glimmers in the eyes of their creators. Instead, the computing environment had grown from mainframes to include a whole series of "mini" computers such as the DEC PDP, PRIME

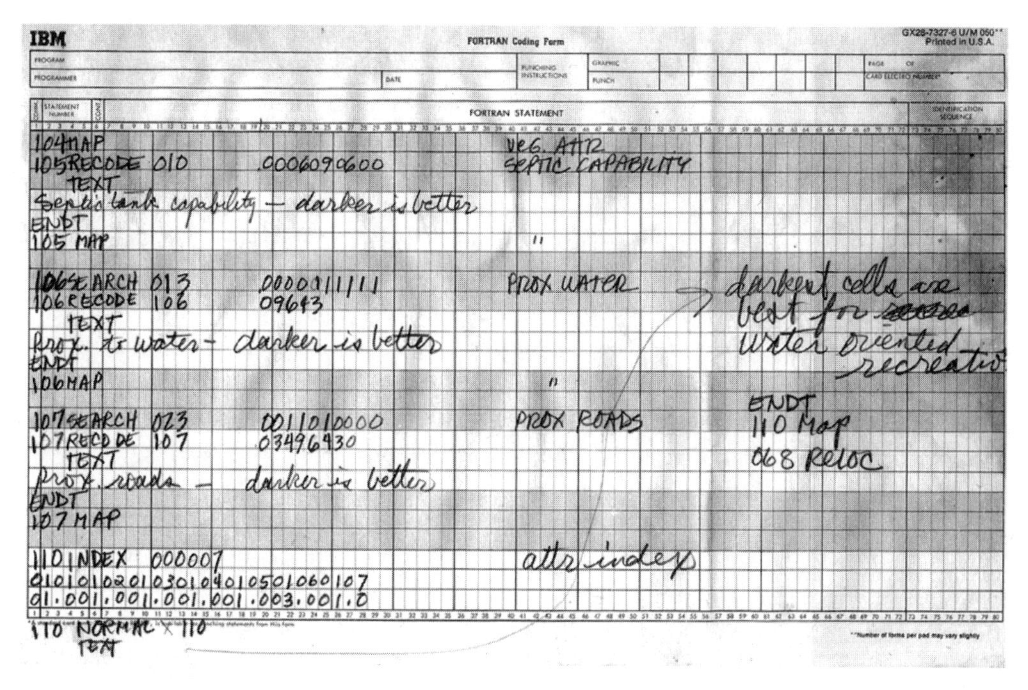
IBM FORTRAN Coding Form GX28-7327-6 U/M 050** Printed in U.S.A.

FORTRAN STATEMENT

104MAP
105RECODE 010 0006090600 VEG. ATTR
SEPTIC CAPABILITY
TEXT
Septic tank capability — darker is better
ENDT
105 MAP "
106SEARCH 013 00001111111 PROX WATER
106RECODE 106 09643
TEXT
Prox. to water — darker is better
ENDT
106MAP "
107SEARCH 023 0011010000 PROX ROADS
107RECODE 107 03496930
TEXT
Prox. roads — darker is better
ENDT
107MAP
110INDEX 000007 attr index
0101010201030104010501060107
01.001.001.001.001.003.001.0
110 NORMAL 110
TEXT

darkest cells are best for water oriented recreation
ENDT
110 Map
068 RELOC

Figure 6.2 Early IMGRID coding form used by graduate students at Harvard.

50 series, Hewlett-Packard's 1000/3000, Data General's Eclipse series, and other units from Perkin-Elmer, Interdata, SEL, and the like.

King of all of these minis was Digital Equipment Corporation's VAX series. With awe-inspiring 1 MIP performance, proprietary VMS virtual operating system, huge capacity 205 megabyte removable disk cartridges that resembled a stack of Domino's pizzas, and freight train-like cabinet enclosures weighing several hundred pounds, the VAX-11/780 was the scientific machine of choice for heavy-duty number crunching. The VAX enjoyed immense popularity among its users as being reliable, well-supported, and a standard that nearly every manufacturer of peripheral displays, printers, scanners, and commercial applications software could interface with. Key punch cards had disappeared and in their place were VT-100 terminals, digitizing tablets, and cartridge or 9-track tapes. Chain printers were replaced by electrostatic and pen-based plotters. In its remote sensing configuration, the VAX was frequently found driving a COMTAL, Gould DeAnza, or an International Imaging Systems Model 75 image processing subsystem, along with a Dicomed color film recorder for hard copy output.

Most universities, government agencies, research labs, and large corporations had no difficulty in providing the required raised-floor computer room environments for superminis like the VAX along with paying for these systems. When compared to the cost of an IBM mainframe, a fully configured VAX seemed a bargain for "only" a few hundred thousand to one million dollars. The prevailing attitude in the late 1970s was, "The answer is VAX, now what was the question?" This hardware expense was nonetheless overwhelming to many small businesses and entrepreneurial start-ups who were anxious to develop GIS and Remote Sensing products, services, and systems without going deeply into debt or becoming victims of the venture (VULTURE) capitalists of the day.

Significant progress had been made in the mid 1970s by Nick Faust at the Georgia Institute of Technology on adapting and optimizing the IMGRID software and other NASA-based Remote Sensing capabilities to lower cost minicomputers, such as the Data General Nova. The capabilities of these minicomputer systems were validated in a highly successful statewide land-cover GIS project cosponsored by the Georgia Department of Natural Resources and NASA's Earth Resources Lab in Bay St. Louis, Mississippi, under the leadership of another visionary, Wayne Mooneyhan. Several of Mooneyhan's principal investigators and collaborators at ERL would later become key players in the commercial Remote Sensing industry, among them a then young Tim Foresman.

We now had easy-to-use (by previous standards) software and moderately priced mini-computer hardware in the $75,000 to $150,000 range. However, the price was still too high for many of the users who wanted to get into GIS with a much smaller budget, say under $25,000 or so. Further, the costs for the mini went up substantially when maintenance contracts were acquired. Minis also still required an electrical engineer or other scientific type to develop the hardware and software interfaces for connecting peripherals to the system. This dependency on a technician to keep the system up and running served as a reminder that, like the mainframe days, we still had a long way to go to gain the independence we thought we had achieved through the earlier revolt and put the user directly in control of the technology.

Microsystem Development

According to the "Way-Back Machine," microprocessor technology was becoming more prevalent in the marketplace in the late 1970s, outgrowing its early popularity among hobbyists with do-it-yourself packages from makers such as Heathkit. However, the term "PC" for personal computer would not become a familiar buzzword until the early 1980s. A company called Cromemco had a commercially available Zilog Z-80 based microprocessor that was built on an S-100 bus architecture and was generally regarded as a solid piece of equipment. Unlike the minis, the micros were compact, lightweight, and could easily sit on top of a desk and operate in room temperature environments. Disks were 8" single-sided, single density floppies with 1/4 megabyte capacity. RAM memory was a whopping 48Kb, and the operating system was either CPM or the increasingly popular CDOS (DOS). The user terminal was something like a Hazeltine CRT, and a light-pen pointing device was originally included but later replaced by a joystick or trackball. (A trackball was really just an upside-down mouse.) Image display generators used were multicard units such as the CAT-400 with 256 by 256 by 16 bits resolution and later 512 by 512 by 24 bits. Originally developed for TV weather broadcasts, these display generators handled both digital and video signals. Therefore, a regular TV set such as a Sony Trinitron could easily be converted and connected to the system for use as an image display screen. During the day, analysts could perform digital image processing and GIS tasks, then at night flip a switch on the monitor and watch the CBS Evening News with Walter Cronkite. (Saturday mornings were always reserved for Bullwinkle!) Best of all, the micros were inexpensive for the time, with systems priced in the few thousand dollars range. Of course, their major drawback was lack of speed and computing power, with processing tasks often running for hours. No one pretended that micros could or should attempt to compete directly with something as powerful and expensive as a VAX. However, many users of larger computers who were required to submit batch runs and wait in a queue were surprised to find that corresponding microcomputer results would often be available before the time-staggered batch run was completed on the larger system.

Building on previous successes by Faust and others, a streamlined version of IMGRID was integrated with Landsat processing software and adapted to this low-cost micro system by a newly formed company, Earth Resources Data Analysis Systems (ERDAS), Inc., in 1978. The complete rewrite of the software by ERDAS that was necessary to make it run on such a small computer created new opportunities for the user interface. A multidisciplinary team of employees with backgrounds in landscape architecture, planning, geography, remote sensing, electrical engineering, computer programming, and mathematics all collaborated to contribute to the new design. The "look and feel" was enhanced to include intelligently-computed default answers, a batch feature for long overnight runs, a basic tablet digitizing capability for GIS data capture, and interactive updating on the color display using imagery. At that time, Landsat imagery was available only on 9-track tape, so some creativity was required to get the data on floppies and into the system. To maintain mathematical precision when using the complex algorithms required for remote sensing, a Floating Point Processor Card (forerunner to the PC's Math Coprocessor Chip) was added.

For mass storage, an optional 96Mb (80 fixed and 16 removable) CDC Cartridge Module disk drive was added. This device, which resembled a washing machine (and shook like one during sequential disk accesses), preceded the sealed Winchester-type disks, and it had air filters that needed to be cleaned regularly. As a brief aside, ERDAS's first major hardware failure on the micro occurred with one of these disk drive units under what can only be described as unusual circumstances. The company originally began operations in a renovated garage (we were a *true* "garage startup"), complete with a small, cute gray mouse who was highly evasive and who refused to leave the premises. We caught the mouse in a wastepaper basket several times and escorted him out to the woods, but somehow he always found his way back. We eventually gave up trying to capture the mouse and instead designated him as the official company mascot, figuring that the office was his home first. During the winter months, the micro's CDC disk drive doubled as an excellent source of heat, and it kept both programmers and the mouse warm. On one cold winter day, the system refused to boot up, and the diagnostics indicated a head-crash on the disk. Disassembly and removal of the disk platter confirmed our worst fears: not only a head crash, but worse yet, small tufts of fur and little pieces of mouse feet and tail clogged the filter! The look on the face of the CDC service technician truly was unique. Because of our fondness for the mouse, no one dared to make any smart remarks about "building a better mousetrap" or "going to the big bit-bucket in the sky. . . ."

With the hardware repaired and software enhancements completed, the microsystem was fully operational, handling input, display, analysis, and output tasks all on the desktop in a friendly and conversational manner. An important guiding principle to this system's design was its fundamental concept of operation that remote sensing and GIS were no longer treated as two separate unrelated fields like they were on larger systems at the time. Instead, the philosophy was that remote sensing was an integral part of a geoprocessing system. Integration was implemented in the micro's software by taking the results of Landsat landcover classifications and merging them directly into the raster GIS—along with soils, geology, hydrology, and other layers—for further analysis.

Another positive factor that influenced user acceptance of the micro was the overall lack of hardware intimidation. The micro, unlike the mini, was easy to reconfigure or expand by non-technical personnel and without requiring the use of oscilloscopes, soldering guns, and wire-wrap tools. The S-100 bus cards were small and simple to insert, so swapping out a bad display card or adding a printer did not require a service call or any backplane rewiring. This open card cage allowed easy connection of peripherals, something that could not be done with a closed box such as the early Apple computers. Trouble-shooting was typically no more complicated than turning the machine off and reseating the cards, especially after shipment. Applications specialists using the system now felt a sense of empowerment and control over the technology, with the ability to turn out meaningful analyses inexpensively and in a stand-alone mode. We finally had an easy-to-use system for under $25,000 that integrated remote sensing and GIS and could actually generate useful products, even if it was slow.

Turnkey Solutions

The microsystem, along with a rented Data General mini, was originally used by ERDAS only for consulting services. Applications projects such as flood control studies, environmental assessments, and recreation master plans for the U.S. Army Corps of Engineers were successfully performed, and the results were delivered in the form of analytical maps, statistics, and a report. When clients came by the office to pick up the maps and report, their first question was, "Where's your computer room and the VAX?" This was always followed by a period of long, uneasy silence. When they saw that the project was done on an inexpensive micro with "conversational" software, initial disbelief soon turned into significant interest in purchasing the entire system. Clients wanted not only hardware and software but also a bundled digital data base of maps and analyses for their project area, as well as training and support services. What they really wanted was the same thing "revolting" Harvard students wanted five years earlier—independence. At first, we felt reluctant to release our technology secret, but it soon became apparent that a market was speaking to us.

Figure 6.3 Photograph of ERDAS 400 System which in 1979 was first low cost personal workstation.

By transferring the entire technology to the user, a turnkey solution could be provided (at a profit) for less than the maintenance cost alone of the larger systems. The result was the ERDAS 400 System, and in 1979 it became the first low cost commercial image processing and raster GIS system in the market place—a personal computer before the advent of the PC, (Figure 6.3). This fundamental shift in business focus from a services-only company to turnkey systems would be instrumental to our future global business success.

Once in the field, the micro's applications software was significantly improved through the interaction with key early users. Remote Sensing capabilities were greatly expanded with the valuable input from Bill Campbell and Bill Alford at NASA. A van-mounted mobile technology display and classroom, created by NASA Goddard's Eastern Regional Remote Sensing Applications Center (ERRSAC) group featuring the microsystem, brought the benefits of remote sensing and GIS out of the lab and into the field for local organizations. Another van was built for the Corps of Engineers to facilitate public input through GIS for the largest public works project in U.S. history, the Tennessee-Tombigbee Waterway. The original "student revolt" came full circle when Professor Steve Sperry, himself a member of the GSD uprising in 1973, acquired the first ERDAS 400 system for teaching landscape architecture (of all things) at Ohio State University in 1980. System sales really took off when the software was ported to the "new" IBM-

PC/XT and AT systems and renamed the ERDAS-PC. These systems found an immediate and popular home not only in other universities but also internationally and in local governments and small private firms who faced real world budget limitations. Versions of the software were also created for the Micro-VAX and the new SUN3, along with other minis, and the PC was networked to these as an intelligent workstation. Perhaps the most significant advance for the PC from a business perspective was the integration of the system's raster processing strengths with ESRI's vector capabilities. The result in the mid 1980s was the ERDAS-ARC/INFO "Live Link," the first commercial product on a PC that integrated remote sensing and topologically-structured vector data interactively.

With all the advances in newer generation computer chips and emerging workstation technology, the original PC quietly faded into the background, replaced but nonetheless fondly remembered.

A quick spin of the fast forward dial on Mr. Peabody's Time Machine now propels us ahead to the year 1997. With a lot more gray hairs and no 370s in sight (very few VAXs, for that matter), we find ourselves in perhaps the most exciting time of all for GIS, at the leading edge of a new revolution for on-line access to everything. Our installed base of Imaging GIS systems is the world's largest, with nearly 10,000 systems in more than 90 countries, many of them PCs. The new PC is built on Microsoft's Windows NT and Windows 95 and uses high performance chips like the Pentium, which easily outperforms yesterday's mainframes. The software, now called ERDAS IMAGINE®, has ESRI's ARC Data model built in, and real time Virtual GIS fly-through of high resolution 1 meter commercial satellite imagery is just around the corner. Better yet, Bullwinkle is now available on CD-ROM!

The other day one of our younger technical support people came in and described a project currently underway at a customer's site. She said the client was using an older version of the system and wondered if we knew where to get 8-inch floppy disks. "By the way," she asked, "what is a 400 anyway?" We both smiled—and began looking for the Geritol.

PART IV

Applications Sector Perspectives

William E. Huxhold , Chair & Associate Professor,
University of Wisconsin-Milwaukee

I cannot remember the first time hearing the term, GIS, but I first heard of this technology before the term was used. I can even remember the day—Aug. 23, 1974—when I attended a paper session with Milwaukee's Data Processing Manager, Frank Bayer, at the Twelfth Annual Conference of the Urban and Regional Information Systems Association (URISA) in Montreal. Joel Filler (now "Orr"), Data Systems Manager for Metropolitan Nashville, presented a paper entitled, "LAMP: Metropolitan Nashville's Location and Mapping Program," in which he described the process of designing and implementing an "interdepartmental locational information system" for the various departments in Metro Nashville. (As an aside, his study back then found that 90% of all files in Metro contained some form of locational information.) The heart of LAMP's graphic capabilities was from M & S Computing, Inc. of Huntsville, AL.

Frank Bayer and I went back to Milwaukee and found funds from the first year of the Community Development Block Grant Program to contract with Joel for a similar feasibility study in Milwaukee. On Tuesday, Sept. 21, 1976, the City of Milwaukee issued its Request for Bids for an "Interactive Graphics System." The specifications were sent to the following companies: AUTO-TROL Corporation, CALMA Interactive Graphics, M & S Computing, SYNERCOM Technology, Inc., APPLICON, Inc., Bendix Computer Graphics, Computer Graphics Company, Computer Research Corporation, Computervision Corp., Gerber Scientific

Instrument Co., Digital Graphics, Inc., Instronics, and H. Dell Foster. M & S Computing (now Intergraph Corporation) won the bid, and the system was implemented in 1977.

William E. Huxhold has been involved with urban applications of GIS. His book, An Introduction to Urban Geographic Information Systems *(Oxford University Press) was one of the first textbooks in the field.*

Peter Croswell,Executive Consultant, PlanGraphics, Inc.

My first exposure to GIS occurred when I was an undergraduate in Geography at the State University of New York-Albany in 1974-75. We used the New York State Land Use/Natural Resources (LUNR) System which encoded land cover and land use data by 1-kilometer grid cells. Using SYMAP and some other spatial processing software, we were able to perform some interesting geographic analyses and produce maps (line printer generated, of course).

Peter Croswell has been instrumental in the implementation of GIS technology for hundreds of local, state, and federal agencies. He also serves as president of the URISA.

CHAPTER 7

Land Information Systems: Development of Multipurpose Parcel-based Systems

D. David Moyer and Bernard J. Niemann, Jr.

Introduction

Land parcels have been a major focus for land information for over 1,000 years. Yet, in the last 25 years, with the advent of computerized information systems, parcel-based systems have become even more important. We expect this trend to continue.

We glimpse here the roots of land information systems (LIS) in Europe and the Middle East. We look at current activity, especially politically at the federal level, to place the discussion in context and to emphasize the current importance of LIS. In the bulk of this chapter, however, we review the major factors and trends that have impacted the development of LIS systems over the last 25 years—particularly as they have been affected by state (including educational institutions) and federal interests. We discuss the roles of the Harvard University Graduate School of

D. David Moyer is an adjunct associate professor in the Institute for Environmental Studies and in Land Information and Computer Graphics Facility in the School of Natural Resources at the University of Wisconsin-Madison. He also serves as the Wisconsin State Advisor for Land Information and Geodetic Systems with the National Geodetic Survey.

Bernard (Ben) Niemann is a professor of Landscape Architecture at the Institute for Environmental Studies at the University of Wisconsin-Madison since 1964. He has directed the Land Information and Computer Graphics Facility since 1985. He holds a master's degree in Landscape Architecture from Harvard University. He is presently coeditor of the *URISA Journal*, coeditor of the GIS Innovator column in *Geo Info Systems*, and a member of the Wisconsin Land Information Board. *Authors' Addresses:* D. David Moyer, 1050 WARF Bldg., 610 Walnut St., Madison, WI 53705. E-mail: moyer@mail.state.wi.us. Bernard J. Niemann, Jr., B-102 Steenbock Library, 550 Babcock Dr., Madison, WI 53706.

Design (GSD), U.S. Department of Agriculture (USDA), American Bar Foundation (the research arm of the American Bar Association), Institute for Land Information, National Research Council (of the National Academy of Sciences), and the University of Wisconsin-Madison. We outline the development of the multipurpose land information system (MPLIS) concept and provide examples of applications and analyses. We also review a number of outreach efforts that have been used to facilitate the adoption and use of LIS systems by state and local governments.

Many of the examples in this chapter, particularly at the state level, are from Wisconsin, a product of the authors' close involvement in LIS development there. Many other states have been active in the development of LIS systems and, like Wisconsin, have struggled to advocate and implement a broader role for LIS.

In this chapter, we use the term Land Information Systems (LIS) as a means to distinguish our focus: that interests and rights in real property are an explicit database component. We also suggest that the real explosion and exploration of spatial information technology use has yet to occur; it has not yet been diffused and adopted by the local governments and citizens who have responsibility and interest in land and land rights.

Background

For a long time, professionals in many land-use disciplines have recognized the need for land record systems that are accurate, complete, and up to date. This need is related to a variety of functional uses, but land ownership title, taxation, and physical attributes have been the most prominent. Some of the earliest parcel-based systems came out of the Middle East when ancient Egyptians developed systems to serve the need of reestablishing boundaries after the annual floods of the Nile River. Similar systems to provide ownership and taxation data followed in the Roman Empire and in England. *The Domesday Book* of 1085-86 is often cited as the first time that land parcel and owner records were linked. Once this connection had been made, the ability to provide data and information to owners and managers of land became possible.

Warren Manning, a Boston-based landscape architect, called for such a connection in his land-use plan for Billerica, MA, in 1909 (Manning 1913). These three issues (title, taxation, and physical attributes) depend to a large extent on the land *parcel* as the unit of analysis for which these data are needed. Even in cases where some data are not based entirely on the parcel (e.g., when other polygons provide the boundaries for resources), the parcel has been and continues to be the primary unit for analysis and decision-making. Frank Popper, a planner, called for the inclusion of property interests as part of any realistic land use plans (Popper 1981). He asserts interest in property is the missing dimension in contemporary land use planning. Recent research by Chowdhury (1994) further documents the importance of real property interests by both county and township citizens and professional planners.

With rapid economic expansion and specialization in the twentieth century, the need for and value of systematized and locally sustained land information systems has exploded across America. Rights in real property and the stewardship responsibilities of these natural and cultural resources remain a central issue for rural America although some 20 years have now passed since work began in earnest on automated LIS systems.

Current Events

Recent political events suggest that this transfer of information technology, focused on local government needs and interests in real property rights and the stewardship responsibilities, will accelerate as we move into the next century. Increase in interest in LIS as related to land ownership was elevated to a new level in 1994, when Congressional interest in protecting real property rights was articulated as elemental in maintaining the quality of life in rural America, with its inclusion in the Republican Party's "Contract with America" (Gillespie and Schellhas 1994). A primary tenet of the 1994 election was the promise of the "end of government that is too big, too intrusive, and too easy with the public's money." As a result, major action items in the contract are provisions to "roll back government regulations . . . and to protect individual Americans from overzealous federal regulations . . . [to] make sure that private property cannot be taken away without just compensation. . . . [It is proposed that] private property owners [would] . . . receive compensation [up to 10 percent of fair market value] from the federal government for *any* reduction in the value of their property" (emphasis supplied).

It would be naive to think that modern parcel-based land information systems and the data they contain won't become central to this debate over rights and compensation. Prior experience in such a "taking" in the coastal cascades of Washington suggests otherwise. In this case, the U.S. Forest Service was authorized by Congress to condemn 24,400 acres of land owned by a private timber company to expand the Alpine Lakes Wilderness Area in the state of Washington. The U.S. Forest Service appraisers based their assessment of fair market value on the value of its marketable timber, or about $13.5 million. The owners, with assistance of a resource-based LIS, demonstrated that the company's non-timbered lands such as their glaciers, lakes, and alpine meadows enhanced the fair market value to $25 million. The owners and the U.S. Forest Service agreed to an out-of-court settlement of near that amount (Chenoweth and Niemann 1989), strongly suggesting that information technology will be used by either or both parties to disagreements as to fair market value.

Other informational aspects of the "Contract with America" are also of particular interest to rural America. On November 11, 1994, three days after the election of a Republican majority in the U.S. House of Representatives, Newt Gingrich (R-GA) laid out five changes that America must address "to effectively get to the twenty-first century." The first of these is that "we have to accelerate the transition from a secondary-wave mechanical, bureaucratic society to a third-wave information society. . . . There's no objective reason that institutions of government have to be two or three generations behind the curve in information systems and management, but they are. And that means, for example, if we're really serious about distance learning and about distance work, we could revolutionize the quality of life in rural America and create the greatest explosion of new opportunity for rural America in history."

This is not to suggest that the modernization of land information systems and its benefits are limited to rural areas. Quite the contrary: Urban areas and those areas facing urbanization and infrastructure management pressures are the heavy investors in LIS/GIS technology. Those of us who focus our interests on the long-term management and sustainability of natural systems and environmental integrity have another focus. That is to give rural Americans the capacity to assess and manage these natural and cultural systems through GIS (in light of Alvin Toffler's "third-wave information technology"), which could "revolutionize the quality of life in rural America in history."

Developments in the Late Twentieth Century

Some innovations in our society can be traced to a single location, event, or person. However, in the case of geographic and land information systems, a number of parallel and eventually convergent efforts beginning in the 1970s were major factors in the LIS systems that have emerged in the 1990s. While it is not possible to discuss all of the relevant developments here, major ones include work at the Harvard University Graduate School of Design (GSD), the U.S. Department of Agriculture (USDA) (including work organized by the North Central Land Tenure Committee), the American Bar Association and Foundation, and the University of Wisconsin-Madison.

Landscape Architects at Harvard Graduate School of Design

In the late 1960s, landscape architecture faculty and students were beginning to explore the use of computer technology for resource management and facility location. One place that was particularly active was GSD. Building on pioneering transportation-planning work (ROMTRAN) by Edgar Horwood from the University of Washington, Howard Fisher went to GSD and designed and implemented a computer-mapping program called SYMAP (Niemann and Niemann 1994a), and, with a Ford Foundation grant, established the Laboratory for Computer Graphics (later, the Laboratory for Computer Graphics and Spatial Analysis) at GSD. Initially, faculty and students in Landscape Architecture at GSD worked synergistically with Fisher and others at GSD (Steinitz 1993b).

Carl Steinitz from the School of Architecture and Urban Planning at Massachusetts Institute of Technology (MIT) completed his work there on urbanization and form using a modified version of SYMAP. Upon his graduation from MIT, he and his graduate student, David Sinton, further modified SYMAP into GRID, a raster-based computer-mapping and analysis system, for a variety of GIS-related research studies.

One such study was of computer applications to investigate processes of urbanization and change. This included a major grant from the National Science Foundation (NSF) program for Research Applied to National Needs (RANN) (Steinitz 1993a). In concert with this research, Steinitz and Sinton brought GRID into the classroom at GSD. The first effort was an environmental assessment of the Delmarva Peninsula in Maryland using GRID. The "Delmarva Studio" brought together a mix of disciplines including a civil engineer, Donald Belcher from Cornell; a geographer, Angus Hills from Canada; and two landscape architects, Ian McHarg from the University of Pennsylvania and Philip Lewis from the University of Wisconsin–Madison. As a result, GSD attracted a number of students and scholars who would later take the lead in various aspects of LIS and GIS development.

In retrospect, the emphasis of landscape architects on resource analysis and the emphasis of the geographers and cartographers on spatial geographic integrity caused each group to go its separate disciplinary way even though they were housed in the same building only one floor apart at GSD. Nevertheless, the roots of GIS innovation can be traced to two developments at GSD.

One was David Sinton's and Dana Tomlin's 1971 user-friendly interface for GRID called IMGRID (Niemann and Niemann 1993c; Sinton n.d.), a response to a student "revolution" against the complexities of GRID.

Another was a significant contribution to the evolution of vector-based GIS. With funding from the National Science Foundation, a group of geographers and computer scientists set out to design and implement a vector-based GIS solution using topological concepts as the basic data model. This initial group consisted of Nick Chrisman, Denis White, Geoffrey Dutton, and Scott Morehouse. A major breakthrough of this effort, known as ODYSSEY, was its ability to conduct real topological overlay efficiently. Until this breakthrough, the ability to maintain the spatial integrity of ownership information and overlay or merge that with natural resource information was not technically feasible (for a discussion about the origins of ODYSSEY, see Niemann and Niemann 1994c).

Sinton eventually moved to Intergraph Corporation and helped establish its GIS capability and reputation. Other notable GSD graduates in landscape architecture rendezvoused in Atlanta and created the Earth Resources Data Analysis System (ERDAS) Corporation, known for its robust raster-based remote sensing and resource analysis capabilities. This group, including Lawrie Jordan, Bruce Rado, and Steve Sperry, had also participated in the IMGRID revolution (Niemann and Niemann 1994a).

Another IMGRID revolutionary, Bruce Rowland, moved to the Tennessee Valley Authority (TVA) and helped design and implement a system capable of managing both TVA's ownership database and its natural resource and recreational management responsibilities (Niemann and Niemann 1994b). Tim Murray, a GSD landscape architect, accepted the responsibility for implementing GIS for electrical transmission corridor planning at the Bonneville Power Administration (BPA) in the Pacific Northwest. Eventually this evolved into the overall management and assessment of BPA's 15,000 miles of transmission lines (Niemann and Niemann 1993b).

Also in this group of GIS innovators was Jack Dangermond who later founded the Environmental Systems Research Institute (ESRI). Dangermond, a landscape architect, benefited from having access to faculty from both landscape architecture and the lab. Beginning with GRID- and SYMAP-based software, he helped introduce GIS technology to the resource management world. With his eventual hiring of Morehouse from GSD, he set about in 1981 to create a topological vector-based GIS now known as ARC/INFO (Niemann and Niemann 1995). ESRI has become a major force in the GIS field, developing software and distributing it around the world. They provide major support for many users, converting data to computer systems, and technical assistance to their customer base.

Other GIS innovators included Bernard J. Niemann, Jr., one of the authors of this chapter, who has spearheaded the development of LIS in Wisconsin and been active in the conceptual design of LIS in general, and Dutton, who has remained active in the field in a variety of ways, including efforts to harness LIS to monitor the ownership of U.S. land by foreign investors.

USDA—Economic Research Service

Another group that developed an active program in what was to become LIS was the USDA Economic Research Service (ERS). That development, under the direction of Gene Wunderlich and with the creative work of D. David Moyer, one of the authors of this chapter, began in the early 1960s, the result of a number of converging forces with which ERS researchers were faced.

ERS is the research arm in USDA that provides assistance to Congress and USDA policymakers regarding a wide variety of issues facing the farming sector and rural sectors of America. To provide this assistance, a complete, accurate, up-to-date database is necessary. A major portion of these data came from the Bureau of the Census, which provided data every five years. In the early 1900s, both the Bureau of the Census and USDA began to supplement these census data with a variety of sample surveys. As the U.S. economy continued to grow and become more complex, researchers were constantly seeking ways to improve the databases they needed. Improvements were essential to assisting in the development and evaluation of increasingly complex policies concerning ownership of and responsibility for the management of the vast resources of rural America.

By the early 1960s, ERS was facing a number of challenges regarding its land-related research efforts: The programs that ERS was developing and evaluating were becoming more complex, mirroring the U.S. economy and society in general; more precise data, updated more frequently were needed; primary data collection was becoming more expensive and therefore more difficult to fund. Concurrently, relevant databases, particularly at the local government level (county or county equivalent) were being collected for a variety of purposes. However, while many of the needed data items existed, they were stored in systems that were incompatible, difficult to access, and generally ineffective in meeting the needs of many local data users, let alone federal researchers.

At this point, ERS began the development of a coordinated research and development effort to address the land data problems at all levels of government. One of the earliest efforts was the Comprehensive Unified Land Data System (CULDATA) project. CULDATA was to explore the feasibility of developing a model land data system that would serve the needs of a variety of data users. Robert N. Cook, University of Cincinnati Law School, was awarded a two-year contract from ERS in 1964 to spearhead the first phase of this work.

A 1966 CULDATA Conference was the first effort to bring together many parties from a variety of disciplines, all with a common interest in developing or using data that a comprehensive land data system could provide. Included in the conference proceedings were papers by lawyers, economists, planners, engineers, and government officials from the local, state, and federal levels. These authors addressed issues covering specific needs and uses for land data and the possibility of serving these needs and uses with emerging computer technology in fields such as mapping, surveying, indexing, and land records management.

Cook presented the CULDATA model as one possible approach to coordinating the collection and use of land data (Cook and Kennedy 1967). He pointed out that the conference brought together people with an interest in modernizing land records to help determine "where we are, where we ought to be, and how we might get there." The result was that this group of economists and lawyers saw for the first time land data and land records as a rich resource that could be used to serve a much wider user community than had been the case up to that time. However, the technology needed to make the CULDATA idea become a reality did not exist in 1966 nor were participants cognizant of the work being developed at GSD and vice versa. At the very least, a clearer vision began to emerge of what the future could bring to the land information field.

USDA continued to pursue LIS modernization, one of the major efforts being through the North Central Land Economics Research Committee (NCLERC). NCLERC was a continuing effort between the USDA and the land-grant universities in the north central United States. In 1968, NCLERC and several other participating institutions from the CULDATA conference organized a second conference held at Mackinac (pronounced *mak´-i-naw*) Island, MI. Participating institutions included ERS, NCLERC, the U.S. Department of the Interior–Bureau of Land Management (BLM), the National Archives and Records Service, the American Bar Association, American Congress on Surveying and Mapping (ACSM), American Land Title Association (ALTA), and several universities in the United States and Canada.

The Mackinac Conference gave particular attention to land title and records, focusing on two major issues. The first was an evaluation of legal requirements for marketable land titles, recording, notice, title insurance, and other elements necessary for the easy transfer of various property-related land rights (e.g., mineral, timber, water, and air). Special attention was given to improved procedures in a number of technologies such as computing, photogrammetry, and land surveying. The second issue focused on the relationship between land title data and the various functions that require and rely on land data. These functions include property tax assessment and collection, community planning, highway and utility location, zoning, enforcement of an array of codes and regulations, and a wealth of research activities.

In a summary of the conference, Neil Harl of the University of Iowa set out five "parameters of the ultimate system as a pattern for law reform to assure that alterations will tend toward the desired state" (White 1969). Harl's presumption was that the land title system was much more likely to undergo evolution, as opposed to a sudden, major revolution.

The five basic areas in which Harl saw a need for change were:

- recordation of instruments affecting land titles,
- storage and retrieval of information affecting land titles,
- efforts to develop an accurate, simple, and uniform method for the identification and designation of individual tracts of land,
- the definition of marketability and legislation to cut off ancient and unstable claims, and
- organizational changes in title and information service agencies.

Canadian Maritime Provinces

At about the same time, activity was also picking up in eastern Canada. New Brunswick, Nova Scotia, Newfoundland, and Prince Edward Island joined together under the Maritime Council of Provinces to develop an ambitious program to modernize the land records in those four provinces. The Maritime program was unveiled in Fredericton, New Brunswick, in late 1968 (*Canadian Surveyor* 1969). Speakers and commentators provided insights into and details of the existing and potential needs, as well as of barriers that would need to be overcome.

The model proposed for the Atlantic Provinces Surveying and Mapping Program contained four phases (Roberts 1968). Phase I was "the precise establishment of monuments at predetermined intervals in each province." Phase II was "the establishment of a provincial large-scale topographic map series at scales of 1:24,000 to 1:6,000." Phase III was "a proposal to implement a computer-based land titles system, based on the Torrens principles." Phase IV was "the implementation of a data bank."

The Atlantic Provinces program continued, with several modifications, for over 25 years. It was generally viewed as a top-of-the-line system and struggled for continuing support from the provincial governments throughout its life of operation. Much progress was made, but recently the quasi-government corporation, Land Registration and Information Service (LRIS), seems to have gone out of business. Further research on how this system evolved and the future direction of land records handling and modernization in the Maritimes would seem to be a fruitful area to explore.

CLIPPP Conference

NCLERC and the American Bar Association (ABA), along with ERS, continued to discuss the need for land records system reform and how such reform might most effectively be approached. They concluded that the single most important thing that could be undertaken and most likely to produce results fairly quickly would be to pursue Harl's third point, the need for a simple, accurate, uniform method for identifying parcels or tracts of land. Therefore, in 1972, the American Bar Foundation, in cooperation with a number of other agencies and associations, hosted a conference in Atlanta entitled Compatible Land Identifiers—Problems, Prospects, and Payoffs (CLIPPP) (Moyer and Fisher 1973). The CLIPPP conference brought together people from a wide variety of disciplines and, in this case, focused on a specific task: developing a recommendation for a model system that provided a unique, compatible identifier for each land parcel.

Conference participants, working in small groups to address and understand the needs of all users and all functions, developed the criteria for parcel identifier evaluation and provided guidance for making the final selection of an identification system. The model identifier system chosen based on the State Plane Coordinate System (SPCS) and provided coordinates for the visual centroid of each parcel (Moyer and Fisher 1973).

Several jurisdictions, including the state of North Carolina, adopted and are still using SPCS-based identifiers for all land parcel-related documents, map indices, etc. It appears, however, that the more important product of the conference was the set of criteria by which to judge an identifier system. In the nearly 25 years since that conference, many jurisdictions have adopted new parcel identification systems that meet the criteria established in the early 1970s. These include:

- Simplicity (which, in turn, requires an identifier that is easily understood and reasonably permanent);
- Uniqueness (meaning a one-to-one relationship and implying ease of maintenance);
- Accuracy (in that spatial information is contained in the identifier itself and therefore provides means for statistical sampling, as well as linkage of parcel numbers to data about parcels);
- Flexibility (to be compatible with current operating systems and to incorporate advances in future information systems);
- Economy (in initial implementation costs and in system operation costs); and
- Accessibility (availability of information needed to assign and interpret parcel identifiers).

The benefits of using a parcel identifier based on these criteria are substantial. For instance, as Moyer and Fisher (1973) note, "The choice of a [parcel identifier based on a] universally compatible coordinate system rather than an arbitrary local one will provide substantial savings in the long run without increasing the technical difficulty of implementation and use."

MOLDS Conferences

During the remainder of the 1970s, activities at the federal level were relatively low key. The North American Institute for the Modernization of Land Data Systems (MOLDS) organized two conferences (in 1975 and 1979) on the topic of land data system modernization. These conferences kept the issue of LIS modernization alive in the public arena and helped facilitate continuing discussions among those in local, state, and federal government, as well as in the private sector. Other results are difficult to measure. The primary shortcomings of MOLDS activities were probably largely due to the volunteer nature of the institute and the resulting lack of sufficient financial and personnel resources by the federal agencies and professional organizations involved. MOLDS has subsequently changed its name to the International Institute for Land Information (IILI) and continues to host seminars.

Many of the voluntary federal members of MOLDS/IILI are now involved in the Federal Geographic Data Committee (FGDC) of 14 federal agencies, which is bringing new energy to the land and spatial data system modernization task at hand.

National Research Council

By the late 1970s, many agencies and groups concerned with LIS improvements were looking for a way to rejuvenate their modernization efforts. These parties were able to interest the National Research Council (NRC), the research arm of the National Academy of Sciences, in giving LIS improvements some attention. NRC took on the project and relabeled LIS the multipurpose cadastre (MPC). A group of experts from various agencies and associations was assembled and, after a year of study and debate, they produced a report documenting the substantial need for a multipurpose system that would serve a wide variety of users and uses for

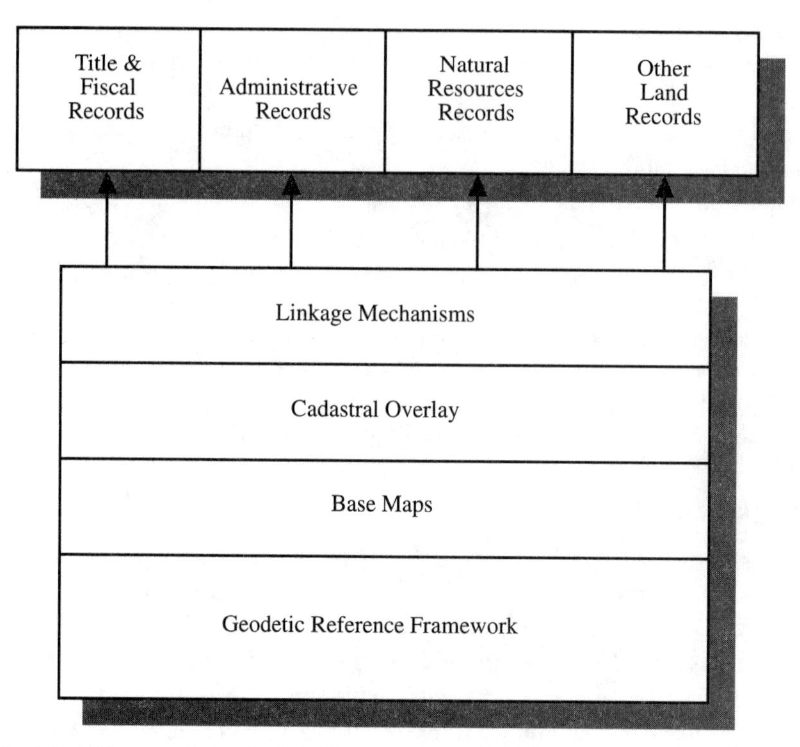

Figure 7.1 The NRC panel members laid out a conceptual diagram of the components of a multipurpose cadastre.

land information (NRC 1980). The NRC panel members also laid out a conceptual diagram of the components of an MPC (Figure 7.1).

The proposed MPC model contained five parts:

- A geodetic reference framework to define the spatial location of all land-related data;
- A large-scale, current, accurate base-map series;
- A cadastral overlay that delineates all parcels;
- Linkage mechanisms, of unique parcel identifying numbers, to link the graphic cadastral parcel map with data files and registers; and
- Data files and registers of land data, each with a parcel index for linkage to cadastral map information.

It should be noted that this model relied exclusively on the parcel as the basic building block and unit of analysis for both maps and data files.

Three years later NRC published a second report, Procedures and Standards for a Multi-purpose Cadastre (NRC 1983), outlining procedures and standards that would be required for the design and implementation of multipurpose cadastre systems at all levels of government. The report was "intended to assist both local governments wishing to pursue the development of cadastral records systems for their own counties or equivalent districts, and also the many other regional, state, and federal agencies, as well as private businesses, whose participation will be needed" (NRC 1983).

The 1983 NRC panel reiterated the basic 1980 MPC components but made one fundamental and essential modification: the panel recognized the important distinction between cultural LIS data (i.e., the cadastral boundary overlay noted in 1980) and natural resource LIS data (which they added in 1983) (Figure 7.2). This conceptual modification provided a recognition of the importance of these two types of data and also provided the means to integrate and analyze these data in one system. The underlying geodetic reference system provided the key to merging these two types of LIS data.

Much of the subsequent work in the conceptualization of Multi-Purpose Land Information Systems (MPLIS) is built on this foundational concept. This conceptual change resulted in a major shift in the technological requirements of a modern LIS. No longer could the data model be simple CAD vector graphics. The data model needed to include topological relationships so that analyses such as topological overlay could be conducted.

This need for an intelligent data model resulted in the need to merge vector renditions of land ownership parcels with vector renditions of more sinuous natural resource patterns such as soils and forests. The net result was the need for software and hardware systems capable of both LIS and GIS functionality—therefore, the emergence of the expression "GIS/LIS" or "LIS/GIS." Other work with NRC was a panel sponsored by the Committee on Integrated Data Mapping, which called for the modernization of the Public Land Survey System (PLSS) (NRC 1982).

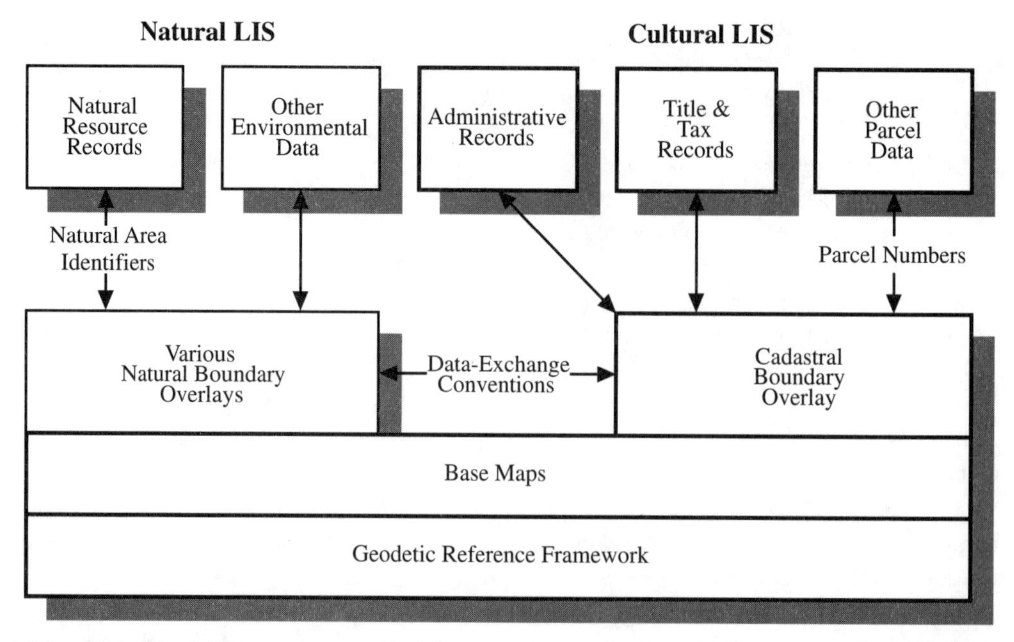

Figure 7.2 The NRC panel recognized the important distinction between cultural LIS data (i.e., the cadastral boundary overlay noted in 1980) and natural resource LIS data which they added in 1983.

University of Wisconsin-Madison

This merger of LIS and GIS modernization branches came together at the University of Wisconsin-Madison (UW-Madison) in the late 1960s. The USDA ERS was continuing its research and development work on improving information systems to compile information about land and related resources. This work included cooperative research activities with a number of universities and involved several academic disciplines. It became apparent that the ideal location to pursue work on modernization of LIS should involve universities with certain characteristics, including researchers interested in improving information about land, expert in disciplines critical to research and development regarding LIS improvements (e.g., economics, law, computer science, and engineering), and willing to enter into a cooperative agreement to pursue jointly (with ERS) a mutually agreed upon research program. After discussions with a number of land-grant universities in the northwest, south, and midwest, ERS selected the UW-Madison as the center for work on LIS improvement.

Researchers were faculty members who had been active in NCLERC, including William Loomer, William Lord, Raymond Penn, Don Kanel, and Stephen C. Smith, along with Walter Raushenbush of the Law School, who was conducting a cooperative project with the American Bar Foundation related to land title system modernization, and faculty in Agricultural Economics, Civil and Environmental Engineering, and Landscape Architecture. The latter faculty, in fact, played an active role in cooperative research from 1969 on, some of which continues.

Initial research focused on land title systems. This was consistent with the emphasis in ERS on land ownership, land tenure, and land taxation policy and data. Research projects included studies at the local, state, and national levels to ascertain the status of land title data and land title recording systems. In some of this work, cooperators included the Bureau of the Census and the National Association of Counties (NACo). The 1974 study, *Land Title Recording in the United States,* which involved ERS, the Governments Division of the Bureau of the Census, NACo, and UW-Madison (Behrens et al. 1974), was a statistical sample survey for the U.S. It was later expanded to provide a detailed census of land title recording practices, costs, and other matters for the State of Wisconsin. The results of these two studies became an integral part of the study of Moyer's Ph.D. dissertation and an annotated bibliography (Moyer 1977; 1978).

While Moyer's work in the 1970s concentrated on land title systems, faculty from Landscape Architecture (Niemann) and Civil Engineering were beginning to explore the use of GRID-based computerized systems for analysis of natural resource problems and the development of natural resource policy. Two of their earliest efforts involved development of alternative power line siting routes and the evaluation of options for siting of a new freeway in eastern Wisconsin (Miller and Niemann 1972).

Because of the emerging citizen interest in environmental and farmland preservation issues in the state, the Wisconsin Highway Commission (now the Wisconsin Department of Transportation [DOT]) sought the assistance of Philip Lewis and others in the UW Department of Landscape Architecture in selecting a corridor for the newly authorized highway between Milwaukee and Green Bay to be Interstate 57 (I-57). Lewis had been involved in the GSD Delmarva Studio exercise. Niemann and others at UW-Madison had taken part in GSD-sponsored SYMAP seminars and workshops at Harvard. SYMAP was being used in classroom exercises in the Department of Landscape Architecture. Given this local exposure to technological innovations, DOT was amenable to using a computerized approach instead of the traditional mylar overlay popularized by McHarg (1969) and Lewis (1964). With assistance of former GSD researchers Steinitz, Sinton, and Murray, faculty constructed a database of 38 natural and cultural resources based on a grid of 1-square-kilometer cells. GRID was used to model and display nine different interstate location corridor determinants (Miller and Niemann 1972) (Figure 7.3):

1. Greatest engineering suitability;
2. Least cost of construction;
3. Least cost of acquisition;
4. Projected traffic generation;
5. Least destruction to the cultural system;
6. Least destruction to the ecological system;
7. Conservation of quality agricultural land;

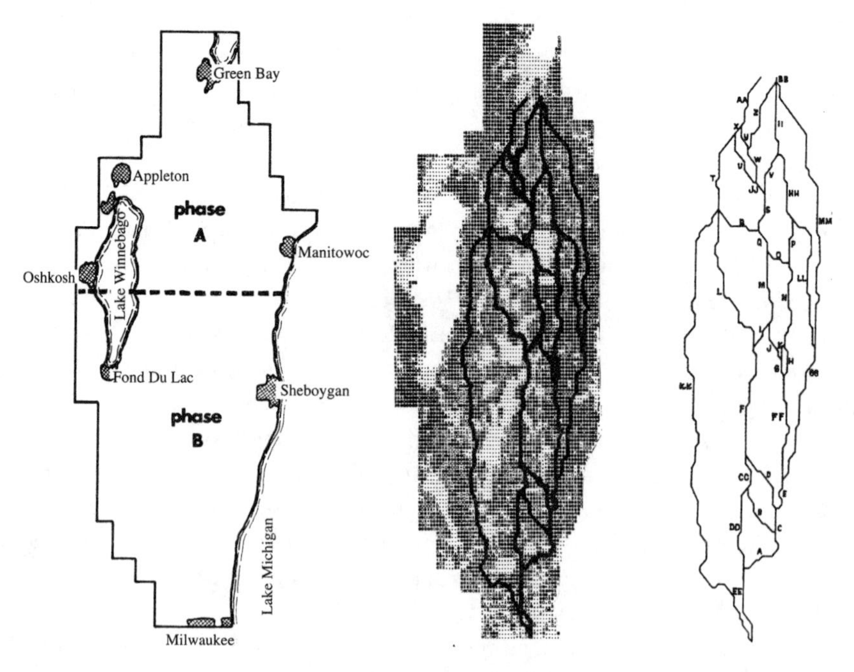

Figure 7.3 GRID was used to model and display ten different interstate location corridor options.

8. Greatest scenic potential; and
9. Least destruction of potential recreation and conservation land.

The database and the various models were presented at public hearings and forums, and alternative routes were suggested. Public sentiment in 1972, however, was not to build an entirely new interstate corridor (I-57) but to upgrade a partially existing corridor (I-43). This was not the first choice of DOT and the highway construction lobby, but public sentiment prevailed and the I-57 database became the basis for comparing the I-57 and I-43 alternatives. Eventually, the I-57 alternatives were rejected in favor of the I-43 solution. It remains only speculation as to how much the UW-Madison computerized effort influenced the selection of the I-43 alternative. What was clear for those involved at UW-Madison was that the flexibility offered by a computerized system could not be ignored.

Nevertheless, the raster- or grid-based GIS approach had two major limitations. The first was a lack of spatial integrity. Various scales of map data had been used to construct the database. Even though data had been verified and even though each cell was one kilometer square, certain phenomena, when spatially overlaid, did not occur within their correct cells. What was thought to be a database with an assumed spatial accuracy of 1 kilometer turned out to be a database of ±1 km (in essence, a 3-by-3-kilometer block) so points fell somewhere within a nine-cell block). This may be acceptable for observing broad suitability patterns, but it is not acceptable when using a linear optimization approach. As a result, the recognition of the need for spatial integrity of an LIS began to emerge.

The second limitation was the inability to address the issue of property boundary severance. In farmed landscapes, this is of major concern to property owners. Because of inherent surveying errors in the original establishment of the PLSS, a uniform-sized grid cell was chosen —a UTM projection. In retrospect, any grid or cellular solution was technically incapable of managing the spatial integrity of a property boundary and its associated rights. This problem of property severance and the difficulties of dealing with a grid-based data model were reaffirmed in subsequent computer applications for electrical transmission corridor applications (Miller and Niemann 1972). It was concluded that it was time to find or develop a vector-based data model that could maintain the spatial and linear integrity of real property boundaries and the sinuous integrity of natural resource boundaries.

Economics of Land Information Systems

Although the benefits of automation are obvious to those who champion it, major issues about the cost of automation always emerge. Certainly, the cost of building *ad hoc* computerized databases for each application would be economically prohibitive. It became apparent that it would be necessary to enlist in the automation process those government data and records collection agencies mandated to maintain these databases on a transactional basis. Initially, these thoughts were met with considerable skepticism. Costs of automation were seen as prohibitive. With the assistance of funding from the U.S. Geological Survey program for Resource and Land Investigations (RALI), the Wisconsin Department of Administration (DOA) and UW-Madison set out to answer two questions:

1. What factors inhibit the ability to gain access to and integrate land data and records?
2. What does it cost each Wisconsin citizen to sustain the non-automated collection and management of land data and records?

In response to the first question, technical and institutional inhibitors were identified:

- Lack of accessibility;
- Lack of availability;
- Duplication;
- Inability to aggregate;
- Inability to integrate;
- Confidentiality; and,
- Institutional restrictions.

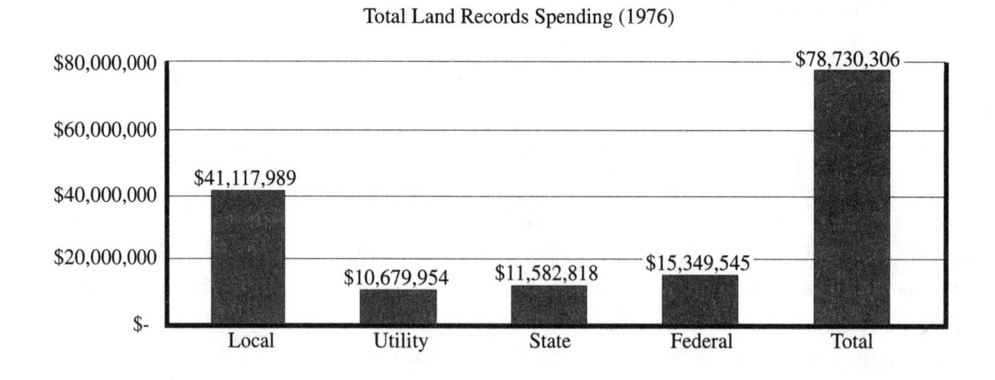

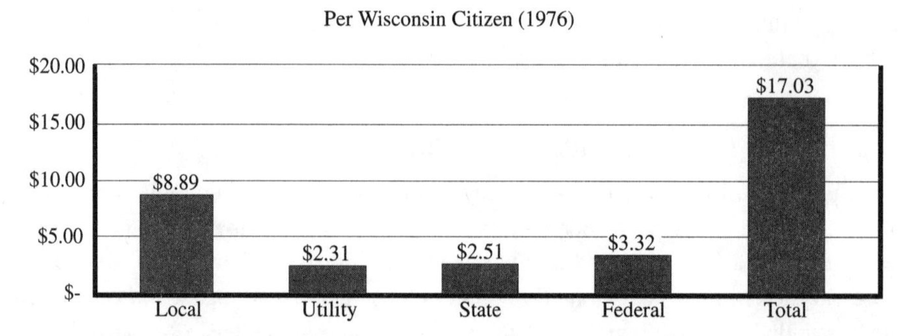

Figure 7.4 Larsen et al. (1978) found the cost to the Wisconsin citizens was about $78 million *annually* in the collection and management of land data and records. The major component (52%) in the cost of collecting and maintaining land data and records was not from federal or state programs but from local governments.

In response to the second question, two major findings emerged (Figure 7.4). First, Wisconsin citizens were spending about $78 million *annually* in the collection and management of land data and records (Larsen et al. 1978). Since this was an annual expenditure, it was concluded to be non-trivial, both because of its magnitude and because of the factors that inhibited integration and use for analysis of data and records. Second, quite contrary to what was initially theorized, the major component (52%) in the cost of collecting and maintaining land data and records was not from federal or state programs but from local governments (Figure 7.4). These are the government units mandated to assume land-use and planning responsibilities, to provide access to federal farm commodity programs, to dispatch E-911 vehicles, to assess and collect real estate taxes, and to maintain all data on property ownership and location. Clearly, what emerged was the need for a model (e.g., automation) that could address this significant annual cost.

Westport Project

By 1978, a joint research by UW-Madison faculty (including ERS's Moyer, who was adjunct to the university) was feasible and appropriate. The participants agreed that the *inter*disciplinary (as opposed to multidisciplinary) project would focus on the development and implementation of a multipurpose land information system (MPLIS) for one 36-square mile PLSS township in Dane County, WI. Disciplines represented included economists, civil engineers and surveyors, computer scientists, geographers, planners, and landscape architects.

The Westport Land Records Project (WLRP) involved collaboration among the UW-Madison and ERS researchers along with the Dane County Regional Planning Commission, the Wisconsin State Cartographer's Office, and township government. The general thrust of the Westport Project was to develop an automated and integrated database for the township. In this case, integration called for the ability to spatially integrate cadastral information with natural resource information. Specific objectives were to examine the usefulness, reliability, required accuracy, and cost of a survey-based land records system. That land records system would incorporate mandated natural resource factors such as regulated wetlands. The process would examine the feasibility of developing such a system within a local government setting and would determine public attitudes about land information among the residents of the study area. Specific project tasks of WLRP included the actual development of compatibility among various layers of resource data (i.e., NR 115 wetlands) by integrating these data with an existing survey control network (i.e., tax parcel data) and improving the techniques used for handling these spatial data (Figure 7.5) (Moyer et al. 1982).

The results of WLRP were encouraging, indicating that, even with the economic limits within which local government must operate, the use of a survey network to integrate data layers provided a significant increase in the quality and quantity of information available for decision-making. Based on these results, WLRP was expanded, first to two additional townships to assure that the techniques and procedures developed were suitable under a variety of topographic, population density, and other local conditions.

Dane County Land Records Project

In 1983, the Dane County (Wisconsin) Land Records Project (DCLRP) was undertaken to modernize the land record system. It included all 35 townships, encompassing an area of 1,234 square miles. The four-year research project involved many county departments, the most active of which were Land Conservation, Land Regulation and Records, Regional Planning, and the Division of Systems and Data Processing in the Department of Administration. Other cooperators included UW-Madison, Wisconsin Departments of Agriculture, Trade and Consumer Protection, and Natural Resources, and U.S. Departments of Agriculture, Interior, and Commerce. Within the university, departments from the Colleges of Agricultural and Life Sciences and Engineering were involved.

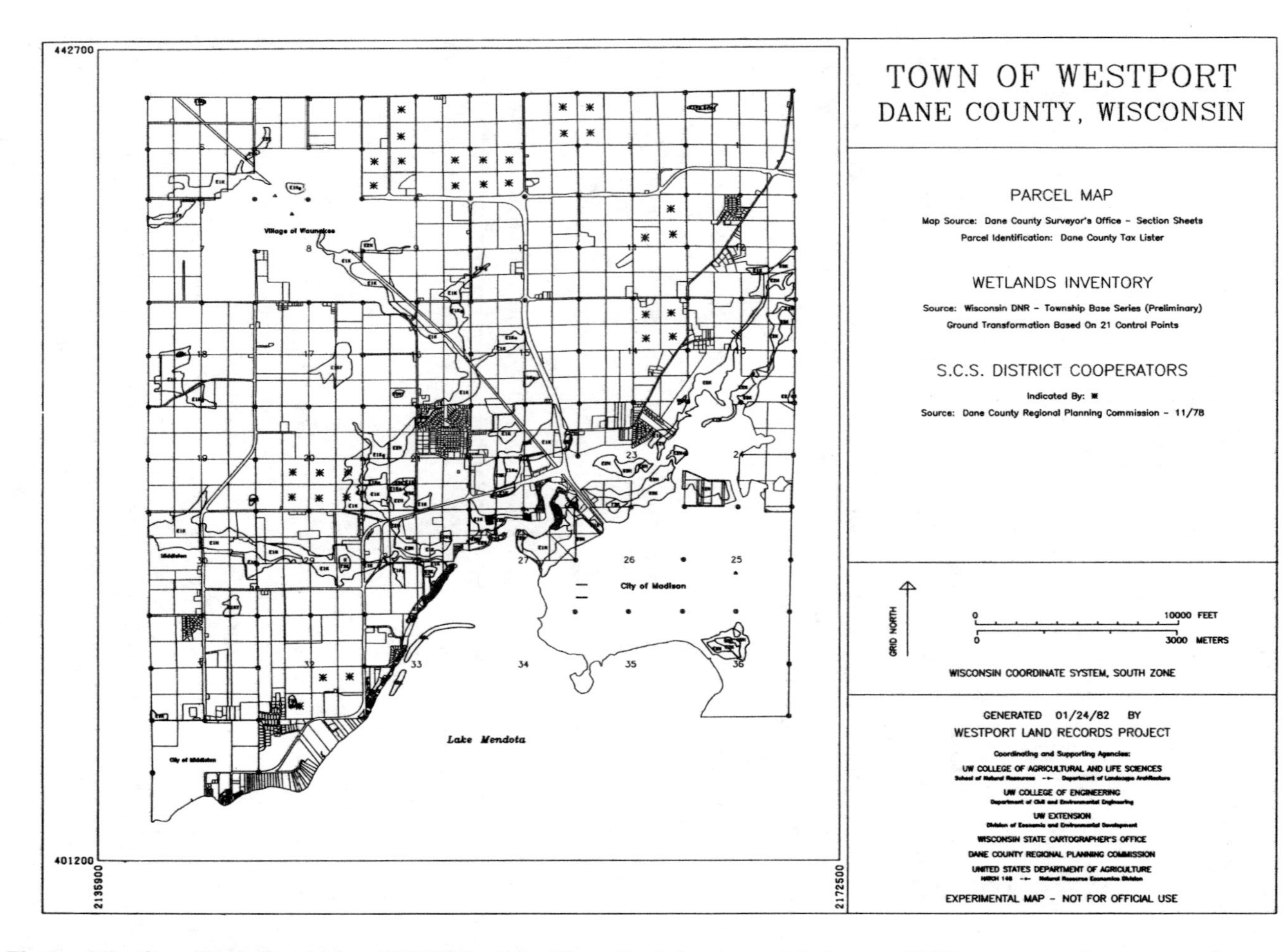

Figure 7.5 Specific project tasks of WLRP included the actual development of compatibility among various layers of resource data (i.e., NR 115 wetlands) by integrating these data with an existing survey control network (i.e., tax parcel data).

The cooperators focused on four areas that had been determined to be critical in any land-record-reform effort: institutional reform, cost effectiveness, technological innovation, and product delivery. The broad goal was to build a multipurpose land information system (MPLIS) based on the four-part foundation. The decision was made to develop, evaluate, and test the MPLIS concept, relying on individual data layers maintained by legislatively mandated agencies. These data layers were to be integrated using a common mathematical reference system.

By the end of the four years, DCLRP had investigated the means to improve efficiency of data collection and handling using scanners, the Global Positioning System (GPS), and remote-sensing imagery. The project directed attention to ways to improve interdisciplinary and interagency cooperative efforts, using cooperative agreements, weekly project meetings, and user-training sessions. Among the assumptions underlying DCLRP were that existing records could be improved to better serve the needs of land planning, assessment, and managing; improvements in land record systems could be made incrementally; the use of computing and geopositioning technology would provide the means to integrate land data layers into a comprehensive MPLIS; and a cost-effective MPLIS system could be developed, gradually transferring data and functions from the manual systems to an automated system; and that this could be done by upgrading and retraining incumbent personnel and implementing institutional reforms as needed.

Multipurpose Land Information System Concept

From the outset, the development of an MPLIS as part of DCLRP was based on two premises regarding LIS improvement. First, each agency that has a legal mandate to collect and store a particular set of spatial information should also be responsible for maintaining this specific data layer in a digital form. Second, a mathematical reference framework (such as the National Spatial Reference System [NSRS], formerly known as the National Geodetic Reference System or NGRS) should be used to provide the linkage between individual layers of data (Chrisman et al. 1984; Chrisman and Niemann 1985). All of the data layers shown were developed as part of DCLRP (Figure 7.6), beginning with the three townships that were the pilot areas for the earlier Westport Project.

Each data layer was the responsibility of a specific government agency. Custodians for those layers included county, state, and federal agencies that took the lead in automation of the initial data set and remained responsible for updating (maintaining) those respective data layers over time.

Many of the data applications of land information addressed in DCLRP involved ownership information. Although deed records are most often thought of as the best source for ownership information, there was no mathematical description of parcels in the Register of Deeds Office. Because of this lack of a coordinate description for land ownership parcels, DCLRP used tax assessment parcels for the applications described below. (Later studies investigated the feasibility of using coordinate geometry [COGO] to convert the textual deed descriptions to coordinate descriptions.) Therefore, the system developed in DCLRP, in which parcel data description was a key component, was clearly distinguishable from GIS systems with their focus

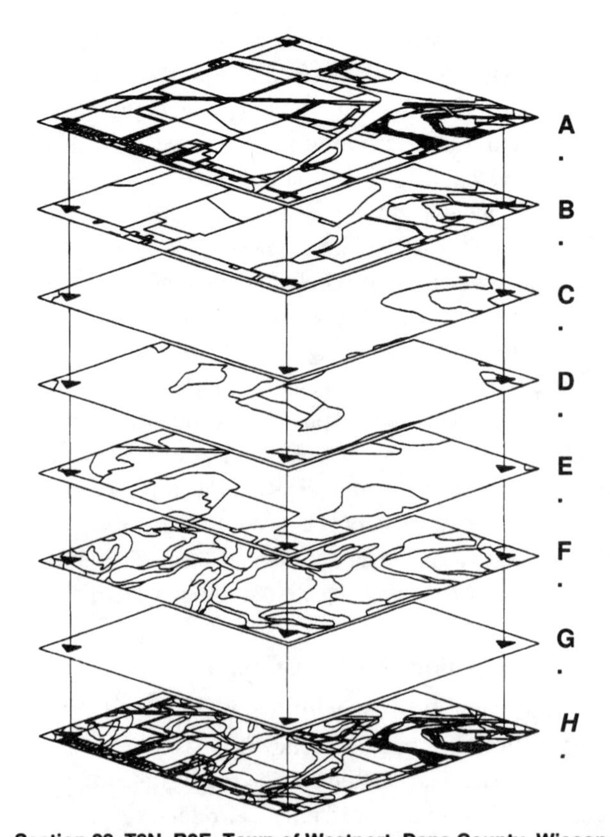

Section 22, T8N, R9E, Town of Westport, Dane County, Wisconsin

Data Layers:	Responsible Agency:
A. Parcels	Surveyor, Dane County Land Regulation and Records Department
B. Zoning	Zoning Administrator, Dane County Land Regulation and Records Department
C. Floodplains	Zoning Administrator, Dane County Land Regulation and Records Department
D. Wetlands	Wisconsin Department of Natural Resources
E. Land Cover	Dane County Land Conservation Committee
F. Soils	U.S. Department of Agriculture- Natural Resource Conservation Service
G. Reference Framework	Public Land Survey System corners with geodetic coordinates
H. Composite Overlay	*Layers integrated as needed; example shows parcels, soils, and reference framework.*

Figure 7.6 The development of an MPLIS was based on the premises that each agency that has a legal mandate to collect and store a particular set of spatial information should also be responsible for maintaining this specific data layer in a digital form and that a mathematical reference framework should be used to provide the linkage between individual layers of data (Chrisman et al. 1984; Chrisman and Niemann 1985).

on spatial data in general. The research team also investigated several techniques to establish the geoposition reference system for the various data layers.

Geopositioning of data layers depends on the linkage to two systems: the legal system (i.e., the PLSS section corners) and the mathematical system (i.e., the NGRS survey network stations). Several techniques were used to place mathematical coordinates on PLSS section corners, including traditional terrestrial (ground) surveying, Doppler surveying, macrometer (a forerunner of GPS), and inertial surveying. These tests revealed order-of-magnitude cost reductions for non-traditional techniques compared to traditional terrestrial surveying techniques. Improvements in GPS hardware and software since DCLRP have made the differences even more dramatic.

Land-value assessment is a function important to most local governments, for property tax revenues are linked closely to the accuracy and efficiency of value-assessment efforts. As a local government function, land valuation is highly visible in that each landowner wants to be assured that his parcel is properly assessed and that the values for comparable parcels provide for equity in property taxes levied. Using the data automated in DCLRP, researchers were able to combine data for two or more layers (Figure 7.7), such as soils, zoning, floodplains, and assessment land classes (from the Wisconsin Department of Revenue). On the basis of these analyses, maps were produced showing by ownership parcel such things as the value of agriculture land per acre and the value of wasteland per acre (Figure 7.8) (Sullivan et al. 1985). These data were then used by assessors to spot inconsistencies in assessment and to make corrections in the underlying database layers.

The power of the MPLIS concept provided the most dramatic application results with regard to erosion-control planning. In 1982, Wisconsin adopted (under Administrative Rule Ag 160) a requirement that all counties develop and implement an erosion-control plan by 1987. To develop such plans required the integration of a wide array of resource and property data. Some dozen land-use categories had to be evaluated for erosion potential: agriculture, developed and developing non-agricultural, roads and rights-of-way, forests and woodlots, streambanks, lakeshores, and other categories required by county conservation departments. These analyses had to be quite specific in that, for the first time, plans had to identify the location of the land and the owner of that land. This meant that the power of MPLIS was the only effective way to develop such an erosion control plan. It also meant that an automated system was the only effective mechanism by which to update plans, whether required by changes in state policies, statutes, and regulations, or by changes in land use and land cover.

The responsiveness of an MPLIS in this application was clearly demonstrated in one particular facet of DCLRP. In Oregon Township, MPLIS was used to demonstrate how automated information could be used to evaluate soil-erosion control alternatives. One alternative was farmers' use of "conservation tillage techniques" to reduce soil erosion. MPLIS was able to illustrate which land parcels (and owners) under traditional farming practices would not meet the legislative intent of Ag 160 (Niemann et al. 1995). It could also illustrate which land parcels (and owners) would meet the intent of Ag 160 if the landowner were to employ conservation tillage practices. This comparison highlights two aspects of the overall power and utility of an

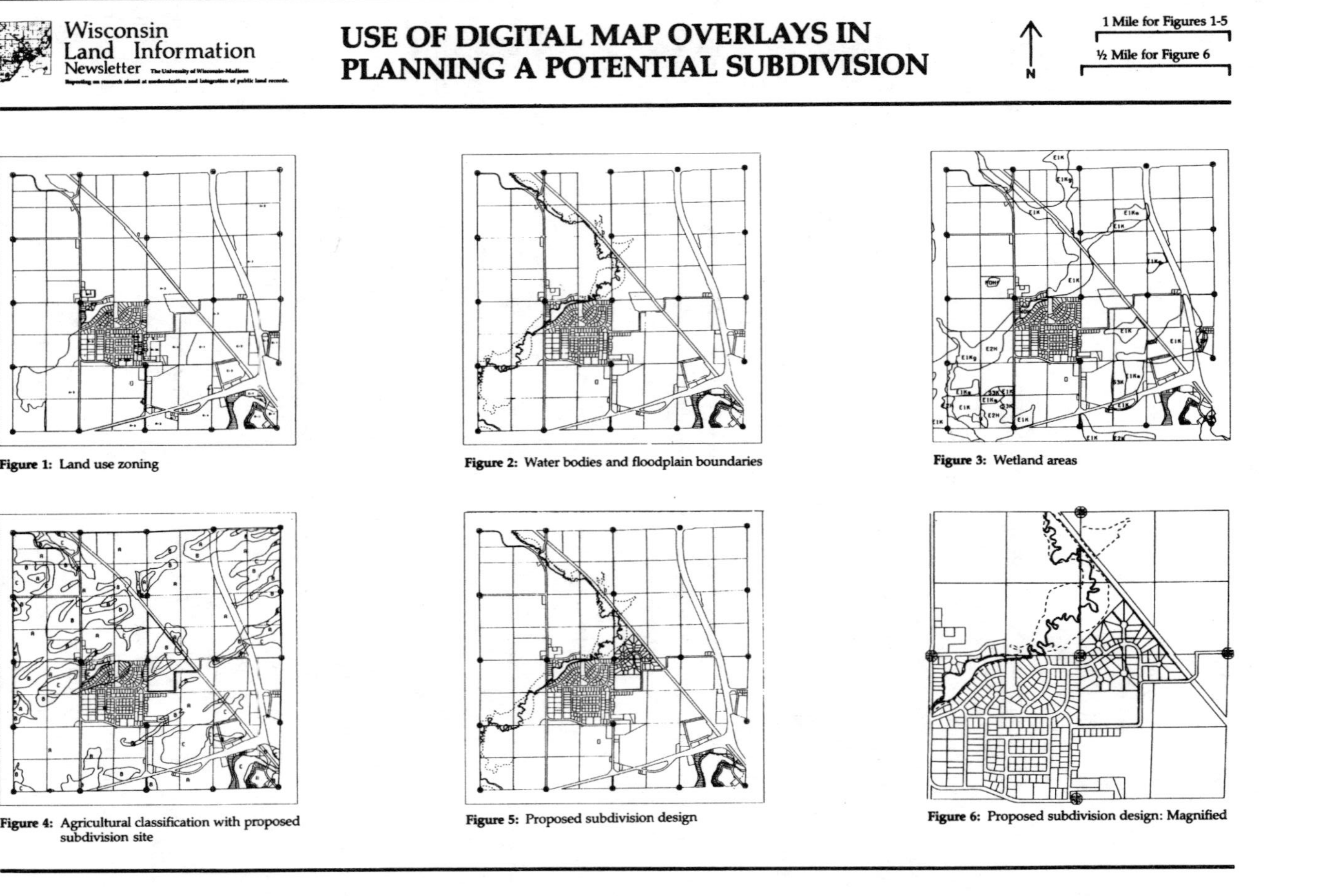

Figure 7.7 Using the data automated in DCLRP, researchers were able to combine data for two or more layers such as soils, zoning, floodplains, and assessment land classes (from the Wisconsin Department of Revenue).

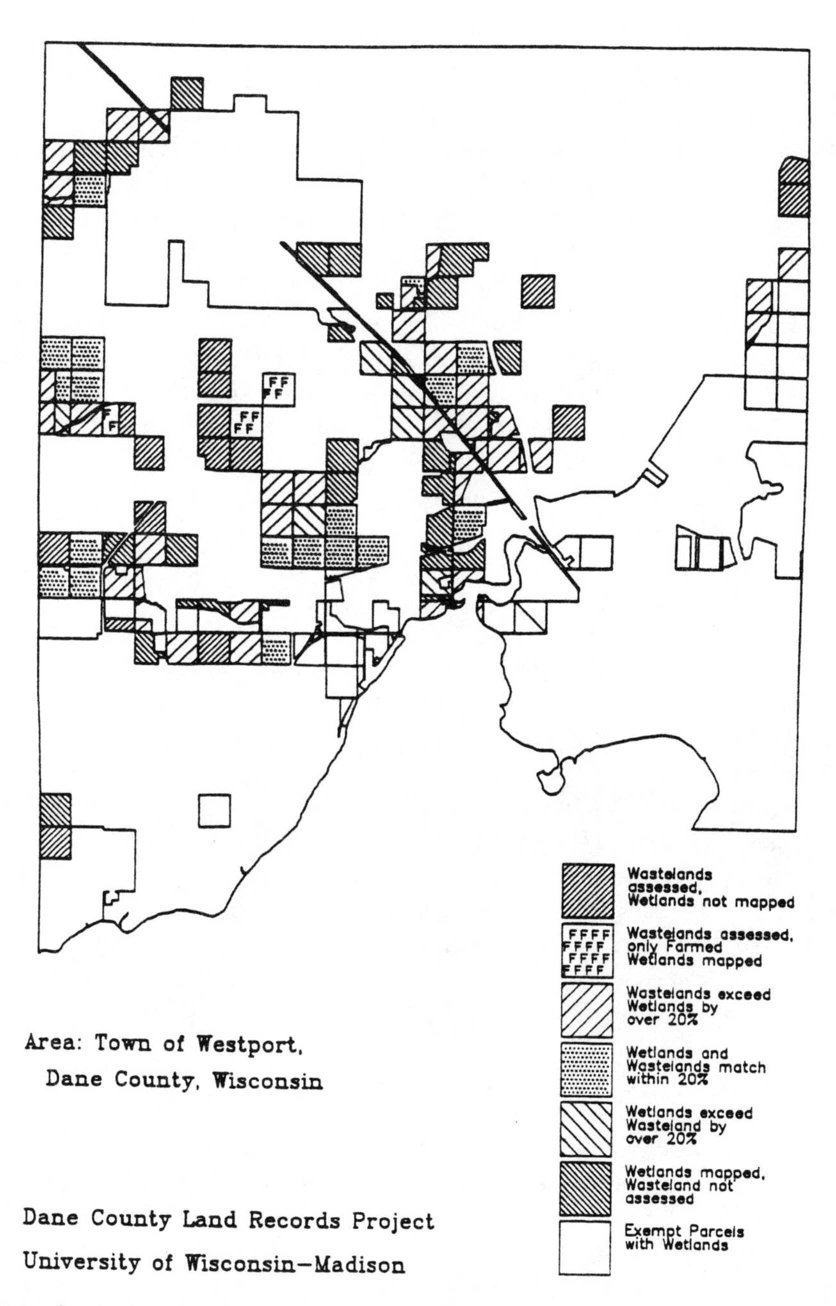

Figure 7.8 On the basis of these analyses of combined data, maps were produced showing, by ownership parcel, such information as the value of agriculture land per acre and the value of wasteland per acre (Sullivan et al. 1985).

LIS using GIS analytical functionality. It demonstrates the ability to conduct "what if" analyses for public policy: What would be the impact if farmers were required to incorporate conservation tillage techniques as the first step to reduce soil erosion? These analyses then could form the initial planning basis for most farm conservation treatments in Dane County. Furthermore, the analysis resulted in what was called by landowners a more equitable process. The Universal Soil Loss Equation (USLE) and its natural resource variances were applied across every land ownership parcel without bias or prejudice.

Additional conclusions about an MPLIS, as shown in DCLRP, are worthy of mention. For example, each government unit (e.g., county department) collects and maintains only the land data needed to meet its mission or legislative charge. Compatibility among land-records-keeping units will improve the availability of land data and lower its cost through reduced duplication of effort. Because of the significant existing investments in land records, modernization should proceed on an incremental basis—that is, agencies should make use of existing land records wherever feasible and should gradually improve the accuracy, reliability, and compatibility of land information as systems evolve.

Several recommendations were also made as a result of DCLRP (Chrisman et al. 1984; Moyer 1987). General recommendations were four:

- Land data should be the responsibility of the lowest level of government that has the technical ability to collect it accurately and efficiently;
- The role of higher levels of government is to develop standards to ensure compatibility among jurisdictions;
- All land data must be based on a mathematical reference framework that will allow the aggregation of local detailed data for more generalized needs; and
- To facilitate providing accessible and usable data at all levels of government, one entity at each level of government should be designated the lead agency, with responsibility for land records and their transactional maintenance.

Specific recommendations were made for local government agencies:

- County government should be the primary access point for publicly held land records; and
- Within each county, all land record activities should be organized into a network using standards, data compatibility and integrability, and modern technology as the key mechanisms upon which data organization is based.

At the state level, it was recommended that:

- The state serve as the focal point for land records and the modernization of land record systems in the state;

- The state should serve as a review authority for state and federal agencies that wish to conduct land data collection or mapping within the state; and
- The state should promote compatibility of land information by encouraging the use of effective, efficient, and compatible land records systems among state and local government agencies.

DCLRP thus built on earlier work of NRC, developing a prototype system that would handle ownership (cadastral) data as well as resource and cultural data and would provide for the integration of all land information in a system that could be used for the development of policy, the evaluation of policy, and the management of the complete range of land-related resources.

Town of Burke

In the Town of Burke (Dane County), major land-use issues had come on the heels of the development of a new corporate headquarters for a major insurance company. A number of municipalities and planning groups were engaged in annexation struggles. As a result, alternate land-use plans and scenarios were being suggested along with evaluations of their environmental consequences.

One such consequence was the impact on water quality in one of the adjoining Madison lakes made famous by Henry Wadsworth Longfellow— Lake Monona—where the Monona Terrace Convention Center, originally designed by Frank Lloyd Wright, opened in 1997. At the Burke site, researchers explored the ability to predict water quality impacts of selected development scenarios. An empirically tested and validated water quality model called Sediment Loading and Management Model (SLAMM), developed by the Wisconsin Department of Natural Resources, was linked to an MPLIS. Given the land-use alternatives, researchers identified the ownership of parcels that would contribute the most pollution to Lake Monona (Pickett et al. 1989). The Burke project demonstrated the ability to employ the analytical functionality of GIS mathematical modeling and the specificity of an LIS as a basis for evaluating land-use planning alternatives. It further confirmed the function of GIS in identifying, by ownership, those areas most susceptible to water quality impacts and most in need of on-site mitigation.

Town of Middleton

Use of an MPLIS for land use planning in the Town of Middleton is another example. Middleton was facing some of the most aggressive urban growth pressures in Wisconsin. The Town Planning Commission requested assistance in preparing land-use planning maps, included parcel maps that noted extraterritorial limits, land use, and existing zoning, and land tenure status of the farmed lands. Other maps depicted productive agricultural soils and sensitive ecological areas, all associated with ownership parcels (Figures 7.9 and 7.10). With these maps, the commission proposed land-use plans whereby each landowner could identify the proposed changes with respect to his/her individual holdings. This application demonstrated that it is tech-

Town of Middleton Project
Dane County, Wisconsin

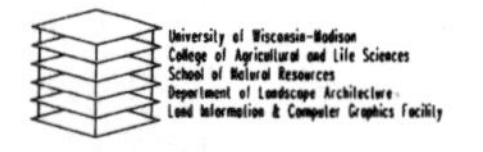

Soil Productivity on Agricultural Lands

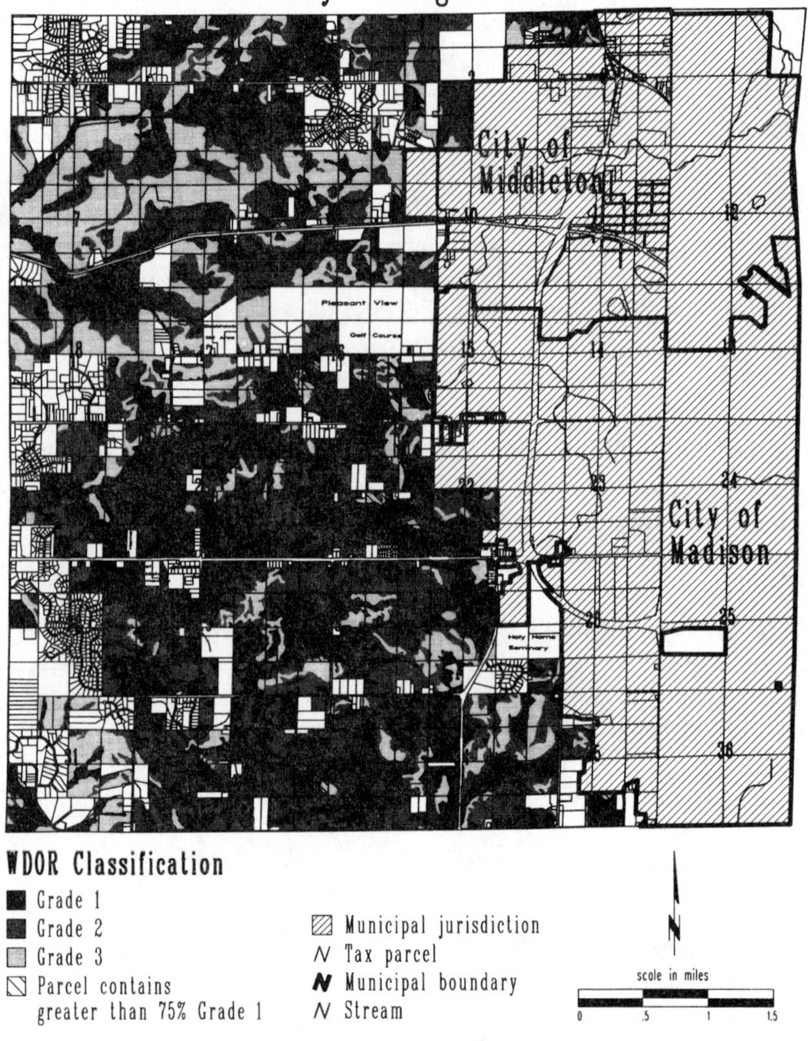

Figure 7.9 The Middleton Town Planning Commission used MPLIS in preparing land-use planning maps including parcel maps that noted, among other attributes, soil productivity grade for each parcel.

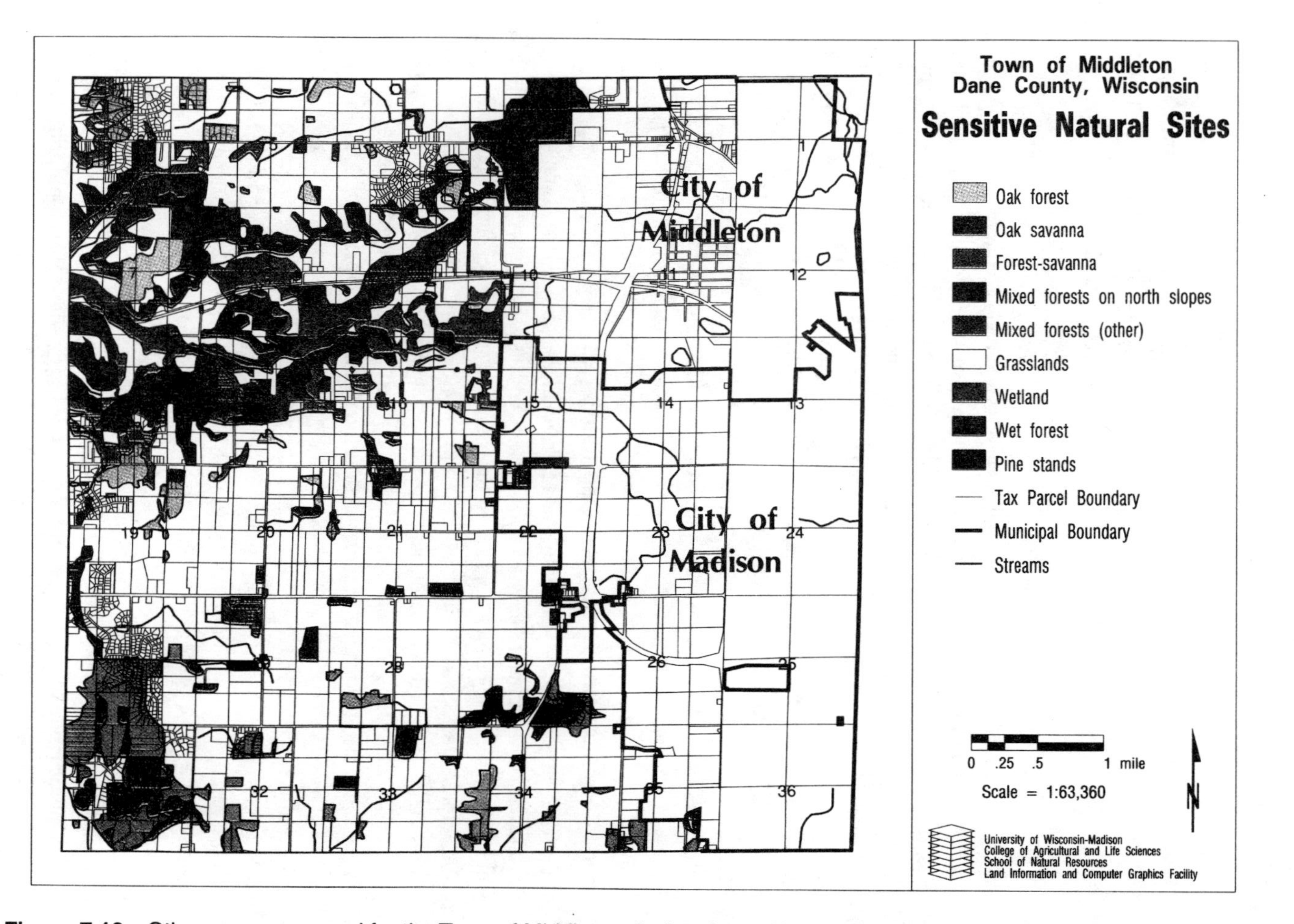

Figure 7.10 Other maps prepared for the Town of Middleton depicted sensitive ecological areas associated with ownership parcels.

nically possible to produce land-use planning maps and information useful to ordinary citizens. It demonstrates the ability to integrate and display parcel-based information with natural phenomena such as soils, wetlands, and sensitive ecological areas. This MPLIS demonstrated another land-use dimension usually left out of the planning process: land tenure probability or landowner intent. Some planners suggest that land tenure is the missing dimension in land use planning (Popper 1981). MPLIS are fully capable of maintaining such variables.

UW Outreach and Wisconsin Initiatives

UW-Madison has continued to refine and expand its work on LIS improvements. Outreach has been and continues to be a major part of UW-Madison's LIS modernization effort throughout the state. UW-Madison has an active outreach program through UW-Extension to communicate findings and to engage the general population in dialogue concerning many issues. Many researchers pursue this outreach objective informally as well. A review of some of these efforts is helpful in understanding the evolution of the LIS modernization program in Wisconsin and in understanding how Wisconsin's LIS program has impacted other parts of the country.

Since the 1970s, the UW-Madison Department of Landscape Architecture has worked with groups throughout the state to publicize research activities and to elicit reaction to findings and recommendations that flow from research. Once the LIS/GIS efforts began to gain steam in the 1980s, the interest in and demand for additional information increased dramatically.

Land Information and Computer Graphics Facility

At that time, the College of Agricultural and Life Sciences approved the establishment of the Land Information and Computer Graphics Facility (LICGF) as a focal site for outreach in LIS, along with research. With colleagues from across the campus, LICGF researchers and faculty were able to concentrate on LIS issues. Early efforts emphasized the interdisciplinary nature of land information systems, reflected in dynamic and far-reaching intercollegiate efforts (see below).

Among LICGF's more recent projects are two major efforts at drawing together the resources of local, state, and federal agencies in developing LIS: CONSOIL and LOCALIS.

CONSOIL (Conservation of Natural Resources through Sharing of Information Layers) built on the results of DCLRP. The linkage of resource polygon and ownership parcel data was particularly important in analyses that provided input to Congress as the 1985 and 1990 Farm Bills were prepared. Work on the conservation and quality improvement of soil and water resources was also a major part of the CONSOIL project.

Project LOCALIS (from the Latin for *local*) has provided a mechanism for interactive research and development between university researchers and local governments as counties respond to the objectives of a recently created statewide land information program. Problem-solving workshops, joint software development, and critical review of national initiatives are some of the activities that have characterized LOCALIS I and LOCALIS II as they have brought university and county LIS experts together.

Intercollegiate Efforts

Beginning in the early 1980s, UW faculty from a variety of colleges and programs held several workshops, seminars, and other special programs, focusing on multipurpose cadastre, LIS, and GIS issues. Two of these efforts are worthy of particular note.

In the spring of 1984, Niemann and his colleagues organized a semester-long seminar to focus on the "technical and institutional issues associated with the creation of modern land information systems" (Niemann 1984). Land information experts from the United States, Canada, and Europe participated. Speakers, representing federal, state, and local agencies as well as the private sector, included surveyors, lawyers, planners, economists, landscape architects, and computer scientists.

More than 1,500 people attended the seminar series. Enthusiasm generated by the seminar led to the formation of a statewide land information coalition of those anxious to modernize LIS in Wisconsin. This group spearheaded efforts that resulted in the governor's appointment of the Wisconsin Land Records Committee (WLRC) in 1985 (WLRC 1987).

In 1986, a second noteworthy educational effort by UW-Madison was the Workshop on Multipurpose Cadastre Information Systems. Twenty speakers from a variety of disciplines had been invited to "present ideas, discuss their experiences, and offer recommendations on how to modernize our land records systems to produce multipurpose land information of mutual benefit to the public and its supporting institutions" (Niemann and Moyer 1988). Since the time of that workshop, UW-Madison has used the term "multipurpose land information system" (MPLIS) rather than "multipurpose cadastre."

Wisconsin Land Records Committee

WLRC was one of the outgrowths of the efforts of the *ad hoc* coalition that organized itself in Wisconsin to promote LIS modernization. The 32-member committee had broad geographic and professional diversity, including non-WLRC members rounding out the 12 subcommittees.

WLRC was asked to "[e]xamine and address the immediate needs of state and local agencies regarding land collection and management" and to "[d]evelop recommendations on how Wisconsin should approach the long-term issues of land records modernization" (WLRC 1987).

After two years of work, WLRC presented to the governor and state legislature a final report outlining a program for LIS modernization in Wisconsin. The WLRC recommendations were five:

1. Establish a Wisconsin Land Information Board (to establish program policy and provide overall direction);
2. Establish an Office of Land Information (to administer program developed by the board);
3. Establish a Grants-in-Aid Program (to help fund development of local and regional MPLIS systems and provide technical assistance);

4. Provide incentive for each county to designate a County Land Information Office (as a specified point of contact between the state and the county); and
5. Encourage coalition members to establish a private, nonprofit, nongovernmental Land Information Association.

Wisconsin Land Information Program

The first four WLRC recommendations were adopted by the legislature in 1989, with a few minor modifications (Holland 1992), creating the Wisconsin Land Information Program (WLIP). Since then, about $30 million has been directly generated statewide to help Wisconsin's 72 counties start the process of modernizing their local land records systems and to begin the countywide implementation of MPLIS. Over the same period, another $30 million has been generated from non-WLIP sources such as utilities, regional, state, and federal agencies and additional revenues from local governments. Using these figures, experts estimates that, by the end of the decade, more than $100 million will have been invested in countywide MPLIS in Wisconsin (Niemann et al. 1995).

Wisconsin Land Information Association

The fifth WLRC recommendation was pursued by the earlier *ad hoc* coalition, and the Wisconsin Land Information Association (WLIA) was formed. WLIA has continued to be a active part of the development of parcel-based land information systems in Wisconsin. Membership totals almost 700 after only seven years.

National States Geographic Information Council

Because of Wisconsin's early and comprehensive work on LIS, efforts in this state have had numerous impacts beyond its borders. The reports of the WLRC and its 12 subcommittees have been requested by GIS and LIS committees, boards, and commissions in many states. Several states have adopted specific parts of the structure of the WLIP for their own use.

Further, once the WLIP was in operation, program staff provided leadership in launching the National States Geographic Information Council (NSGIC). William S. Holland, the first executive director of the Wisconsin Land Information Board, served on the inaugural board of directors of NSGIC and then served as its president for 1994-1995. NSGIC, an organization of the geographic information leaders and managers in state governments, has responded to a large number of requests for guidance and technical assistance from many states and European countries. A number of Wisconsinites continue to play active roles in NSGIC, particularly in regard to its major effort to organize and help implement the National Spatial Data Infrastructure.

National Center for Geographic Information and Analysis

The National Center for Geographic Information and Analysis (NCGIA) was established by the National Science Foundation after competition among a number of universities and groups of universities. A consortium of the University of Maine at Orono, the State University of New York at Buffalo, and the University of California at Santa Barbara was named the center and funded for an eight-year period to develop curricula in GIS and to identify research issues in geographic information and analysis.

The Wisconsin view as to the importance of both land ownership data (usually parcel-based) and natural resource data (typically polygon-based) was not chosen as the approach for NCGIA to pursue, in terms of the selection of the center. The importance of land parcels, natural resource polygons, and LIS generally, however, was recognized and represented on the first NCGIA board of directors by Moyer, who served from 1988 to 1993. In addition to the research conducted on the NCGIA center campuses, NCGIA also supports multistate cooperative research projects. UW–Madison LICGF is currently participating in one such NCGIA-funded project to study patterns of innovation for LIS improvements

Summary and Conclusions

The importance of combining cultural *and* resource data was recognized by the NRC in its 1983 report on the multipurpose cadastre. The concept is as important today as it was then. MPLISes that combine ownership and resource data are the only effective way to carry out the analyses and the management of land and related resources in a cost-effective manner.

Efforts to develop automated LIS have resulted from forces and needs that have appeared or expanded in recent years. Those efforts began with several institutions and involved a substantial number of individuals. Over the years, state and federal involvement and support have waxed and waned as money, politics, and individuals have entered or exited the stage. But the big picture reveals progress. A part of that progress has been the awakening of institutions at every level to an understanding of the potential of LIS, and a recognition of the degree of involvement and the willingness necessary to take advantage of multipurpose, parcel-based land information systems.

The needs in LIS are still expanding in terms of databases required, in terms of the sophistication of software programs required for the analyses called for by decisionmakers, and more. Recent advances in software have reduced the need to distinguish LIS and GIS as separate areas of expertise. The history of LIS suggests that the differences will become even fewer in the future. The goal of both LIS and GIS is to provide data for decision-making—data that are accurate, suitable for the analyses in question, available in a timely manner, and affordable, given the tasks at hand. MPLIS can meet these objectives, provided that the organizations that stand to benefit most can make the institutional changes necessary to maximize its potential.

Bibliography

Behrens, J. O., D. D. Moyer, and G. Wunderlich. 1974. *Land Title Recording in the United States: A Statistical Summary.* Washington, D.C.: U.S. Department of Agriculture and Bureau of the Census, U.S. Department of Commerce, 38.

Canadian Surveyor. 1969. "Proceedings of the Symposium on Land Registration and Data Banks." *Canadian Surveyor,* 23 (1 & 2).

Chenoweth, R., and B. J. Niemann, Jr. 1989. "What Price Beauty?" *Landscape Architecture,* 75 (5): 78.

Chowdhury, M. N. 1994. "Perceived Importance of Land Ownership Information for Rural Land Use Planning in Dane County, Wisconsin: A Case Study." Unpublished Ph.D. dissertation, University of Wisconsin–Madison.

Chrisman, N. R., and B. J. Niemann, Jr. 1985. "Alternative Routes to a Multipurpose Cadastre: Merging Institutional and Technical Reasoning." *Proceedings of AUTO-CARTO 7.* Falls Church, VA: ASPRS, 84–94.

Chrisman, N. R., D. F. Mezera, D. D. Moyer, B. J. Niemann, Jr., and A. P. Vonderohe. 1984. *Modernization of Routine Land Records in Dane County, Wisconsin: Implications to Rural Landscape Assessment and Planning.* URISA Professional Paper Series No. 84-1. Washington, D.C.: Urban and Regional Information Systems Association, 44.

Cook, R. N. and J. L. Kennedy, Jr. eds. 1967. *Proceedings of the Tri-State Conference on a Comprehensive, Unified Land Data System (CULDATA).* Cincinnati, OH: University of Cincinnati College of Law, 253.

Dutton, G. 1979. Chapter 10. In *Monitoring of Foreign Ownership of U.S. Real Estate, A Report to Congress.* 3 volumes. Washington, D.C.

Economic Research Service, U.S. Department of Agriculture. 1979. *Monitoring of Foreign Ownership of U.S. Real Estate, A Report to Congress.* 3 volumes. Washington, D.C.

Forsyth County, North Carolina. 1974. *Land Records Information System.* Winston Salem, N.C., 65.

Gillespie, E., and B. Schellhas. 1994. *Contract with America. The Bold Plan by Rep. Newt Gingrich, Rep. Dick Armey and the House Republicans to Change the Nation.* New York: Times Books, 196.

Institute for Land Information. 1987. "The Economics of Land Information: A Symposium of the Institute for Land Information." *Technical Papers of the 1987 ASPRS-ACSM Annual Convention.* Falls Church, VA: ASPRS, 78.

Larsen, B., J. L. Clapp, A. Miller, B. J. Niemann, Jr., and A. Ziegler. 1978. *Land Records: The Cost to the Citizen to Maintain the Present Land Information Base: A Case Study in Wisconsin.* Madison, WI: Department of Administration, 64.

Lewis, P. H., Jr. 1964. "Quality Corridors." *Landscape Architecture,* 54 (1): 100–104.

Manning, W. 1913. "The Billerica Town Plan." *Landscape Architecture,* 3 (3): 108–118.

Maritime Council of Provinces. 1968. *Proceedings of a Symposium on Land Registration and Data Banks.* Fredericton, New Brunswick: Maritime Council of Provinces.

McHarg, I. 1969. *Design with Nature.* New York: Natural History Press, 197.

Miller, A. H., and B. J. Niemann, Jr. 1972. *An Interstate Corridor Selection Process: The Application of Computer Technology to Highway Location Dynamics. Phase 1.* Madison, WI: Environment Awareness Center, Department of Landscape Architecture, University of Wisconsin–Madison, 240.

Moyer, D. D. 1977. "An Economic Analysis of the Land Title Record System." Unpublished Ph.D. dissertation. University of Wisconsin–Madison, 625.

Moyer, D. D. 1978. *Land Information Systems. An Annotated Bibliography.* Washington, D.C. and Falls Church, VA: Natural Resources Economics Division, Economics, Statistics, and Cooperatives Service, U.S. Department of Agriculture and North American Institute for the Modernization of Land Data Systems, 195.

Moyer, D. D. 1987. "Multipurpose Land Information Systems in Wisconsin: Content and Process." *Assessment Digest,* 9 (September/October): 13–18.

Moyer, D. D., and K. P. Fisher. 1973. *Land Parcel Identifiers for Information Systems.* Chicago, IL: American Bar Foundation, 600.

Moyer, D. D., J. Portner, and D. F. Mezera. 1982. "Overview of a Survey-Based System for Improving Data Compatibility in Land Record Systems." *Computers, Environment, and Urban Systems Journal,* 7 (4): 349–358.
National Research Council (NRC). 1980. *Need for a Multipurpose Cadastre.* Washington, D.C.: National Academy Press, 112.
National Research Council (NRC). 1982. *Modernization of the Public Land Survey System.* Washington, D.C.: National Academy Press, 74.
National Research Council (NRC). 1983. *Procedures and Standards for a Multipurpose Cadastre.* Washington, D.C.: National Academy Press, 173.
Niemann, B. J., Jr., ed. 1984. *Seminar on the Multipurpose Cadastre: Modernizing Land Information Systems in North America.* IES Report No. 123. Madison, WI: Institute for Environmental Studies, University of Wisconsin–Madison, 320.
Niemann, B. J., Jr., and D. D. Moyer, eds. 1988. *A Primer on Multipurpose Land Information Systems.* Report 113. Madison, WI: Institute for Environmental Studies, University of Wisconsin–Madison, 176.
Niemann, B. J., Jr., and S. S. Niemann. 1993a. "An Innovator in Pursuit of Synthesis: Ken Dueker." *Geo Info Systems,* 3 (6): 62–67.
Niemann, B. J., Jr., and S. S. Niemann. 1993b. "GIS Diffusion at Bonneville Power Administration: Tim Murray." *Geo Info Systems,* 3 (8): 57–61.
Niemann, B. J., Jr., and S. S. Niemann. 1993c. "Lines of Code and More: David Sinton." *Geo Info Systems,* 3 (10): 58–62.
Niemann, B. J., Jr., and S. S. Niemann. 1994a. "Revolutionaries in Action: Nick Faust, Lawrie Jordan, Bruce Rado, and Steve Sperry." *Geo Info Systems,* 4 (2): 53–57.
Niemann, B. J., Jr., and S. S. Niemann. 1994b. "The Whole Nine Yards at the Tennessee Valley Authority: Bruce Rowland." *Geo Info Systems,* 4 (4): 47–51.
Niemann, B. J., Jr., and S. S. Niemann. 1994c. "Nicholas Chrisman: Innovation with Affect, Parts 1 & 2." *Geo Info Systems,* 4 (7): 20–30; 4 (9): 46-52.
Niemann, B. J., Jr., and S. S. Niemann. 1995. "Scott Morehouse: 100 Percent Overhead." *Geo Info Systems,* 5 (10): 20–25.
Niemann, B. J., Jr., D. J. Tulloch, and C. H. Kirk. 1995. *Overview of Program Accomplishments (WLIP) 1989–1995.* Madison, WI: Wisconsin Department of Administration, 35.
North American Institute for Modernization of Land Data Systems (MOLDS). 1975. *Proceedings of the North American Conference on the Modernization of Land Data Systems.* Falls Church, VA: ACSM.
North American Institute for Modernization of Land Data Systems (MOLDS). 1979. "Land Data Systems Now." *Proceedings of the Second MOLDS Conference.* Falls Church, VA: ACSM.
Pickett, S. R., P. G. Thum, and B. J. Niemann, Jr. 1989. "Using a Land Information System to Integrate Nonpoint Source Pollution Modeling and Land Use Development Planning." *Proceedings of the URISA Conference.* Washington, D.C.: URISA, 373–388.
Popper, F. 1981. *The Politics of Land Use Reform.* Madison, WI: University of Wisconsin Press, 321.
Roberts, W. F. 1968. "Some Basic Features of an Environmental Integrated Data Bank." *Canadian Surveyor,* 23 (1): 30–33.
Sinton, D. (n.d.) "Reflections on 25 Years of GIS." *GIS World* (Supplement): 8.
Steinitz, C. 1993a. "GIS: A Personal Perspective." *GIS Europe,* 1 (5): 19–22.
Steinitz, C. 1993b. "Geographical Information Systems: A Personal Historical Perspective, the Framework for a Recent Project, and Some Questions for the Future." *Papers of the European Conference on Geographic Information Systems,* 7.

Steinitz, C., P. Parker, and L. Jordan. 1976. "Hand-Drawn Overlays: Their History and Prospective Uses." *Landscape Architecture,* 66 (5): 444–445.

Sullivan, J. G., N. R. Chrisman, and B. J. Niemann, Jr. 1985. "Wastelands vs. Wetlands in Westport Township, Wisconsin: Landscape Planning and Tax Assessment Equalization." *Proceedings of the URISA Conference.* Washington, D.C.: URISA, 73–85.

White, J. P., ed. 1969. *Proceedings of a Workshop on Problems of Improving the United States System of Land Titles and Records.* Indianapolis, IN: University of Indiana Press.

Wisconsin Land Records Committee. 1987. *Final Report of the Wisconsin Land Records Committee: Modernizing Wisconsin's Land Records.* Madison, WI: Institute for Environmental Studies, University of Wisconsin-Madison, 53.

CHAPTER 8

AM/FM Entry: 25-Year History

Keith McDaniel, Chuck Howard, and Hank Emery

Introduction

Many people ask the question, "What is the difference between AM/FM and GIS?" (AM/FM stands for automated mapping/facilities management.) If the question were addressed to most software vendors, they would respond that their system was both. The authors believe that in 1997 at the GIS package software level, the differences are negligible. But historically the differences were very real, and today they can still be measured at the application requirements level. This chapter tells the history of the discipline that came to be known as AM/FM. The reader can judge for himself or herself if there is any real difference other than application of the technology.

This chronology starts with the basic requirements under which utilities and public works departments (hereafter referred to collectively as infrastructure organizations) had to keep records for their installed plants. It progresses to the early and middle 1960s where new demands put a strain on the capabilities of these records systems. This strain is what gave birth to the ideas we called AM/FM.

Keith McDaniel is executive director of Strategic Planning Utility Applications with Intergraph Corporation. He has more than three decades of experience in the utilities industry with the design and implementation of major data conversion for automation of AM/FM and GIS.

Chuck Howard is the president of Geographic Information Technology, Inc., providing infrastructure organizations with consulting and system implementation services.

Hank Emery is a principal of Emery & Associates, providing consulting services to public utilities for automation services. He has 35 years of experience in AM/FM/GIS involving more than 100 major utilities projects.

Author's Addresses: Keith McDaniel, Intergraph Corp., 7400 E. Orchard Rd., Suite 3000, Englewood, CO 80111. Chuck Howard, Geographic Information Technology, 1115 Berkeley Ct., Longmont, CO 80503. Hank Emery, Emery & Associates, Inc., 7462 East Princeton Ave., Denver, CO 80237. E-mail: hemery@sni.net.

The next two sections tell about the early pioneers who contributed ideas. The text focuses on the utility companies which made investments that spurred the technology development along, but it also talks about some of the individuals who made notable contributions. Special emphasis is placed on the work done by Public Service Company of Colorado (PSCo) because it contributed an extraordinary number of enduring ideas and individuals who made real contributions to the development of this important technology.

These authors recognize the years of 1978 through 1986 as the years a real AM/FM industry was formed. Large companies went beyond research and developed products. Utilities that were normally risk adverse started to investigate the technology. Professional organizations were formed to educate and promote the discipline. These were exciting years with many new businesses entering the field, especially conversion service companies.

The last two sections of this chapter outline events in more recent years, when the technology became more robust and the lines between AM/FM and GIS began to blur. The reader may conclude that being able to define the differences is not very important any more. However, we can all celebrate that the competition that existed between these two camps produced an array of choices in today's marketplace that no one could have predicted in the 1960s.

An Aging 100-Year-Old Process: 1860s–1960s

As utilities and cities began to provide their distribution services in the 19th century, the record systems were inconsistent and incomplete by any standard. Initially, they were the minimum necessary to obtain rights of way. Obviously, buried facilities required more documentation than those that were above ground, but most records were merely sketches of what was built, and many of these were kept by the people who actually did the construction and had the maintenance responsibility. Much of what we consider essential information today was recorded only in the memories of those who did the installation work.

Government regulation probably had a role in the early formation of "official records." As privately owned utilities spread their services, they had the need and the right to set the rates they would charge. As competition grew, government saw the benefit to consumers of limiting the amount of installed infrastructure to only what was necessary to provide the needed services. This meant that service territories needed to be assigned to eliminate duplication. From this first level of regulation for consumer protection came the rules for setting rates. To encourage rapid proliferation of services, and the massive investment this took, came the concept of rates to guarantee a return on investment. This required good accounting records of what was installed to assure a proper "rate base."

Most likely, the next step was taxing authorities wanting payment for the utilities' right to exist (and make profit) in their jurisdictions. This added the element of needing knowledge of location of the facilities to just knowing their value.

The utilities had growing information needs, also. As business and industry became dependent on these essential services, they demanded reliability. This created the need for networking to provide alternative service sources. This meant better engineering, design, and maintenance, which demanded better records. Safety issues added their demands, as did other drivers. As the systems became more complex, so did the drawings. Since utilities were noncompetitive, they openly shared business practices, and since they were regulated, they had common reporting requirements. So it was natural that some consistency would develop in how they kept their records. By the late 1930s, maps became a critical recording medium for the complex facilities systems.

Many of the early mapping systems had little cartographic discipline. In many cases, north was some direction other than up, and scales were set at whatever met the current need (some were not scaled). Land information was just a reference to find assets, so projection systems were not even considered. Little was done to resolve mete edge discrepancies.

The 1960s were volatile years with much societal change. People took a deeper interest in the environment, and their attitudes toward business became more aggressive. The movement toward more underground utilities began, as did more litigation over a variety of issues. However, there were still a few utilities that did not yet have manual mapping systems. The regulators intervened in a couple of cases and mandated that some level of mapping be done. Carolina Power and Light was one such utility. This brought about a lot of change in infrastructure organizations, attitudes toward records, and technological developments were showing promise of providing solutions to known problems.

One big technological event occurred in April 1964 while most engineers were using slide rules and tables for problem solving. IBM introduced its 360 line of computers. These computers were significantly different from previous generations in that they were multipurpose. Prior to that event, computers were either "scientific" or "commercial" in design and application. These two computing platforms had different operating systems, languages, and user profiles. System 360 was such a huge success for IBM that other mainframe providers followed its lead.

The combining of these two environments had a dramatic impact that affected infrastructure organizations in two ways. The first was that the engineering organizations gained more access to the company's computing resource. Utilities had always been big users of commercial style computers for billing and financial records. However, they never spent much on scientific computing platforms that would have been more appropriate for the engineers. As utilities purchased these multipurpose platforms, engineers and operations people began developing decision support tools in an effort to modernize.

The second way this new generation of computers helped is that it enabled ideas and encouraged development of the early database management systems. The ideas and discussions that surrounded the anticipation of database managers included the concept of enterprise-wide computing architectures and Management Information Systems (MIS). Although it took many years before technology could support these concepts becoming reality, the ideas were born in the early to mid 1960s, and many forward thinking people started to plan for that idealistic future.

One of the early preparations for automation was the assignment of grid coordinates to facilities as a means of knowing where they were with respect to everything else. Vernon Graphics, a small mapping company in Elmsford, NY, had made a good business from contracting the mapping services for several eastern utilities. These utility maps were made on a cartographically correct base. Bell of Pennsylvania, one of Vernon's customers, wanted to assign a unique identifier to each facility. The idea emerged to use the state plane coordinate as an identifier that would be unique to each facility and be a clear indication of location. In 1963, Bell of Pennsylvania presented the idea to AT&T, which had to approve such things in those days. The idea got a lot of discussion but was not approved because of the expense involved.

Not long after, in 1964, Pennsylvania Power & Light implemented the idea in its mapping system. The map sheets were indexed by the high order digits of the state plane grid number, and each facility carried a number that was the low order digits, down to a 10 foot by 10 foot square. This was all manual mapping, but it was good preparation for what was to come. A few other utilities, such as Metropolitan Edison, followed this lead. Even Pacific Bell dared to do it in spite of AT&T's earlier objections. On grid designators, the Y coordinate was listed before the X coordinate. Some say this was Pacific Bell's defense by being different from the idea AT&T rejected for Bell of Pennsylvania.

Vernon Graphics did some work to automate its process of making utility maps and assigning grid locations. This may have been the first infrastructure semiautomated mapping system developed. However, this work did little to contribute to the body of knowledge of this emerging discipline. Vernon saw its role as providing mapping services and kept its technology development secret.

Computer Aided Drafting (CAD) systems began their emergence in the late 1960s. Some people saw this new technology as a productivity tool to help keep map backlogs under control. However, as more and more software was developed to support engineers, the need for operational models became visible. This need, coupled with the MIS department's quest for enterprise solutions, was all that was needed to convince infrastructure organizations that CAD mapping was a short-sighted solution.

In 1968, Public Service Company of Colorado embarked on a project to determine what was the key identifier that could be used to link all its files together in an effort to have an enterprise-wide database. It was discovered early on that all parts of the utility depended heavily on location information to do their work. That work became so significant to the evolution of AM/FM technology that it has been given a section of its own in this chronology.

PSCo of CINS Project: 1968–1972

It seems ironic that it was almost three decades, until 1995, before an AM/FM International local chapter would be formed in Colorado (the birthplace of AM/FM and a community that still serves as the worldwide focal point for AM/FM/GIS products, services, and education). Could this be a milestone noting the technology evolution towards entering the mainstream?

The newly-formed Colorado chapter had its first meeting in late 1995. The agenda for this first meeting was a round table discussion about the "good old days" by several of the "good old boys," who are, amazingly, still active in the industry. Chuck Howard is with Howard & Associates, Jim Hargis is with UGC, Keith McDaniel is with Intergraph, and Hank Emery is with Emery & Associates. McDaniel and Emery were a part of the original Public Service Company of Colorado (PSCo) Project Team; Hargis, with Tobin Research, was a part of the Cheyenne Pilot; and Howard joined as part of the IBM joint development group—all on what was known at the time as the "PSCo CINS Project." CINS, although it sounds like something decadent, stood for the *Common Identification Number System.*

It is interesting to realize 30 years later that AM/FM/GIS had such a humble birth. A Common Identification Number System: a number system for what? What did this have to do with utilities or anything else for that matter?

There really was a logical answer. Like all utilities of that time, the PSCo Gas and Electric Engineering and Operating Departments were using manual maps as the source of information to do their jobs. Again, like most utilities, the maps which they were using were individual map sheets with no connection between sheets. In an attempt to make the maps more usable, an unknown engineer had placed a 100 x 100 unit grid on each of the electric facilities maps. These grids were primarily used to identify transformer locations on each map. The problem was that the individual grids were placed arbitrarily, and just as there was no edge match between maps, there, of course, was no connection between these map grids.

So the very simple first mission of the CINS Project was to find a numbering scheme or indexing system that could be used to tie all the maps together in an accurate manner across the entire service territory. In other words, find a coordinate system that could be used for an accurate continuous mapping system which could easily be used for identification and location of gas and electric facilities. Soon after the project began, it became obvious that location was the most important item. It was determined that location was somehow involved in 90% of the information that the utility used on a day by day basis. It was not known at the time that this search was really to find a common index for all future data bases.

Today, the solution sounds easy, everyone knows about latitude and longitude, everyone knows about state plane coordinate systems, UTM (Universal Transverse Mercator), and the multitude of mapping projection systems. Back in those days, not many utility people knew about any of these systems. A survey was made, and it was found that one utility (Pennsylvania Power and Light) was using its state plane system on manual maps, with a handful using "homegrown" and local grid systems based on trying to put map sheets together.

So the CINS solution and recommendation to use the Colorado and Wyoming state plane coordinate systems was a major step forward. The project team felt that they were real pioneers.

Another idea that evolved from the CINS study, which may have been just as important as the Number System, was the idea of having a central land and facilities database. Prior to this, utilities were beginning to do computer applications such as Load Flow Analysis, Phase Balancing, Transformer Load Management, Outage Analysis, and so forth. At that time, a facilities database was created and recreated for each of these applications. The result was multiple databases, differently organized but containing basically the same data. The sad part was that as soon as the yearly studies were run, no one bothered to keep these databases up to date, so they were all recreated the next year and on and on. Eighty percent of the data was duplicated in each database, so the question became: why not create a facility core model that contained all the needed data and keep it up to date on a continuous basis?

The second stage of the CINS Project was to do a pilot to test the state plane coordinate systems with real, live data. This second stage became known as the "Cheyenne Pilot" because it was done for Cheyenne Light Fuel and Power Company, a wholly-owned subsidiary of Public Service Company of Colorado. The pilot was large, by today's standards, using some 20,000 customers worth of data.

The pilot involved the acquisition of new rectified aerial photography. Since rectification in those days was only to remove errors caused by tip and tilt of the photography, one can imagine that some strange things were encountered. For example, a pole showed up on the corner of four adjoining photos. First, it had to be determined that it was the same pole on all four maps, and then a determination had to be made as to which map it should be placed on.

The locations of the gas and electric facilities were assigned coordinates. However, the land was manually drawn because it was thought at that time that to digitize the land was too expensive. Long, complicated, and confusing coding sheets were filled out and key-punched for entering the facilities coordinate and attribute information on a "span" by "span" basis, with a record for each point facility (i.e., pole, manhole, valve, cross, and the like). These point records were called the "At" points. For describing linear facilities (i.e., pipes, conduits, conductors, and so forth), it was necessary to define the facility between two of these "At" points, but, because line configuration was critical, it was necessary to understand the direction of the data being described. Thus the "To" point was introduced and all linear facilities could be defined from a given point "At" to the next point "To." Point facility attributes were associated with the "At," and linear facility ones were associated with "At"–"To" with each "At" point assigned a state plane coordinate. It should be remembered that at that time, "interactive graphics" was a thing of the future. Then, the only output was "batch" plotting, which meant submitting "key-punched" cards and waiting several working days for the plots.

The fact that the technology was quite primitive in those early days makes the development and progress that was made all the more interesting. Figure 8.1, tracks the major development of AM/FM for the first 15 years after the modest 1968 beginning.

The Beginnings of AM/FM

The figure shows how important the PSCo CINS Project was to establishing an AM/FM industry or discipline. PSCo seeded the IBM joint project with knowledgeable people who passed on their experiences and enthusiasm to people at IBM. They eventually developed products and entered the market, bringing a lot of credibility and energy. Other people left PSCo to form Computer Graphics Corporation (CGC). CGC grew and was acquired by Butler International and eventually Intellegraphics. People left CGC and took influential roles at M&S Computing (Intergraph) and Kellogg. Kellogg sponsored the early Keystone Conference in 1978 that

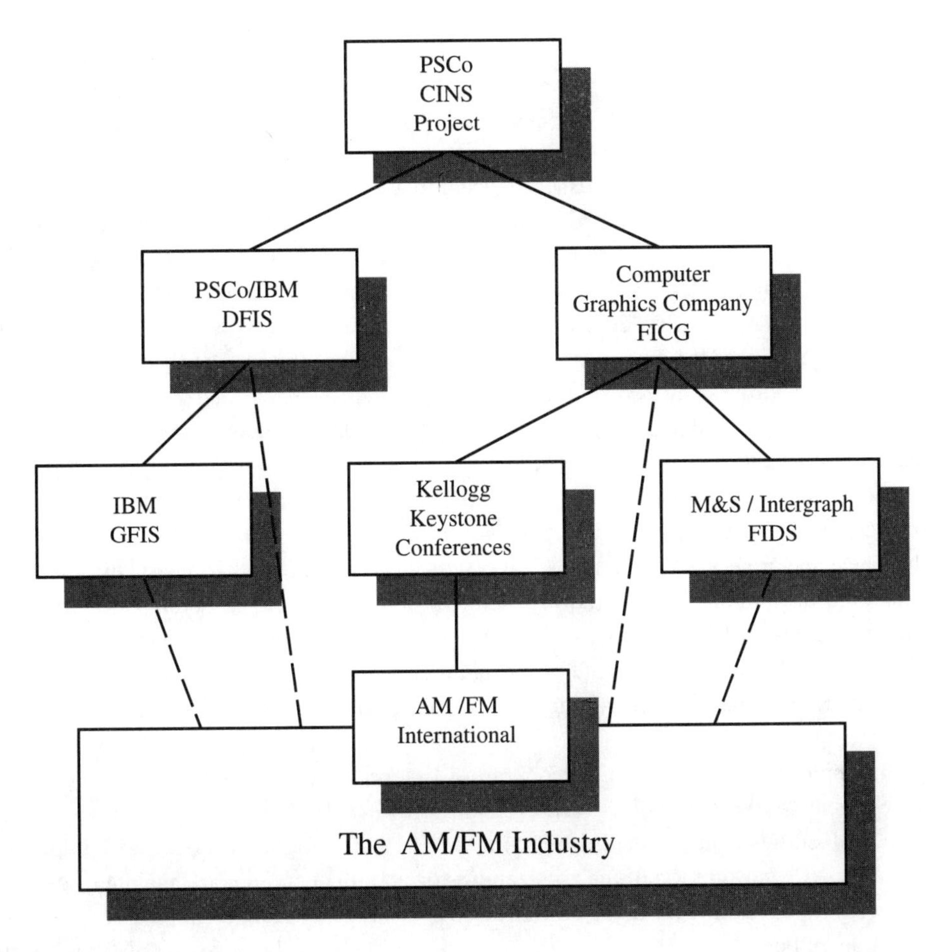

Figure 8.1 The AM/FM industry.

evolved into AM/FM International. It would be hard to name another project that contributed as many people and sound ideas to the formation of this industry.

In 1969, PSCo leaders started to share what they had learned with utility professional associations and technology user groups. In May of 1969, they presented a paper to the Operating Section of the American Gas Association. In November that same year, they presented that report to the Applications Division of GUIDE, IBM's large systems user group. This paper was well received by the utility participants and captured the interest of the IBM Utility Industry Marketing Department.

These industry marketing people took the ideas from PSCo, packaged them into a presentation, and conducted a survey of utilities. They visited more than 20 large utilities, primarily explaining the concepts and asking if they thought there was value in products to support geographically based facility databases and mapping applications. The answer was a resounding "yes."

As the study was being conducted, the IBM Research Division was reporting the belief that computing speeds and storage capacities could be doubled every 18 months for as far into the future as anyone could predict. IBM projected that unless some new types of compute/data intensive applications were developed soon, the entire world's computing needs would be saturated in three to five years. The company embarked on a two pronged approach to solve the dilemma. First, projects were started to improve application development tools and languages. Secondly, some 20 application development projects were funded with IBM customers. This was known internally as the Major Application Expansion Program (MAEP). Some of the first on-line insurance processing and shop floor scheduling software came from this effort. PSCo was selected and accepted a proposal to develop application software for the facility mapping and database market.

The interest was not just confined to IBM. PSCo was asked to make repeated presentations at professional conferences, and more than 70 utilities from all over the world sent personnel to see the Cheyenne project. As primitive as the system shown in Figure 8.2 may seem, it served as a means to discuss and debate the requirements of these systems, long before AM/FM or GIS technology was available. The next several paragraphs describe those requirements. One can see that they went well beyond just spatial or mapping requirements.

At the foundation of utility applications is the facilities model. The facilities model has very different characteristics from either the scientific or application models employed in many other GIS applications. The facilities model must first reflect the specific system connectivity and component configuration of a given installation with a secondary emphasis on geographic location. In facilities management, the FM in AM/FM, this is necessary because the facilities model is the key to integration of the applications for design, operation, and maintenance of the company's physical assets. By contrast, most GIS applications place geographic positioning as the first order of business, because most of the application relationships are spatial. The elements can usually be connected with coexistent points or spatially, in terms of relationships between data sets.

The complex configuration of the utilities and telecommunications systems makes it difficult to capture all of the components and relationships using spatial operators alone. Many refer to the facilities model as the engineering network model, a model which is so complex that literally hundreds of unique detail drawings and schematics have been necessary to depict the complexity of the components, connectivity, and characteristics. At the other extreme of the facilities model, a 50,000- foot view perspective, reside the key structures and the generalized connected pathways of the facilities. Here is where integration with most spatial applications occurs. It is the difference between the generalized 50,000-foot perspective and the engineering network model that has created great controversy and defines the uniqueness of the facilities model.

The applications which compose facilities management are wide-reaching across the enterprise. In the traditional approach, the human work force has been the integrator for all of

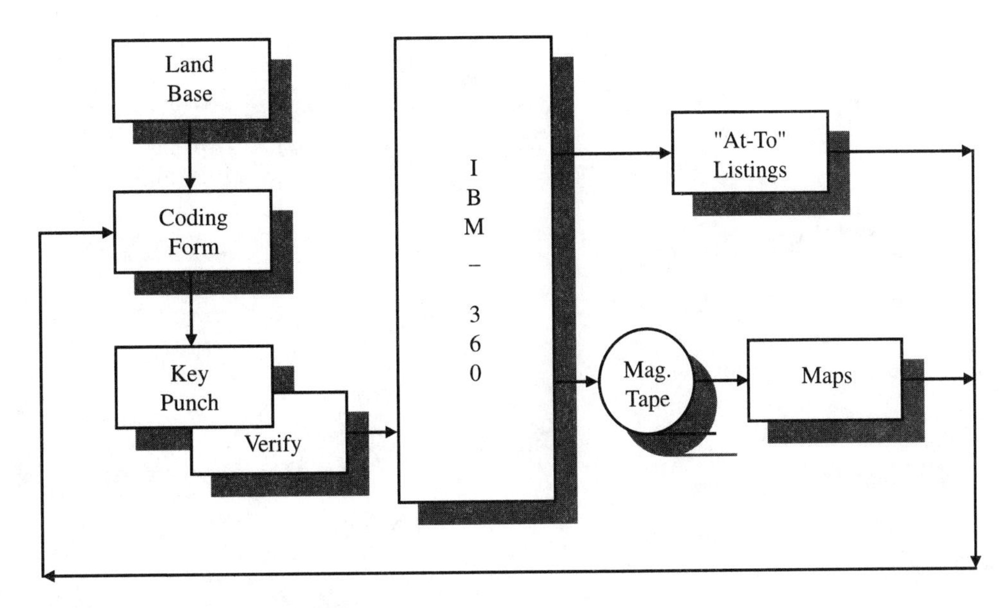

Figure 8.2 CINS project work-flow.

the isolated manual systems and islands of automation. AM/FM introduces the ability to integrate digitally these systems by replacing mapping and drafting with a sophisticated facility model that has both spatial and connectivity properties captured from these records.

Of all the systems shown in Figure 8.3—customer information, forecast and planning, trouble management, network management, material management, work management, design and analysis, and mapping—mapping is typically one of the last to be automated and holds the key to the real integration. Mapping systems developed out of a need, and they depict the model of the distribution system, which includes the geographic location, the configuration of the facilities, and most of the keys to the other systems. This has been the last to be automated by utilities because it is the most difficult to capture and manage in a digital format, but its automation is considered paramount for a truly integrated enterprise.

The following presents a conceptual overview of the traditional facilities information management components and the intersystem data dependencies.

Customer Information System

Although most of the functionality in the Customer Information System can work independently, there are areas that require integrated information and processes. The customer's geographic location, position on the network, and load factors are intersystem data required for trouble management, network analysis, work management, and forecast and planning. As the user interface becomes more refined, it is also expected to provide customer representatives with geographic based information in response to customer inquiries.

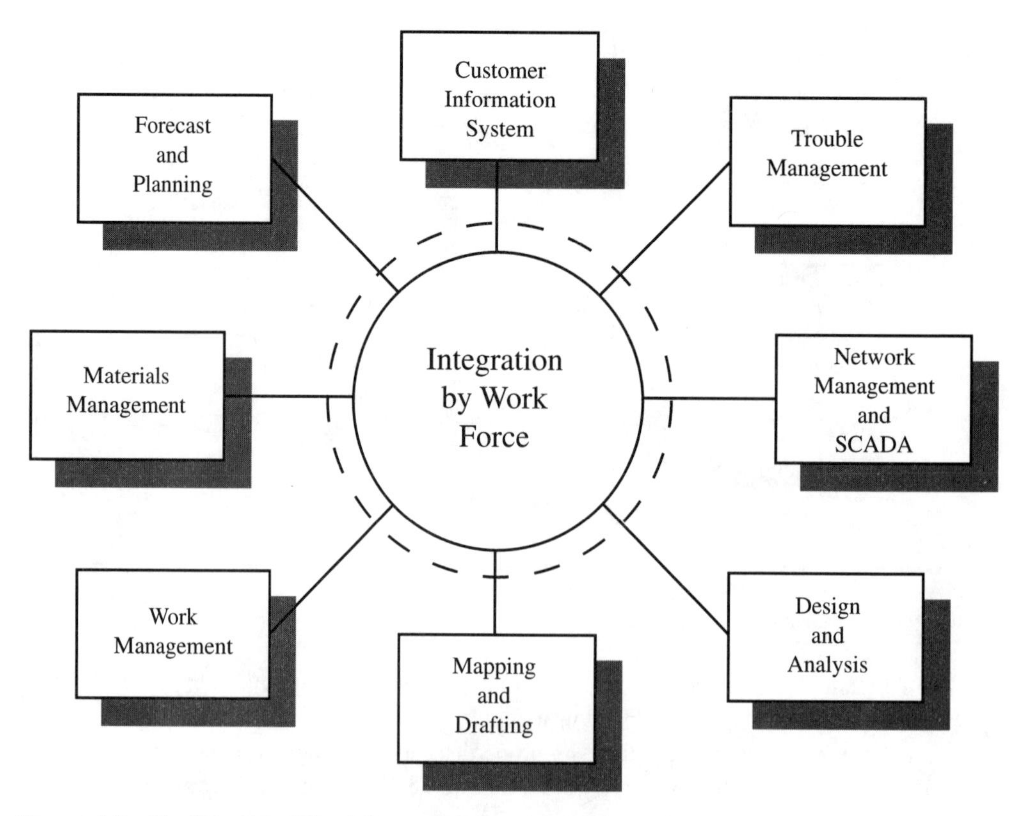

Figure 8.3 Traditional facilities information management.

Trouble Management

This application requires a view of the distribution system that depicts the connectivity that is maintained in the facility model. This is a selected set of data which identifies the relationships with connected upstream devices in the network that are used to identify the probable cause in outage situations. The Customer Information System also has intersystem data interface because it is typically used for call taking and restoration reporting.

Network Management and SCADA

The real time distribution management systems also use a view of the facility model. The network management view is a reduced segment model that eliminates many of the details required for managing the individual facilities. Because of the nature of the real time environment, this may be a redundant set of data. It is, however, derived from the facilities model and should receive updated transactions as they occur.

Design and Analysis

The design function is a short-term process that uses small network analysis data sets. The network analysis function, on the other hand, uses much larger data sets and typically requires additional source of input and a more extended evaluation period. Design and analysis has an extremely tight link to AM/FM. The intersystem data dependencies between Design/Analysis and AM/FM are so extensive that they may almost be viewed as a single unit.

AM/FM

By its very nature, providing the model and the primary user interface for distribution system changes, AM/FM has intersystem dependencies with all of the other modules. This system must formulate the appropriate transactions for inquiries, updates, validation, and applications. The facilities network connectivity in the model has explicit and binding relationships and keys to link to the other modules. These are relationships that must be relayed as they change. The spatial relationships are primarily positional and nonbinding. Many of the relationships are actually generated during the application process. Thus they only require positional accuracy as a management concern.

Work Management

This module is driven by changes being made to the distribution system. There are intersystem dependencies for the request or initiation and for the work order designs. The component list or bill-of-materials generated during the work order design session in the AM/FM module is passed to Work Management where it is used for costing, scheduling construction, tracking, reporting, and closing.

Materials Management

The Materials Management intersystem dependency is primarily with Work Management. The design and build process requires the interaction with Materials Management to insure the availability or procurement of job materials.

Forecasting and Planning

An area of considerable change results from the introduction of GIS tools and the availability of commercial and government data. The applications run from marketing through long-range planning. The key areas for intersystem data exchange are: AM/FM for the spatial positioning of facilities; Design and Analysis for network analysis and capacity planning; Customer Information System for usage, interruption history, and potential; and Network Management/SCADA (Supervisory Control and Data Acquisition) for critical capacity evaluation.

This overview provides a foundation for a conceptual view of the major requirements of an integrated enterprise and helps explain the difference in focus from other GIS applications.

With this background, we will take a look at how technology has played its role in producing solutions.

The Pioneers: 1972–1977

As the 1970s unfolded, the pioneers on the IBM/Public Service Company Study Team faced totally new challenges. The multiple-year study validated the concepts and theories explored in the Cheyenne Pilot Project of integrating corporate data through common keys and geographic locators. However, the team realized that the system was far from a reality for practical implementation. Great controversy erupted within the team over the most critical issues to address. Public Service Company management and IBM elected to pursue the building of the database management system and the enhancements of dynamic database driven graphics. Hank Emery, Duane Gilbert, Keith McDaniel, and Phil Schaffer, project leaders at PSCo, and Jim Hargis, the programming manager at Tobin Surveys, were all convinced that the process would fail unless better conversion and communication tools were developed. This segment of the team began investigating new commercially available interactive graphic products developed around the electronics industry and believed that this technology would play a critical role in meeting the conversion and communication requirements.

Three significant paths were spawned from the basic foundation of the modeling approach: DFIS (Distribution Facilities Information System), the IBM entry, pursuing the database management and facilities files interface problems; FICG (Facilities Interactive Computer Graphics), the Computer Graphics Company entry, based on a Bendix interactive graphics system and pursuing the user interface and data conversion problems; and, later, FIDS (Facilities Interactive Design System), the M&S Computing (Intergraph Corporation) entry, pursuing the building and maintenance of the model using interactive graphics combined with database technology that addressed the legacy data structure interface problems. Each of these approaches had a common goal of building, managing, and maintaining a distribution facilities model with interface capability to corporate databases, spatial applications, and producing the required maps and reports.

The IBM database proponents, under the project direction of Lonnie Martin and the technical direction of Dick Martin, began the development of DFIS. This was an extension of the AT-TO approach developed by Jim Hargis, enhanced to use a database as the repository for storing the full distribution model. The program concentrated on the process of refining the database model to make it more complete and more efficient. One of the key advances in these early years was the introduction of point connectors in the database for modeling the point and line segment facilities back to the X,Y coordinates. The coordinate values were the key to the database structure, with point, line, and facility information being stored by geographic location. The concept was to integrate all of the facilities information throughout the corporation with the use of the coordinate values keys as the common thread.

The concept of database driven graphics also became a reality during this early effort. This used a refined and enhanced set of plotting routines conceived in the CINS Project but using a database rather that an unmanageable card deck for the storage media of the facilities model. This effort enhanced the CINS concepts and demonstrated that more complex connectivity could be captured in a database.

When PSCo and IBM began their joint study in 1972, it was not the only facility mapping/database project IBM funded under the MAEP program. Brooklyn Union Gas (BUG) had wanted to do the same thing but had not progressed as far as PSCo. IBM agreed to work with BUG in parallel with PSCo so long as there was a basic difference in approach. The difference chosen was that the BUG project could not use a grid coordinate as a file key in its solution. BUG spent about three months hypothesizing different approaches and decided it could not succeed under the conditions set. IBM agreed to share all the information learned at PSCo with BUG, and their joint project was terminated. BUG, however, was determined to automate its mapping processes and hired Raytheon Corporation to develop and implement a system. All of Staten Island was mapped using aerial photography, and there were some attached attributes to the CAD file maps. This pilot showed some promise, but BUG determined it was too expensive to do the whole company and decided to wait for the technology to develop.

There were several other utilities that agreed to act as validation sites for the requirements being developed at PSCo. They included: Northern States Power, Texas Power and Light (TP&L, now a part of Texas Utilities), and San Diego Gas and Electric (SDG&E). Most of these companies had activities of their own that related to the long-term requirements. TP&L was developing a continuing property records system (alphanumeric, not graphic) that used grid coordinates as the key. SDG&E bought a Computer Aided Drafting (CAD) system from Calma Corporation to reduce the backlog of underground electric mapping. IBM at one point believed it would never produce graphics hardware of its own and worked with SDG&E to link the Calma mapping system to the database produced at PSCo. Dick Martin developed the Interface Format File (IFF) as a means for any CAD system to link to an IMS (Information Management System) database. This ended up being a very useful tool for conversion vendors to load data after IBM released its Geographic Facilities Information System product set. IBM changed the name of its product set from Distribution Facilities Information System (DFIS) to Geographic Facilities Information System (GFIS). This was done to accommodate waste water customers who wanted to avoid any notion they would distribute sewage.

Meanwhile, during this same period, the conversion and communication proponents struck out on their own. Emery set off to find investors for a new venture using the new interactive graphics technology and found John Russell at American Appraisal willing to back the effort. Russell had a vision of providing comprehensive products and service to municipalities and utilities for managing their assets from many aspects, operations to property tax assessment. He would eventually pull together under one roof Chicago Aerial Survey for photogrammetry, Cole-Ayre-Trumble for tax mapping and valuation assessment, and the Emery team for facilities management. This would be the first of many spin-off companies in this industry and was started when Emery, McDaniel, and Schaffer left PSCo and were joined by Hargis from Tobin Surveys to form Computer Graphics Company.

The team developed an early form of today's benchmark for evaluating two competing vendors, Computer Vision and Bendix. Bendix won the benchmark because of an ability to associate graphics elements with attributes, a feature considered an absolute necessity when combined with spatial components for capturing the facility model. The system configuration in Figure 8.4 provides an interesting perspective on that system with its 8 bit architecture and 32K RAM, compared with today's systems of 32 and 64 bit architecture and ever growing RAM. One would probably question just what a system this simple could actually do. Considerable, after the team developed significant advancements with techniques that links attribute and graphic components, "parent/child" relationships, and off-page connectors. Models created for such companies as Detroit Edison, Portland General Electric, Southwestern Bell, and Public Service of Indiana were created using this system and have since migrated forward to today's current technology.

Plotting from graphic files was also introduced, greatly reducing the through-put time and demanding only a fraction of the computer resources. The big difference was the storage of the graphics rather than their generation for each plot. Another advancement was the introduction of graphic symbology and text manipulation to produce multiple-scale and multiple-purpose maps.

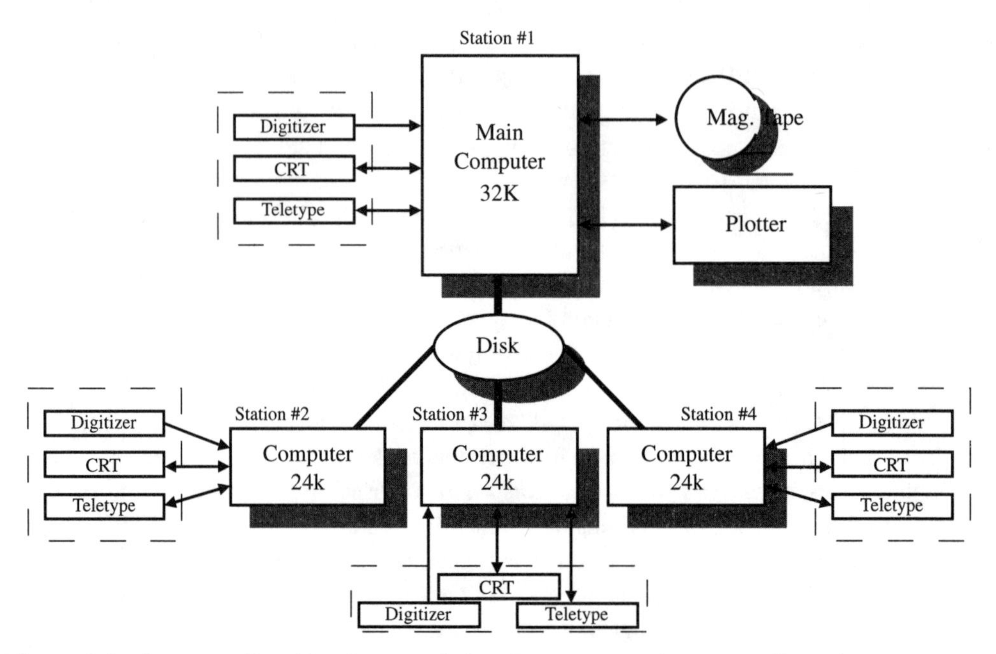

Figure 8.4 Computer Graphics Company's Bendix system equipment configuration.

Although this approach was radically different from DFIS, its purpose was rooted in the same set of requirements. Interactive graphics were very effective for converting and updating facility information in a digital form while achieving corporate mapping standards. The one drawback in this methodology was that no database model actually existed. Graphics were structured appropriately for constructing a database facilities model from coexistent points and graphic file pointers at a future date.

Introduction of interactive graphics revolutionized the conversion process. The original method used the point-to-point approach with only the points being given global coordinate values. All of the graphics were defined at or between these points through detailed alphanumeric coding. Software then used this data to generate all of the graphics dynamically from scratch each time to produce plots. This original method went dormant until the 1990s when a hybrid approach of the technology resurfaced. Interactive graphics permitted the data entry personnel to input data in a format which very closely resembled the source document. This greatly reduced the cryptic encoding and interpretation of information between the dissimilar formats and provided instant response for visual verification of entry.

There was an obvious dramatic impact from the user interface perspective but a less obvious impact on the data structures and validation processes. The original IBM method grew up out of the tightly controlled data processing environment where data structures were designed with data validation incorporated in each step. Interactive graphics by contrast introduced a more freeform computerized drawing method. The challenge was to take this freeform entry method and add the necessary data management constraints to ensure data integrity. A challenge which eluded most of the commercial and homegrown systems CAD approaches of the time, but the Computer Graphic Company (CGC) team had this as the primary focus. While others also produced quality pictorial representations, most failed to maintain data integrity, and some errors were not discovered until years later when users attempted to perform other applications. They found picture perfect maps but invalid facility models.

Developing the capabilities of using simple graphics input to produce structures that created and validated a complete facility model was perhaps the first use of expert system technology; however, the user still provided most of the expertise. Although limited and without any form of inference engines, this was definitely an interpretative process of inferring structures from the graphic relationships of data and coexistence points. The early stages of such processes forced the user to provide all of the relationships by the point and click methods which were very susceptible to error. However, with time, these became very sophisticated and allowed the user to build picture perfect graphics that produced valid facility database models.

Independently, in 1973, M&S Computing (later Intergraph Corporation) entered the picture with two map based projects—one the City of Nashville mapping project, under the direction of Dr. Joel N. Orr, and the other the Bell Laboratories' Outside Plant Engineering Records Automated Interactive Drafting project (OPERAID). The City of Nashville mapping project employed a new diminution in the use of double precision 8 bit values (16 bit coordinates). This mapped true coordinates rather than the single precision 8 bit values used in most interactive graphic systems of the time, where only limited coverage was required for individual map or drawing applications. The use of 16 bit precision opened the door for seamless mapping and interchangeable map projection calculations.

The Bell Laboratories' OPERAID project was conducted at the New York Telephone Company using M&S Computing software and data that was converted by Western Electric Company using an Autotrol Technology Corporation system. The unique focus of this project was on the preparation of work prints and study maps with a strong emphasis on the integration processes of posting and preposting the change order information to the plant location records database. Although this project never materialized into a fully deployed system, it was a critical initial attempt at incorporating interactive graphics to assist in the work order sketch and mapping database management process.

Other commercial entries into the market in this time frame included Computervision, Synercom Technology, Inc., and Calma with each of their variations on the themes and a few simply using a graphics-only Computer Aided Drafting (CAD) approach for reproduction of maps and drawings. The early CAD vendors all saw utility mapping as a big opportunity and launched sales campaigns. A few utility companies, such as Southern California Edison, Utah Power and Light, and several European and Asia Pacific companies, bought systems primarily for mapping, but most took a bigger view that included facility database and enterprise architectures.

The Birth of an Industry: 1978–1986

If the AM/FM discipline (or industry) is defined as that which provides products and services utilizing spatial analysis and processing, tightly coupled with network modeling functions, then it is fairly safe to say there was no such industry prior to 1978. CGC had all the concepts, some of the tools, and a few customers, but they were realizing less than $10 million revenue annually. Synercom had a few customers and rudimentary networking logic, but it was also small and relatively unknown. IBM had done some research but had not yet committed to a product. Other large technology companies had pieces of the puzzle and either had not focused on infrastructure applications or simply offered CAD tools to be applied to parts of the problem. No company with market making power had it all together.

In 1978, three unrelated events occurred that marked the beginning of a legitimate industry. McDaniel and Hargis took the CGC concepts to Intergraph and combined them with already strong cartographic and CAD tools. IBM's utility industry group decided to produce a product. And Hank Emery, then with Kellogg Corporation, started a publication and a conference where ideas could be freely exchanged. Over the next few years, the presence of two large, well-known competitive vendors erased many doubts about the viability of the technology. Even if the infrastructure buyers did not see all the necessary tools, they knew they would come eventually.

McDaniel and Hargis, satisfied in 1978 that they had broken through the major conversion barrier problems by eliminating the coding sheet and keypunch process through the use of interactive graphics, set out in search of a solution using interactive graphics for real time communication with the database model. Intergraph with its mapping and engineering based solutions and earlier experience with the City of Nashville and the Bell Laboratories' OPERAID project offered the greatest opportunity. It also happened that Jim Meadlock, president of Intergraph, believed very strongly in the same principals of interactive graphics driven database models for engineering. This combination and the commitment of resources to solving the problem changed the user interface for managing the facilities model forever. Intergraph immediately began acquiring new utility and municipality customers—such as Detroit Edison, Portland General Electric, the City of Edmonton, Consolidated Edison, Pacific Gas and Electric, and many others—with its new industry workshop series called the Project Guide and System Implementation Program, a program that helped clients develop their unique systems using structured techniques for design, development, and project management. But the major breakthrough occurred with the Southern Bell Project and a visionary named Dick Snelling.

"You can't face the challenges of tomorrow with yesterday's technology" was Dick Snelling's quote as he launched a deal with Intergraph, AT&T, and Bell Labs that would reshape the way Southern Bell managed its facilities information systems and influenced the world. AT&T and Bell Labs (before divestiture) were acting on behalf of all of the Bell operating companies to provide technical direction to the project. The single most important factor for all of the participants was the building of a real time continuous facilities model that could be accessed and driven from either graphic or alphanumeric terminals. Intergraph offered a Database Management and Retrieval System (DMRS) link to its Interactive Graphic Design System (IGDS). This combined system was used to model the telephone outside plant, a very intricate system, one which modeled circuits down to the individual pair count level—a dynamic system that demanded the ability to reconstruct components of the model interactively as upstream reconfigurations changes occurred and rippled through the system. The model could not be composed of spatial linkages; it had to stand alone as a character database with appropriate data pointers, similar to the PSCo model but interactively driven by graphics. Circuit configuration changes could be launched from alphanumeric terminals, and the results needed to be dynamically reflected in the graphics, again similar to elements of database driven graphics found in the original PSCo project.

This was seen by McDaniel and Hargis as the project to test the theories of graphic driven database and database driven graphics combined into a single solution. The specification and a test bed called the King Street Wire Center were completed in 1981, and the rest is history. Under the direction of Joyce Rector and Frank Castleberry, the project became the most comprehensive and aggressive in the industry, still unsurpassed after a decade. With more than 1,200 users of the technology, the project became a milestone in AM/FM system achievements and earned the second prestigious AM/FM International Pioneer Award.

There was a belief that since Bell Standard Operating Practices had been enforced during the project, and that Bell Practices were followed by the Bell Operating Companies, and that a large percentage of the telephone systems around the world followed Bell Practices, now a generic package could be produced. Divestiture hit, and the telephone companies went mad—"don't share at any cost." And, even though Intergraph went on to develop industry generic packages for gas, Gas Facilities Interactive Database Systems (GFIDS), electric (EFIDS), and telephone (TFIDS), industry standards were in reality a concept. An interesting bit of information is that the telephone product built on the Bell Standard Operating Practices was actually more successful in the international market and with non-Bell companies in the U.S., such as GTE, than it was with the Bell companies. The Intergraph products and services helped open the market to the more pragmatic utilities, telecommunication companies, and municipalities. This widely expanded the use of AM/FM technology to more than 150 companies worldwide, but the persistent problem of customization to meet each individual company's practices and legacy system interface requirements continued.

Convinced that the industry was not interested in establishing standards and not ready for off-the-shelf products, Intergraph redirected its development to the next generation of product called Facilities Rulebase and Model Management Environment (FRAMME). In 1982, Intergraph made a major commitment to object-oriented technology and began the development of a product called Topologically Integrated Geographic and Resource Information System (TIGRIS), which was used primarily by government and military mapping agencies for spatial applications. It was not practical to use TIGRIS for AM/FM because of the large database requirements of the facilities model, but many of the object-oriented techniques were incorporated into the new FRAMME development. FRAMME used a rule base as the central control mechanism which pulled graphic and character data from the respective databases and managed the elements as objects in memory, then produced transactions updates to the appropriate AM/FM or other enterprise databases and support systems. Certified in 1986, this product was designed to take full advantage of all of the similarities "standards" in the industry while providing the flexibility to customize components for individual companies. The significant contribution of this product to the industry was its ability to incorporate the corporate standards and practices into a rule base system that managed the graphic and character components in a conflict resolution environment previously only seen in character base management systems. Rules could be stored in either the master function library or as unique functions for each company. As more and more similarities were discovered, they could be incorporated into standard library routines for everyone to use, thus simplifying the customization process and reducing overall development, a system designed to evolve as standards emerged.

IBM had a long way to go in 1978 to get a viable product to market. Their only graphic hardware was aimed at high-end CAD users for the aerospace and automotive markets. They owned the preeminent hierarchical database management system for utilities, but there were no spatial or cartographic functions. The prototype software developed at Public Service Company of Colorado was installed at San Diego Gas & Electric to assure it was transferable. It would take two years for IBM to get its first AM/FM software product to market and another three to get acceptable graphics terminals and overall product support.

Chuck Howard had been a part of the Joint Study with PSCo, acting as the project officer and chief technical writer. His role in the early part of this time frame was to prepare the market for when products would be ready. He published a three-volume set entitled *Design Guides for a Distribution Facilities Information System.* He used these to educate the IBM utility sales force and distributed several hundred sets to utility people attending IBM workshops and conferences. For those companies that showed more than a casual interest, he would conduct a needs analysis and feasibility study. Some of these efforts resulted in sales for IBM. A few resulted in sales for competitors. However, by the end of this time period, the availability and usefulness of the technology was well known to the utility industry. IBM had almost 100 customers worldwide, with more than 20 U.S. utilities and several city public works departments as customers.

IBM's main technical contribution of this time period was its database management product. Geo-Facilities Database Support (GDBS) was built on IMS, a standard database manager, and provided a robust set of spatial functions as well as extensive networking tools. Its distinguishing feature was an element called the point connector. This allowed any number of connections (nodes) to be present at any spatial location. It met the utilities requirements, and they bought it. However, the life of GDBS was short because it was hierarchical when relational was becoming the standard, and it was mainframe bound with AM/FM being a natural distributed system served better by client server environments. IBM's entry into the AM/FM market did a lot to instill buyer confidence, and its technical contributions set new standards for spatial database products.

Upon leaving CGC in 1977, Hank Emery joined Kellogg Corporation, a construction consulting firm, and led an automated mapping division. In his first year, Hank held a meeting in Keystone, CO, with representatives from every utility company that was showing an interest. This was a rather informal exchange of ideas, but it became clear that everyone wanted a way to network and share ideas and experiences. In 1978, Hank started the *Automated Mapping Chronicle*, a newsletter, and he held the second annual Keystone conference that would later evolve into AM/FM International and its conferences.

Shortly after starting the *Chronicle,* Emery invented a character that was half robot and half cartoon. It had a globe for a head, a T-square for a body, and a slide rule for arms. He called it "AM/FM" for automated mapping/facilities management. Its purpose was to make editorial comments Emery would rather not say himself. The name was partially a play on the film industry robot R2D2 who was everyone's sweetheart in those days. However, the name expanded the notion of automated mapping into all those other applications for facilities information. All product and service providers prior to this time had varying names for their own offerings; GIS had not yet entered the vocabulary of this segment of the industry; and the name was descriptive of what infrastructure organizations wanted to do. So, the name of a cartoon robot became the name of a discipline that was maturing in parallel with what others called GIS.

Kellogg continued to sponsor the Keystone Conferences through 1981. The conferences grew in size until the expense became an unreasonable burden for one company to carry.

Besides, all the vendors were now attending and benefiting from the gathering of potential customers. In 1981, Emery suggested that if the sharing and networking were to continue, it was time to form a professional organization. During that conference, two committees were formed: one to put on the program for the next year's conference and another to form the new organization. This was the birth of AM/FM International. Kellogg did not stay in this business much longer, but its people had a major impact on the newly forming industry. Emery left Kellogg to form Emery Data Graphic.

The formation of AM/FM International did a lot to add credibility to this new scientific discipline and brought more people's energy to bear. Chis Harlow, the vice president of information services at SDG&E, headed the committee for the first year's conference and later became the organization's first president. Bill Folchi, marketing vice president of Synercom, pledged the organization's first $2,000 and quickly got Intergraph and IBM to do the same. He became the first treasurer and later was the fourth president. Warren Ferguson, who owned a small map and printing company in San Antonio, served as a founding director of the organization and donated the first two years' printing needs to the new organization. His company later became one of the premier conversion service companies under the name of "Ferguson Cartographic Technologies" and later "Cartotech." The mission statement was very simple: to promote the discipline of automated mapping and facilities management. Education was the primary vehicle chosen to meet the mission. The organization grew quickly and by the end of the time period was holding two conferences a year and had more than a thousand members.

Coming of Age: 1987–Present

As this era opened, the scene had dramatically changed from the start of the previous era where only a few vendors and visionary users participated. The industry was now showing an acceptance of the technology and experiencing an atmosphere of growth. AM/FM International had more than 85 sponsors and contributor vendor companies, a European division had been formed, and world conference attendance had grown to nearly 900. The seeding of many of the new entries came from most of the pioneering companies as one would expect in a growing industry. This also makes it more difficult to define individual contributions as multiple organizations provided similar advances within relatively the same time frames.

During this time period, we have seen the established GIS vendors like GeoVision and ESRI, along with such new entries Enghouse and Smallworld, following the lead of IBM and Intergraph in recognizing the modeling needs of infrastructure organizations, and consulting firms, such as PlanGraphics and UGC, having an impact on new start-up projects with need assessments and feasibility analysis. Companies like McDonnell/Douglas have taken a run at the business, and large consulting firms and integrators like Anderson Consulting and EDS showed keen interest. Large utility projects such as Bell South and PG&E gave birth to the data conversion industry. Vendors and research companies strove to develop technology to reduce the time and cost impacts of data conversion, notable achievers being Intergraph, Coherent Research, and

ExperTech. The most significant change, however, has been the entry of the application providers and the ability to achieve the benefits of an AM/FM/GIS central model through the integration of applications. This signaled the breakthrough that AM/FM was not simply another isolated system but the enabling technology for enterprise integration as promised for all of the proceeding years. Although the individual and company achievements are harder to distinguish in the more current years, the significant technology advances have increased at a much faster pace than previously experienced. In the following paragraphs we present some of the more important issues and their impact.

Document Management

The utilities and telecommunications industries create and manage hundreds of different types of documents for describing the facilities location, attributes, connectivity, configuration, and operations and maintenance information. There are thousands of these documents in every company, with larger organizations possessing hundreds of thousands. Xerox Corporation, Formtek, SysScan, and others entered the market from the document management perspective. This presented a radical departure from the previous approaches. One of the critical driving forces behind the use of this technology was the reduced costs for conversion, the concept being that the documents could be scanned quickly and inexpensively. These could then be made available electronically across the organization. Accessing the documents was only a part of the problem, as they fall into two categories. The first classification of documents is the one used for maps and record keeping. A great percentage of the time when these are accessed, they require updating for changes in the facilities. The second classification of documents is the static variety. These are images that will have no change, such as photographs, archived construction information, service card sketches, manhole drawings, permits, and the list goes on.

The second group proved to be ideally suited for the document management environment where all of this historic and static data could be made available to the work force throughout the organization. Although the access for the first group was still very beneficial, the update and change management proved to be a problem. Numerous solutions were developed from raster editing to the integration of CAD. Another problem was that these systems handled the documents as individual units and did not provide for a seamless mapping environment where users could easily work across the document edges. The conclusion drawn from projects using this technology was that document management could be highly effective if it were integrated with the AM/FM capability. This integrated technology is now on the horizon.

Scanning and Raster/Vector Integration

Scanning entry into the market actually appeared many years earlier and brought with it extremely high expectations for easy and automated conversion of existing records. Solutions were perceived for converting scanned raster images into interactive graphic vectors, then recognizing the line, symbol, and characterization from the vectors, and finally, automatically building the facilities database model. This was comically referred to as "drawings-in" and "database-out." It would be many years of development before these expectations would begin to be realized and then only for clean and accurate records. When scanning was introduced, it was probably the most overrated technology in the history of AM/FM and actually slowed the progress for many projects while they waited for the miracle. The problem was that it was easy to scan and produce the raster image in a digital format, and this presented the illusion that the data had intelligence.

The realization of raster capabilities soon settled into practical applications. A real breakthrough did occur with the integration of raster and vector images into an overlaid configuration. First came the overlay of images for an individual drawing or map, followed by the more sophisticated image warping and global positioning processes. This opened a whole new world of techniques for the integration of existing drawings and maps with aerial photo images and interactive graphics vector data. The development occurred primarily as a result of the industry's inability to convert all images easily into an intelligent vector format and the realization that much of the data should remain in a raster format. This technology has drifted quietly into the solution with very little fanfare. This has happened because it follows on the heels of the oversold miracle, but it will provide far greater benefit than the original perceived solution. It is, in fact, one of the greatest contributions for AM/FM solutions in 25 years. It is the foundation for new and incremental conversion processes, heads up digitizing, and the integration of facilities information management with document management.

Facility Objects

The introduction of the digital facilities model into the previously isolated environment of systems for unique purposes was revolutionary to database management as well as to operations and procedures. The facilities model provides the linkages to tie all of the independent systems together. For the first time in facilities management history, the individual facility in the field had the potential of being viewed as a single facility object across databases.

The term "objects" began to creep into the AM/FM vocabulary with the introduction of object-oriented technology. The problem is that the information for the facility object is scattered throughout the enterprise in numerous systems. As a result, the concept of managing an object has been viewed from two perspectives. The first involves managing all of the information, graphic and nongraphic, that exists for a facility in a single system. The second means managing all of the information, graphic and nongraphic, that exists for a facility across the enterprise. There is a tremendous difference between these two perspectives and the systems that manage them. The term "object management" has confused as much as it has clarified because of the incompatibility of definitions used between the general concept and its practical implementation in AM/FM.

Figure 8.5 shows a typical corporate arrangement containing numerous independent files. Several record types may be required to achieve a complete description of the physical asset and

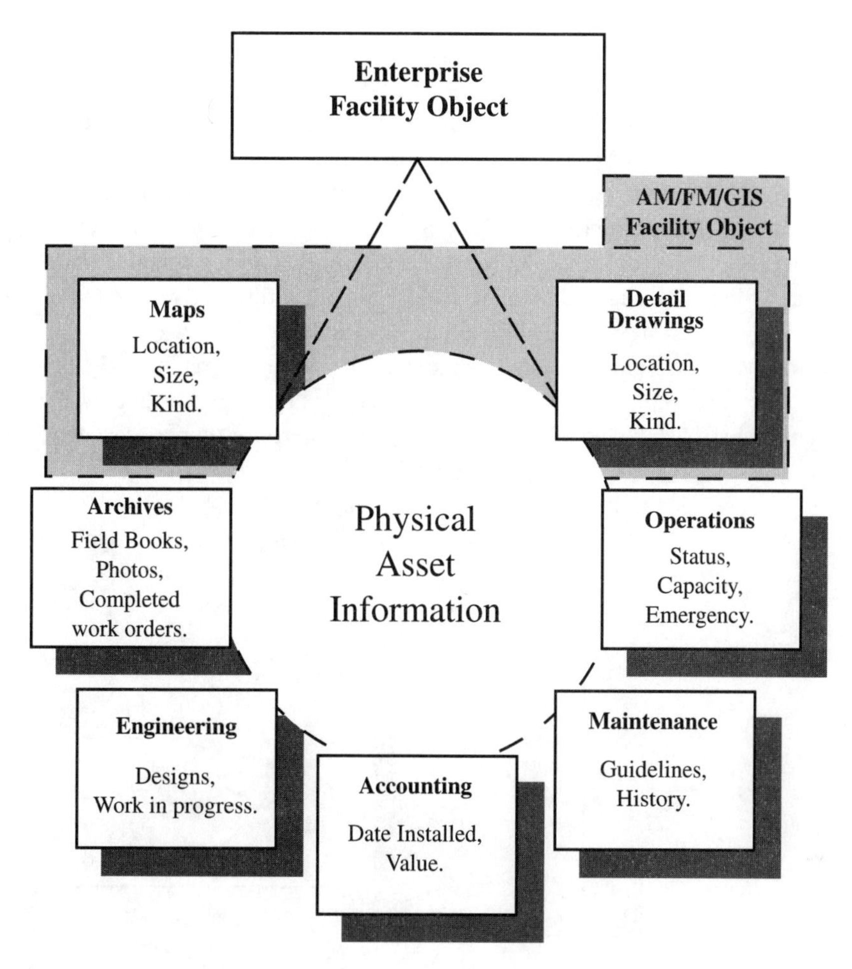

Figure 8.5 Enterprise facilities object information management.

may include numerous graphic representations of the same facility object on different maps, schematics, drawings, and sketches.

The use of object-oriented technology in this industry began with its focus on GIS analysis. A debate has been raging for several years over the use of open relational database models or proprietary object-oriented database models for managing AM/FM/GIS and its integration with other enterprise support systems. Both of these models are borrowed from the general database industry and bring to AM/FM a significant investment in functionality and maturity that is shared across many applications and industries. The technology is still evolving with the current direction headed toward open RDBs with spatial data extensions such as Oracle Corporation's SDO (Spatial Data Options.)

Database Integrity

The last ten years have been dedicated to the integration of graphic processes with database management techniques that could communicate with the other enterprise support systems. Since no system existed to perform the task, numerous individual efforts have produced a variety of proprietary approaches, each having components of the full desired functionality but none possessing the total. These proprietary solutions have been pushed pretty much to their limits, and the magnitude of the problem has now captured the interest of the computer industry giants like Oracle and Microsoft. Figure 8.6 depicts where we are and where we need to be for a true integrated solution which supports enterprise wide database integrity.

Some of today's systems assemble data from many sources, graphic and tabular, put the components into memory, manage them as objects, and then return transactions to the appropriate relational database (RDB) model, CAD files, or other enterprise support systems (ESS) databases. Some work in proprietary object-oriented databases or topology databases and translate the models into a relational database for interface with other ESS. Some are still CAD based with only macro controls. Still others are database warehouse-oriented with dynamic database

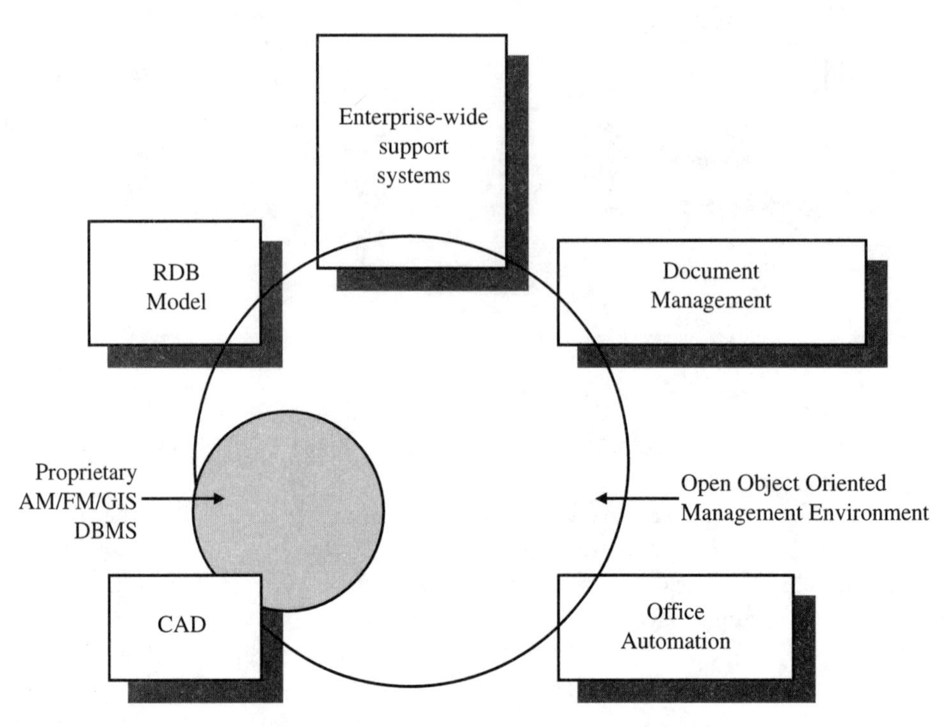

Figure 8.6 True integrated solution.

graphics. Functionality varies with some managing in a live link mode with full facility life cycle control and conflict resolution, while others use posting and temporal methods. There are approaches that incorporate document management and others that integrate office automation. Most of the functionality described above is required in the desired system but probably won't be realized until an open object-oriented management environment is available from the giants like Oracle and Microsoft.

Adding artificial intelligence to the conversion process, the long-awaited miracle of automatic conversion begins to appear again in the 1990s. Development efforts from McDonnell/Douglas, Intergraph Corporation, Coherent Research Inc., ExperTech Corporation, and Information & Graphics Systems, Inc. focused on artificial intelligence. Conversion costs persist as a major blockade to automation. Several development projects were started globally in major companies like Ameritech, French Telcom, Niagara Mohawk, NYNEX, and Bell Atlantic, with the telephone industry providing the greatest interest because of the magnitude of the conversion problem. It is estimated that somewhere in the neighborhood of $100 million has been spent on development in this area alone. Variations on a basic theme where undertaken in these projects, with some working directly from raster and others first converting to vector. The key to these new ventures was the introduction of the inference engine as a tool to convert images and relationships into intelligent facilities models and the embracing of an earlier technology, the old database driven dynamic graphics generation routines from the early 1970s. Many of these had been dormant for 20 years, because of the extremely high computing demands. The new computing power now made this approach a more viable technology.

The one problem with these techniques is that they must be highly customized for the unique source materials and presentation standards. Customization cost may be prohibitive for the medium sized and small companies. This is the primary reason why processes have largely been driven by the large telephone companies. They have a better return on investment because of the large number of records to be converted.

Aerial Photography and Satellite Imagery

The photographic image for background has evolved from rectified photography through orthophotography to the current digital orthophotography. The digital ortho provides an image and manipulation tool that exceed the requirements for most AM/FM projects. As a raster image, it is also integrated with the other AM/FM components. Another image being considered as the background is new, high-resolution satellite imagery. The technology is available to provide quality images with global mapping accuracy as background for AM/FM, but the use has been very limited for this application by utility and telecommunications companies.

Global Positioning Systems (GPS)

GPS has been introduced to numerous application areas for utility and telecommunications companies. GPS shows promise wherever field crews need to reference or record their global position. In the early stages of AM/FM, the use of the coordinate numbers was considered unwieldy because of the extensive amount of digits. GPS on the other hand can make the handling of these large numbers transparent to the user. General acceptance and practical use of GPS for facilities information management have not yet been realized. This technology is, however, projected to grow as applications are assimilated into the field and accuracy in the one-meter range becomes cost effective.

The "OPEN" Revolution

The word open has been abused and misused by this industry for years, but as all of the smoke begins to clear, the concepts are crystallizing into practical applications, applications which would have been difficult to impossible only a few years ago when attempting to integrate the mishmash of proprietary systems. In the early years, the facilities information was viewed as a series of independent processes residing in separate and distinct organizations. Today's perspective is quite different. The competitive pressures are refocusing the process for greater efficiency and forcing the reengineering of facilities management into an enterprise view. Here each of these independent processes is broken down into its functional requirements, viewed as components of the total business demand. These processes can no longer stand alone if utility and telecommunications companies are going to be competitive in this new market.

From GIS data sharing to the integration of enterprise-wide facilities information systems, the movement is away from translations and into integrated data and systems. In this new environment, the open systems approach will provide the avenue to the future. We have already seen the first waves of change in this revolutionary approach, and the momentum is building. We have seen the publishing of data formats which was considered taboo only a few years ago. Vendors feared the release of this proprietary information would lead to the decline of their products, dominance or superiority. This does not appear to be the case, and, in fact, the openness may actually be stimulating product growth as users are able to perform more applications.

Proprietary databases that presented extremely difficult problems for integrating with corporate applications and systems are being replaced with new open Relational Databases (RDB). This single technology shift to RDBs has had a major impact on the real world of reengineering and integrating the enterprise. Computing power has shifted, and we are currently experiencing an operating system shift from the multitude of UNIX based operating systems to the more popular office automation variety, an approach that puts both of these sets of tools at the fingertips of the user at a single device. Organizations like Open GIS are forging new ground, and relational database companies are stepping up to address open object-oriented technology issues. The open movement is revolutionary in changing the commercial protective mindsets and the corporate information management philosophies, but, most critical, it is providing a foundation for the reengineering of the business processes. The open systems approach will have a profound effect on those who seek the full benefits of the integrated enterprise.

The Realization of an Integrated Enterprise

The facilities model from its inception in the late 1960s was always a geographic-based approach, but the greatest emphasis was placed on the relationships with other corporate systems for operations and management of the facilities. The early 1990s brought an additional focus of planning and marketing with relationships to more extensive spatial data. A tremendous amount of controversy developed over the issue of whether these systems were AM/FM or GIS. They are both and always have been. They are a GIS but also have characteristics that go far beyond the spatial components for managing the facilities model and the integration with other enterprise databases and applications. Much of this controversy was introduced as GIS began to gain popularity. It did not change the requirements or the solutions. The availability of GIS data did add greater potential for spatial analysis. Regardless of the labels of yesterday, today, or tomorrow, the system must provide functionality to reside in both worlds (the spatial world and the integrated enterprise facility network model management world).

Although this industry's application development is more than 25 years old, it has only experienced the inception of the truly integrated enterprise in the last few years. No installed system today reflects the total integration of all of the components. However, implementations of combinations of most of the components have been achieved, and many of the key projects have been recognized by AM/FM International in its awards program.

The Pioneer Award has been given to those companies that pushed the envelope of AM/FM technology forward in their companies when risks were extremely high and the territories mostly uncharted. Pioneer Awards were given to Public Service Company of Colorado in 1989, Southern Bell Telephone Company in 1990, San Diego Gas and Electric Company in 1991, City of Edmonton in 1992, and Indianapolis Mapping and Geographic Infrastructure System in 1995.

The Excellence Award is presented to those companies that have achieved significant benefits from fully deployed robust applications. These vary greatly between companies, but they all have a common thread of highly successful implementations with executive support and an appetite for additional integration to achieve greater enterprise-wide benefits. Excellence Awards were given to Wisconsin Public Service Company and Detroit Edison in 1993, Citizens Gas and Coke Utility and Mountain Fuel Supply Company in 1994, Cobb EMC and Commonwealth Gas in 1995, Louisville and Jefferson County Information Consortium in 1996, and Clark County, NV, in 1997.

The Special Achievement Award is given to those companies that are, once again, pushing the envelope of technology in the fast track new world of competition, by integrating commercially available enterprise support systems with AM/FM/GIS, a process that will accelerate the development of standards and promote widespread adoption of the technology. Special Achievement Awards have been given to Illinois Power in 1996 and Texas Utilities in 1997.

The technology is now available for integration and the reengineering of utility and telecommunications companies' processes to meet the demands of the competitive environment.

These systems will continue to evolve, and the open system technology will play an important role in the deployment of the applications to the entire work force.

Bibliography

Engelbert, E. F. 1978. *OPERAID, Report on the Application of Computer Graphics to Outside Plant Engineering Drafting*. Bell Laboratories.

IBM Corporation. 1991. "History of IBM's GIS." *GFIS Newsletter,* 30 (Winter).

IBM Corporation. 1975. "Resource Management for Public Utilities." Data Processing Division GE20-0517-0. Unpublished report.

McDaniel, K. E., and M. A. Stice. 1991. "AM/FM System Implementation Approaches, The Path to Full Facility Information Management." Intergraph DDUM054A0. Unpublished report, Intergraph, Huntsville, AL.

McDaniel, K. E., and E. Downing. 1989. "AM/FM/GIS, History, Differences, and Similarities: Overview of Technology and its Impact on AM/FM/GIS." *AM/FM International EMS Proceedings.* Aurora, CO: AM/FM.

CHAPTER 9

Remote Sensing and GIS in Agriculture and Forestry—The Early Years

Roger M. Hoffer

Introduction

The author of this paper was invited to help document a few of the key events in the early years of the development of remote sensing and GIS for agriculture and forestry. It has been a privilege to do so as well as to have been a part of an exciting series of events as remote sensing and GIS sciences have matured.

This chapter focuses primarily upon the research program and development activities oriented on remote sensing technology in the 1960s and early 1970s. These events formed the creative environment which fosters development of many of today's principal raster GIS algorithms. The importance of the early agricultural and forestry application on raster GIS evolution is further discussed in the chapters by Faust and by Estes and Jensen.

Even before the advent of the airplane, people recognized the many advantages of looking at Earth's features from the air. In 1858, an aerial photograph was taken from a balloon near Paris, France, by the photographer Gaspard Felix Tournachon, later known as "Nadar" (ASPRS 1960). During the Civil War in the United States, the use of balloons to take pictures of the enemy positions clearly showed the value of "the bird's eye view." The advent of the airplane provided a more effective platform from which to obtain aerial photos, and it was Wilbur Wright who took the first aerial photographs from an airplane in 1909. World War I pushed the

Dr. Roger Hoffer is a professor in the department of Forestry and Remote Sensing and is director of the Remote Sensing and GIS Program, College of Natural Resources, Colorado State University. He has been a principle investigator on Landsat, Skylab, Shuttle Image Radar, and other projects involving remote sensing and Earth resources. Author's Address: Dr. Roger Hoffer, Department of Forest Sciences, College of Natural Resources, Colorado State University, Fort Collins, CO 80523. E-mail: RMHoffer@aol.com.

development and use of aerial photography to new levels, and after the war, there were many trained photo-interpreters who could see the value of aerial photos for non-military purposes. Several aerial survey companies were formed in the 1920s and 1930s, and soon the use of black and white aerial photos for topographic mapping was common. Mapping land use, forest cover, soils, crops, geology, archaeologic features, and many other areas of application clearly showed the utility of aerial photographs. The American Society of Photogrammetry (now called the American Society of Photogrammetry and Remote Sensing—ASPRS) was formed in 1934 to advance the science and art of photogrammetry and photo-interpretation, largely through the publication of the journal, *Photogrammetric Engineering* (now called *Photogrammetric Engineering and Remote Sensing*). The advent of color film and then, during World War II, color infrared (initially called "camouflage detection") film provided many new potential uses for aerial photography. In 1956, Dr. Robert N. Colwell of the University of California published a classic article concerning the use of "camouflage detection film" for detecting black stem rust in wheat (Colwell 1956), which led to more interest in the potential use of both camouflage detection and black and white infrared films for detecting diseases in various types of agricultural crops. Much of this early work is well documented in the Manual of Photo Interpretation, published by the American Society of Photogrammetry in 1960. The reader is referred to the latest edition of the *Manual of Remote Sensing* for a comprehensive update of the ASPRS series.

Computer Analysis of Multispectral Data: The Beginnings

Due to concerns about the impact of insects and diseases on crops and forests throughout the United States and the world, the Agricultural Board of the National Research Council recommended in 1960 that a committee be formed to investigate the potential of aerial surveys for monitoring insect and disease infestations in agricultural crops and forests. This committee, called the Committee on Remote Sensing for Agricultural Purposes, was chaired by Dr. J. Ralph Shay, then head of the Department of Botany and Plant Pathology at Purdue University. Dr. Robert N. Colwell of the University of California was well recognized as an expert in aerial photo interpretation, and he played a critical role in the activities of the committee. Dr. Marvin Holter of the Institute of Science and Technology, University of Michigan, also was a key member of this committee, in that he headed a group developing airborne multispectral scanners under a military contract called "Project Michigan." Other members of the committee included such well-known individuals as Robert C. Heller, David M. Gates, Harry J. Keegan, Frank E. Manzer, Victor I. Myers, and several others. From the beginning, the committee considered the potential advantages of multispectral scanners and other sensors, including radar, for obtaining data in portions of the electromagnetic spectrum beyond just the visible and near infrared wavelengths to which photographic emulsions were sensitive. Much of the work of this committee was ultimately documented in the classic book, *Remote Sensing—With Special Reference to Agriculture and Forestry*, which was published by the National Academy of Sciences in 1970.

At about the same point in time that the Committee on Remote Sensing for Agricultural Purposes was initiating work, the National Aeronautics and Space Administration (NASA) was rapidly developing a satellite capability, and as NASA looked ahead to the potential utility of satellites for various types of applications, the possibility of utilizing multispectral scanner systems for agricultural surveys emerged. Thus, three key elements had converged to form the basis, in part, from which has developed the current worldwide system of satellite multispectral scanner systems and applications, namely: (1) The need for various types of agricultural information over large geographic areas to be obtained in a timely manner; (2) the potential utility of multispectral scanner systems, since they could obtain data in discrete narrow wavelength bands and in wavelengths beyond the photographic portion of the spectrum, and (3) NASA's interest in the potential use of spacecraft for various applications of benefit to the citizens of the United States.

These three elements were fused into a proposal to NASA from the Committee on Remote Sensing for Agricultural Purposes to conduct some preliminary feasibility studies to test the potential utility of multispectral scanner systems for monitoring agricultural conditions. Since Purdue University was in the center of the corn belt of the U.S. and had an agronomy farm where a large variety of crops and crop conditions were present under well-known and controlled conditions, it was decided to conduct a series of aerial missions during the summer of 1964, using the University of Michigan multispectral scanner system and the Purdue Agronomy Farm as a test site.

On May 3, 1964, two single-engine L-19 aircraft flew over the Purdue Agronomy Farm with a system of multispectral scanners and cameras. This was the first time in history that a multispectral scanner had been flown for the purpose of gathering information on crops and soils, rather than for military purposes. The scanner system consisted of two double-ended scanners, one of which obtained data in the ultraviolet portion of the spectrum on one end of the scanner and the thermal infrared on the other end. The other scanner collected data in one thermal and three reflective infrared channels, respectively, from the two ends of the scanner. The data were recorded onto analog tapes, which were then used to create imagery for analysis using photo interpretation techniques, since computer analysis capabilities did not then exist. Neither of the scanners obtained data in the photographic portion of the spectrum (0.4- 0.9 micrometer). Data in the photographic wavelengths were obtained using an old Graflex camera, but, instead of a normal lens, there was simply a piece of plywood in which nine holes had been drilled and some small lenses inserted. On top of each lens was a thick packet of filters designed to allow only certain portions of the spectrum to be transmitted. The imagery was recorded on 4 x 5 inch I-N spectroscopic glass plates, each carefully loaded into a film pack on the back of the Graflex camera. The result of the combined scanner and camera system was a set of 18 wavelengths of imagery, representing the entire spectrum from 0.32 micrometer in the ultraviolet to 14 micrometer in the thermal infrared.

Five missions were successfully flown over the Purdue Agronomy Farm during the summer of 1964 under the grant from NASA. The task of interpreting all of the resulting imagery was given to Dr. Hoffer, who had been hired by Dr. Shay late that summer under a grant from the U.S. Department of Agriculture, Economic Research Service. It was decided that the best approach for interpreting these 18 wavelength bands of imagery was to use cardboard gray-scale tone cards and to attempt to quantify the gray tones of different crop types and conditions seen on the imagery. In light of today's technology, such an approach sounds like something out of the Stone Age, but it was about the only thing available back then. This crude method of data analysis quickly led to the conclusion that before we pursued the goal of trying to identify diseases in various agricultural crops, we needed to verify the utility of multispectral data to simply identify the different crop species! The resulting efforts to identify various crop species resulted in a clear indication of the importance of obtaining remotely sensed data at the critical stages of crop development, plus the need to develop a good understanding of crop phenology.

It is interesting to note that, because the scanner had been developed under a military contract, all imagery generated was considered "classified" and had to be kept under lock at all times when not being used. Unauthorized personnel couldn't even look at these images of agricultural crops! One of the early reports of the data analysis showed an illustration of a few crop species that displayed distinct differences in reflectance in the photos obtained in the visible (0.4–0.7u) and near infrared (0.7–0.9u) wavelengths, but only an "artist's concept" sketch by Dr. Hoffer could be used to illustrate the tones of these crops for the thermal infrared image (4.5–5.5u).

The crude early attempts to interpret manually 18 wavelength bands of black and white imagery also led to the conclusion that methods needed to be developed to quantify both the data collection and the analysis processes. One possible approach that was considered involved using a densitometer to measure the photographic opacity of the film for each agricultural field of interest in each of the wavelength bands of data obtained. Fortunately, however, the engineers at the University of Michigan were working on the development of a new multispectral scanner system that collected data simultaneously in 12 bands (ten visible and two near infrared), and these data were recorded onto analog tape. The analog tape could be changed later into digital format via an analog to digital converter. It was this scanner that provided the data for the research needed to develop and test digital pattern recognition techniques, many of which are still in use today.

Meanwhile, in early 1965, the Purdue researchers had learned that faculty in Electrical Engineering were doing work in an area of research referred to as "pattern recognition." Meetings between Dr. Shay, Dr. Hoffer, and an electrical engineer, Dr. Roger Holmes, developed into a plan to form an interdisciplinary team to attempt to apply pattern recognition techniques to this multispectral scanner data. A proposal to NASA from Dr. Hoffer and Dr. Holmes was funded for the purpose of "establishing methods whereby various soil and agricultural crop parameters may be determined remotely, through a program of comparative multispectral sensing. Crop and soil parameters to be studied include species identification, state of maturity, disease conditions, soil types, and soil moisture conditions." A companion proposal to NASA from Dr. Shay for needed computer equipment and the hiring of additional staff provided the remainder of the funding needed to establish, in February 1966, the Laboratory for Agricultural Remote Sensing (LARS). Electrical engineers Dr. David Landgrebe and Dr. K. S.. Fu, who was well known in the pattern recognition field, became the core of the data analysis group. Robert MacDonald was hired from

IBM to become the director of LARS, and Terry Phillips was hired for data processing. Agronomists Dr. Marion Baumgardner and Chris Johannsen brought their agricultural expertise into the group. These individuals formed the initial nucleus of this very interdisciplinary LARS team. The basic goal of their research was to develop techniques to digitally analyze multispectral scanner data of agricultural crops. (It should be noted that in 1969, LARS was renamed the Laboratory for Applications of Remote Sensing to reflect the research activities of the group in forestry, geology, hydrology, and geography as well as agriculture.)

At the same time that LARS was being formed at Purdue University, NASA also funded Dr. Colwell in the Department of Forestry at the University of California to pursue the development and use of remote sensing techniques for forestry applications. In order to minimize duplication of efforts, and in light of Colwell's expertise in photo interpretation, the research efforts at the University of California concentrated on the use of multiband photography for forestry applications. A third grant was given to Dr. Mike Holter at the University of Michigan to continue to develop multispectral scanning systems and to pursue the use of analog methods of data analysis. Dr. Victor Myers, U.S. Department of Agriculture (USDA), Weslaco, TX, was funded by the department to assess the utility of remote sensing for various agricultural applications. Thus, there was a significant amount of research being conducted at several locations, all dealing with remote sensing for agricultural and forestry applications, but each team of researchers was pursuing a different aspect of the overall research activity. The research at these four locations formed the nucleus of the remote sensing research being conducted with multispectral scanners in the late 1960s and early 1970s—research which provided critical data analysis and processing capabilities needed for the Landsat era. It must be pointed out, however, that NASA had also funded a major research effort on the use of radar data at the University of Kansas, under the direction of Dr. Richard Moore. Drs. Fawaz Ulaby, David Simonett, and Stanley Morain, as well as several other researchers, made particularly noteworthy contributions concerning the understanding and use of radar in these early years of remote sensing.

At Purdue University, a set of data processing and pattern recognition programs known as LARSYS was developed and tested on multispectral data obtained by the University of Michigan during three agricultural flight missions in 1966. The LARSYS programs are recognized as a milestone for raster processing and formed the basis for much governmental and commercial GIS software (Chapter by Faust). A LARS Information Note (21567) titled "Automatic Identification and Classification of Wheat by Remote Sensing" by David A. Landgrebe and staff of the Laboratory for Agricultural Remote Sensing was published in March 1967, documenting the first successful application of pattern recognition techniques to multispectral scanner data. The data had been collected over an agricultural area designated as Flight-line C-1, south of the Purdue Agronomy Farm. To display the data, a line printer and different alphanumeric symbols were used to represent different levels of reflectance in a particular wavelength band, thus providing a rough gray-scale map of the area. The application of pattern recognition techniques to the multispectral scanner data resulted in a printout in which only the points classified as winter wheat were displayed, using the letter W. One field near the middle of the flight-line had oats planted in the middle and wheat around the outside portion of the field, and it therefore became affectionately known to many students and visiting scientists in the following years as the "donut field." In the 1968 LARS report, it was pointed out that out of 64,240 total points classified, 5,469 points were classified as wheat, thus indicating the potential for using such computer processing techniques not only for mapping a particular crop species but also for providing acreage estimates, if the size of the scanner resolution element is known. The classification results reported in this publication proved for the first time that the concept involving the application of pattern recognition theory to the analysis and classification of multispectral scanner data was valid. This report is thus considered to be one of the milestones in the development of remote sensing technology and raster GIS.

The Corn Blight Watch

In 1970, a potentially grave situation was developing in the form of a disease known as the Southern Corn Leaf Blight. At that time, essentially all of the corn being grown in the United States had been genetically developed with a type of cytoplasm that was susceptible to this new strain of potent, fast-spreading disease. The blight would first be seen as small brown lesions on the lower leaves of the corn plant. These lesions would spread on the leaf, and leaves farther up on the plant would develop then until eventually the entire plant would be killed. At the request of NASA, LARS conducted some preliminary studies in late 1970 to determine if remote sensing could be effectively utilized to identify areas that had been affected by the corn blight. Some of this preliminary work was carried out in August and September of 1970, and the results showed that one could see differences in fields that had been moderately to severely affected by the corn blight from those where the corn was still healthy. Based upon these preliminary results, NASA and USDA joined forces and proposed a major effort for the summer of 1971 to monitor the rate of spread and severity of the corn blight. This project become known as the "1971 Corn Blight Watch Experiment." One interesting aspect of the study was that because it had not been an anticipated project, it was not funded in the normal sense of having a line item in the budget of either agency—they simply assigned people and resources from other projects to the task at hand. Purdue University was asked to coordinate the corn blight watch program under the leadership of R. B. MacDonald (the director of LARS) and a steering committee composed of agency personnel from the USDA, NASA, Purdue University, the University of Michigan, and various other agencies. A seven-state region, including all of Iowa, Illinois, Indiana, and Ohio as well as portions of Nebraska, Missouri, and Minnesota, was designated as the study area. To cover such a huge area, it was decided to utilize color infrared photography obtained by a NASA aircraft (RB-57) from a height of 60,000 feet and that this photography would need to be obtained every two weeks throughout the growing season.

NASA and the USDA were also very interested in determining the potential for using multispectral scanner data for detecting and monitoring disease such as the Southern Corn Leaf

Blight, so it was decided to have the University of Michigan C-47 fly over an intensive study site in western Indiana every two weeks. This resulted in a set of 30 segments of multispectral scanner data that also had to be collected and analyzed using computer analysis techniques on a rigorous two-week cycle. A team of photo-interpreters located at LARS under the direction of Dick Mroczynski carried out the photo-interpretation analysis throughout the summer every two weeks, while the multispectral scanner data was analyzed by teams of people at the Environmental Research Institute of Michigan (ERIM) (formerly the Institute of Science and Technology, University of Michigan) and at LARS, Purdue University. Both the photo-interpretation work and the analysis of the multispectral scanner data relied on field observations obtained by personnel from various agencies who were local to the various test segments. The results of all these analyses were summarized and transmitted to the USDA every two weeks. An interesting side note was that we were told that those of us involved in the analysis work could not invest in the futures market that summer, under a potential penalty of a $10,000 fine and six months in jail!

Both the color infrared photography and the multispectral scanner data were found to be very effective for monitoring the severity and spread of the blight. It was found that because the blight is first detected by the small lesions on the lower leaves on the plant, farmers and field personnel participating in the experiment knew if a field was infested with the blight long before it could be detected using remote sensing techniques. This pointed out the significant difference between plant diseases that are systemic, in which the entire plant is affected, and diseases such as the blight that do not cause a visible or spectrally observable effect over the entire plant early in the development of the disease. However, even though remote sensing could not detect the disease at the outset of an infestation in a field, both the color infrared photography and the multispectral scanner data were shown to be very effective in identifying fields with moderate to severe levels of blight and therefore were effective in monitoring the rate of spread and the severity of the disease.

As it turned out, the summer of 1971 was not as hot or as humid as normal, and therefore the blight did not develop into as serious a problem as had been anticipated. However, a number of significant results did come out of this experiment. Remote sensing provided the best, most accurate source of information that existed for monitoring the spread and the severity of the corn blight that summer. The ability of two major agencies within the government to work together under a very tight time frame and to coordinate the activities of the hundreds of people that were involved in the program was also significant. It has been stated that the 1971 Corn Blight Watch was the largest agricultural experiment of its kind ever conducted. To indicate the magnitude of this project, it was estimated that NASA flew 38,000 miles of flight-lines, and more than 2,600 man-days were spent by field crews in obtaining the ground observation data during the summer of 1971. Aerial photography for an area of more than one million acres in the seven-state region was obtained, analyzed, and reported upon every two weeks during the three-month period of the experiment. It was indeed a massive undertaking!

Analysis of First Frame of Digital ERTS-1 Data

The launch of the first Earth Resources Technology Satellite (ERTS-1)—later renamed Landsat-1—initiated a new era for resource managers in all disciplines. For the first time ever, high-quality data could be obtained from satellite altitudes at reasonably frequent intervals for nearly any portion of Earth's surface. ERTS-A was successfully launched to become ERTS-1 on July 23, 1972. Two days later, the four-band multispectral scanner was utilized to collect image data over the U.S. for the first time. At that time, LARS was one of the very few facilities in the U.S. having the capability to analyze digital multispectral data, especially for an area the size of a full frame of ERTS data. Therefore, NASA requested LARS to conduct an analysis of the first frame of digital data obtained from the satellite after it was launched. To achieve the most rapid analysis possible, we were asked to send someone to Goddard Space Flight Center to obtain the data as soon as it had been collected. The purpose of this analysis was to obtain preliminary indications concerning the operating characteristics and potential value of this satellite as a data gathering device, as well as to evaluate the quality of the data obtained.

Bill Simmons, one of the computer technicians at LARS went to Goddard, and late in the day on July 26, he was given a black and white hard copy image from Channel 5 (0.6-0.7 micrometers) together with the digital tapes from the multispectral scanner for the full frame of data. He arrived on campus about 11 p.m. and by the next morning had the data displayed on the LARS digital display computer and ready for further analysis. The analysis was conducted by David Landgrebe, Dr. Hoffer, Forrest (Bud) Goodrick, and other members of the LARS staff. The results of this analysis were reported at the First Earth Resources Technology Symposium, held at Goddard Space Flight Center on September 29, 1972 (Landgrebe et al. 1972).

The frame provided to LARS for analysis was taken from the first pass of the satellite across the U.S. and was of the Red River Valley area along the border between Texas and Oklahoma. Preliminary examination of the image on the computer display screen revealed that the frame contained an interesting array of agriculture, rangeland, forestry, cultural, geologic, and water resource features. However, we had no ground observations available or even much general knowledge of the area.

On July 30 and 31, two members of the LARS team—Chris Johannsen and Bud Goodrick—traveled to the area to obtain field observation information. With the knowledge gained from the ground and low altitude aircraft mission, as well as the wealth of information obtained from the resource people contacted in Oklahoma, a second series of classification analyses was conducted.

In summary, the analysis of this first frame of ERTS-1 data provided a detailed and, as nearly as we could determine, a reasonably accurate classification of the various cover types present. The sequence of a preliminary classification based on clustering, followed by the field trip to obtain detailed information about the area, both on the ground and from the air, and then the final detailed classification was shown to be very effective. From the time the ERTS-1 satellite was first turned on until the analysis, including the field trip, had been completed for the entire 12,000+ square mile area involved less than two weeks, thereby demonstrating the potential for a rapid turn-around using such satellite data. Now, 25 years later, it is interesting to note one particular statement in the Summary and Conclusions section of the paper describing the analysis which says: "One of the most difficult problems at the present state of the remote sensing art appears to be the refinement of a straightforward technique to relate the spectral classes present to the significant categories of interest defined by the users." In many instances this is a problem with which analysts still struggle and clearly defines the role of GIS for remote sensing classification!

Data for Mapping Forest Cover in Mountainous Areas

One of the earliest attempts to combine GIS data with remote sensor data involved the use of topographic data (elevation, slope, and aspect) in combination with Landsat data to identify individual forest cover types more accurately (Hoffer et al. 1975). The test site involved a mountainous area in Colorado containing a complex mixture of forest types, rangeland, alpine tundra, water bodies, geological features, and various manmade features. Initial studies in this area had shown that Landsat data and computer classification techniques could be used to identify and map deciduous and coniferous forest cover types with a reasonable degree of accuracy (i.e., 70 to 85%), but they also showed that individual species of trees could not be effectively differentiated and identified using only the Landsat spectral data (Hoffer et al. 1975). However, in the San Juan Mountains of southwestern Colorado, as in most mountainous areas, there is a rather well-defined distribution of forest cover types as a function of elevation. It was hypothesized, therefore, that if the areas of coniferous and deciduous forest cover could be identified and mapped using just the spectral information in Landsat data, topographic data could then be used to differentiate the data into individual forest species groups or cover types. A study was therefore undertaken to develop and test techniques which would utilize both digital topographic data and spectral data to map forest cover types at a greater level of mapping detail and with increased accuracy than was possible using spectral data alone. Elevation data was available for the test site in the form of Defense Mapping Agency (DMA) topographic data which had been derived from 1:250,000 maps. This DMA data was in computer compatible format with each pixel being approximately 64 meters square, which was reasonably close to the 56 x 79 meter ground area represented by each pixel of the Landsat MSS data. A nearest neighbor algorithm was used to overlay digitally, pixel by pixel, the elevation data onto the Landsat data. Slope and aspect channels were generated from the elevation data, using an interpolation process and added to the Landsat spectral-plus-elevation data set. This provided a database in which every X-Y coordinate contained four channels of Landsat spectral reflectance data, a channel of elevation data, a channel of slope data, and a channel of aspect data.

Analysis of this data set indicated that the most effective method for combining Landsat and topographic data was a "layered classification" technique. This technique allows the analyst to classify the data in a hierarchical sequence of steps or layers (Swain and Hauska 1977). Since different types of data could be used in the different layers or steps in the classification process, such a technique was particularly useful in this study in that Landsat satellite data could be used initially to identify coniferous and deciduous forest cover (as well as other major cover types), and then, in the next layer, topographic data corresponding to the same X-Y coordinate on the ground could be used to identify the particular forest cover type. To accomplish this, two different sets of training statistics were required—one for the Landsat data and one for the topographic data. A hybrid technique for developing spectral training statistics, developed by Michael Fleming and originally called the "Multi-Cluster Blocks Technique" (Fleming et al. 1975), was used to generate the spectral training statistics, and a stratified random sample procedure was used to develop a "Topographic Distribution Model" which quantitatively described the distribution of the various cover types of interest as a function of elevation, slope, and aspect (Fleming and Hoffer 1979). The layered classification tree structure shown in Figure 9.1 was used to classify this combined spectral/topographic data. Only the Landsat spectral data was used in the first layer of the classification to identify the major cover types in the area (coniferous forest, deciduous forest, herbaceous, barren, and water). In this study, 15 spectral classes were used to identify these five major cover types. In the second layer, only the topographic data was used to classify the data into the individual cover type classes of interest. As indicated in Figure 9.1, after the classification was completed, the alpine willow and tundra classes were combined into a single informational category called "alpine." It was also discovered during the analysis that some areas were being misclassified as water when only the Landsat data was used because they had very low spectral response values, particularly in the reflective infrared wavelengths. In reality, these areas were generally spruce-fir forest cover in regions of topographic shadow. The topographic data was therefore used in the second layer of the classification process to reexamine each pixel that had initially been classified as water, and only those pixels having a zero slope were allowed to remain classified thus. If the pixel had a slope of other than zero, it was put into the spruce-fir forest cover type.

The results of this study indicated that by using the layered classification technique and the combined spectral plus topographic data, classification accuracies were improved by 15% over those obtained using Landsat spectral data alone (Hoffer et al. 1975). It is believed that this work also represented one of the first cases in which both Landsat multispectral scanner data and a GIS database were combined to obtain a more accurate cover type map over a relatively large (i.e., more than 500,000 hectares) geographic area. This study led to the operational use of combined Landsat data and topographic data to develop a forest cover type map for the San Juan National Forest which was used for planning activities for the forest.

Summary

Over the last 33 years, remote sensing has matured from a crude prototype multispectral scanner flown in a single-engine observation aircraft to multiple highly sophisticated satellite systems capable of collecting vast amounts of data over the entire Earth. Likewise, as computer technology and storage capacities have grown, Geographic Information Systems sciences have developed at a rapid pace and provide tremendous opportunities to today's society. No longer are we limited to working with small data sets over limited geographic areas, but we have amazing

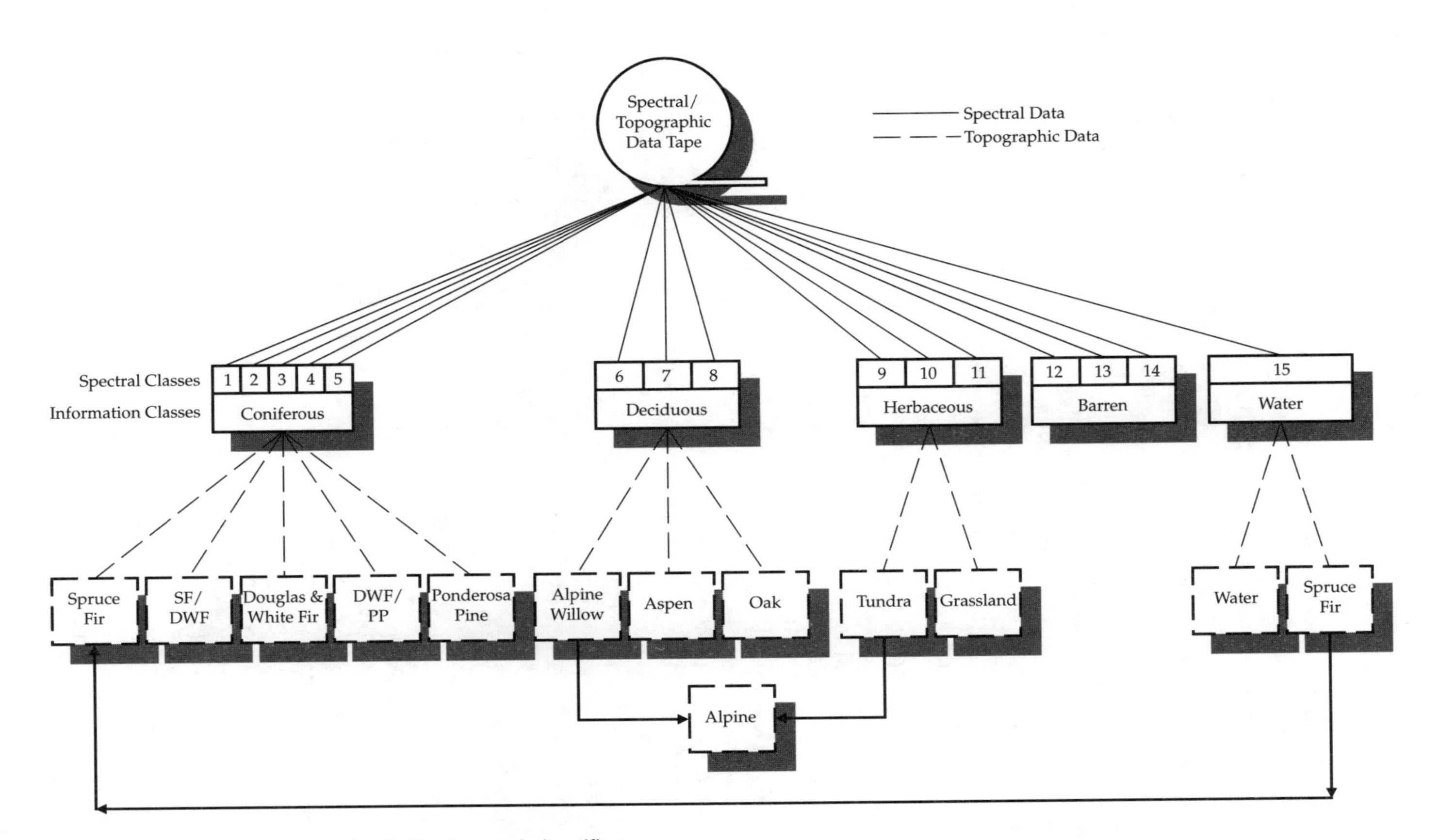

Figure 9.1 Decision tree used with the layered classifier.

quantities of various types of data available for the entire Earth. Current and developing remote sensing and GIS capabilities provide opportunities that could only be dreamed of or hoped for in years past. They make the future look very bright indeed.

There have been many other people and organizations who have been critical to the development of these sciences whose activities could not be adequately acknowledged in these few pages but whose work was also of tremendous significance. However, because of the roles played by the University of Michigan engineers in developing multispectral scanner technology and by Purdue University scientists and engineers in developing computer-aided techniques for analyzing multispectral scanner data, it seems fitting to conclude this chapter by pointing out that in 1976, LARS and ERIM were jointly awarded the Pecora Award by NASA and the U.S. Geological Survey (USGS) for their outstanding contributions to the field of remote sensing. In making the award, the citation stated that:

> "The Laboratory for Applications of Remote Sensing (LARS) and the Environmental Research Institute of Michigan (ERIM) have been pioneers in developing methods for the acquisition and analytical processing of multispectral imagery. Early efforts of these two organizations were aimed at automating agricultural inventories. The same techniques are now being applied to many other discipline activities, including geology, hydrology, and land cover analysis. Many of the Landsat design concepts and analysis techniques are direct outgrowths of these original efforts.
>
> "When Landsat-1 was launched in July 1972, few investigators were prepared to fully utilize the capabilities of the digital data acquisition system. The association between discipline scientists and these organizations has led to a rapid broadening of the user community capable of using data analysis methods.
>
> "The combination of technological expertise, knowledge of user requirements, and comprehension of the total acquisition and analysis systems has made these organizations a unique source for advice and special studies. Their contributions, both as organizations, and as individuals, have been critical to the success of the Landsat program."

Bibliography

American Society of Photogrammetry (ASPRS). 1960. *Manual of Photographic Interpretation.* (R. N. Colwell, ed.). Washington, D.C.: American Society of Photogrammetry, 868.

Colwell, R. N. 1956. "Determining the Prevalence of Certain Cereal Crop Diseases by Means of Aerial Photography." *Hilgardia,* 26 (5): 223–286.

Fleming, M. D., J. Berkebile, and R. M. Hoffer. 1975. "Computer-Aided Analysis of Landsat-1 MSS Data: A Comparison of Three Approaches Including the Modified Clustering Approach." *Proceedings of the Symposium on Machine Processing of Remotely Sensed Data.* West Lafayette, IN: Purdue University, 1B-54 to 1B-61.

Fleming, M. D., and R. M. Hoffer. 1979. "Machine Processing of Landsat MSS Data and DMA Topographic Data for Forest Cover Type Mapping." *Proceedings of the 1979 Symposium of Machine Processing of Remotely Sensed Data.* West Lafayette, IN: Purdue University, 377–390.

Hoffer, R. M., and staff. 1975. "Natural Resource Mapping in Mountainous Terrain by Computer Analysis of ERTS-1 Satellite Data." *Agricultural Experiment Station Research Bulletin 919.* West Lafayette, IN: Purdue University, 124.

Landgrebe, D. A., and staff of the Laboratory for Agricultural Remote Sensing, Purdue University. 1967. *Automatic Identification and Classification of Wheat by Remote Sensing.* LARS Information Note 21567, Research Progress Report 279. West Lafayette, IN: Agricultural Experiment Station, Purdue University, 7.

Landgrebe, D. A., R. M. Hoffer, F. E. Goodrick, and staff. 1972. "An Early Analysis of ERTS-1 Data." Invited paper. *Proceedings of the NASA Symposium on Preliminary Results of ERTS Data Analysis.* Greenbelt, MD: Goddard Space Flight Center, 21-38.

LARS staff. 1968. *Remote Multispectral Sensing in Agriculture.* Laboratory for Agricultural Remote Sensing, Annual Report, vol. 3, Research Bulletin No. 844. West Lafayette, IN: Agricultural Experiment Station and School of Electrical Engineering, Purdue University.

Swain, P. H., and H. Hauska. 1977. "The Decision Tree Classifier: Design and Potential." *IEEE Transactions on Geoscience Electronics,* GE-15 (3): 142–147.

PART V

National Agency Development

John D. Bossler, Director, Center for Mapping, The Ohio State University

I am a relative newcomer to the GIS world, and, while the acronym was vaguely familiar to me, the first time I began to consider GIS seriously was in 1986, when I attended a meeting at SUNY in Buffalo, hosted by Duane Marble. It was at that meeting that we discussed the upcoming NSF solicitation for the NCGIA. Ron Abler attended, along with numerous notables in the GIS field. It was an interesting meeting, and we shared our thoughts about the value and impact of the NCGIA.

I was still the director at the Coast and Geodetic Survey at the time but became the director of the Center for Mapping a month or so later. I convinced the Ohio State University to apply for the NSF grant and also convinced the university to hire four new "GIS faculty." We lost the grant but strengthened the GIS expertise at OSU.

John D. Bossler has a long career in the military development of mapping technology and is a past president of AM/FM International.

Kenneth J. Dueker, Center for Urban Studies, Portland State University

In 1965–66, I was a research assistant to Bill Garrison and Duane Marble on a remote sensing research project at Northwestern University. I sought Duane Marble's advice on a title for my part of the research. We generated titles using several terms—geographic, spatial, data—information systems and models. We discarded geographic information systems for two reasons, even though we liked the acronym, GIS, as used in the Canada Geographic Information System (CGIS). GIS was broader than our remote sensing project warranted, and IBM had just announced its new software package called GIS—Generalized Information Systems. We figured they had preempted the use of GIS. Little did we know about another new term, "vaporware." So my project report, (*Spatial Data Systems,* Department of Geography, Northwestern University, 1966) and my subsequent dissertation (*Spatial Data Systems,* Department of Civil Engineering, University of Washington, 1967) did not reinforce the usage of "GIS" terminology at an early stage in the development of the GIS field.

Meanwhile, others were groping with what to call their work. During this period other terms were more frequently used. GIS developments in the late 1960s and early 1970s were described using such terms as spatial data handling, automated cartography, urban and regional information systems, land management information systems, natural resources inventory systems, and urban geoprocessing. Gradually, GIS became the encompassing concept.

Kenneth J. Dueker continues the academic tradition for GIS development started in late 1950s at the University of Washington.

J. Michael Scott, Biological Resources Division, U.S. Geological Survey

The first time I used GIS was in the summer of 1980. My long-time friend, Dr. Jack Estes, then Chair of the Department of Geography at the University of California at Santa Barbara, was visiting my family in the Hawaiian islands. I had taken Jack down to my field station to show him maps, graphs, and tables detailing the results of an eight-year survey of the endangered forest birds of the Hawaiian Islands. In explaining the results of the project, I expressed my frustration about our inability, despite numerous publications in the refereed scientific journals and one book detailing the results of our survey, to put the information into a format that was easily interpreted by managers and policy makers. He then explained the principles of GIS to me, and we began that night with single-sided sheets of mylar, mapping out the distribution of five endangered forest birds as well as the boundaries of national parks, national wildlife refuges and other areas dedicated to the protection of natural resources. The results were astonishing: when we laid the six different mylar sheets on top of each other, they clearly showed less than 5% overlap between protected areas and the ranges of species we were attempting to protect. When shown to officials in a position to make a difference on the ground (the U.S. Fish and Wildlife Service, the Nature Conservancy of Hawaii Department of Land and Natural Resources), the results were equally astonishing. They said this was the information they needed. One graphic byte had done what hundreds of written pages could not. As a result, the 33,000 acre Hakalau Forest National Wildlife was created by the U.S. Fish and Wildlife Service, and other federal and state forest bird reserves were established.

J. Michael Scott is a Senior Research Biologist with the Biological Resources Division of the U.S. Geological Survey; Leader of the Idaho Cooperative Fish and Wildlife Research Unit, and a professor in the Department of Fisheries and Wildlife, University of Idaho.

CHAPTER 10

Development of Remote Sensing Digital Image Processing and Raster GIS

John E. Estes and John R. Jensen

Introduction

This chapter summarizes remote sensing digital image processing and geographic information system (GIS) development funded by the United States Government, specifically the National Aeronautics and Space Administration (NASA). Much of the history of the development of digital image processing systems is undocumented. Many of the raster-based GIS systems in use today have functions that were derived from these undocumented digital image processing systems. The contributing authors were asked to document the chronology and status of the various digital image processing systems that contained significant GIS functions. The principal authors

---

Dr. John E. Estes is a professor of Geography and director of the Remote Sensing Research Unit at the University of California, Santa Barbara. He has worked in academia and with government and industry more than 30 years in the field of remote sensing and more than 25 years in the area of GIS.

Dr. John R. Jensen is a professor of Geography at the University of South Carolina in Columbia. He leads a research team in investigating advance techniques for integrating remote sensing and GIS for wetland and environmental studies. He is also an author of the leading textbook on digital image processing for environmental applications.

Contributing Authors: Nevin Bryant, NASA Jet Propulsion Laboratory; Leonard Gaydos, U.S. Geological Survey, NASA Ames Research Center; David Landgrebe, Purdue University; Ron J. P. Lyon, Stanford University; Gary Petersen, Pennsylvania State University; Darrel Williams, NASA Goddard Space Flight Center.

Authors' Addresses: John E. Estes, Dept. Of Geography, UCSB, Goleta, CA 93106. E-mail: jestes@geog.ucsb.edu. John R. Jensen, Dept. Of Geography, U. Of S. Carolina, Columbia, SC 29208.

synthesized this diverse collection of heretofore unpublished information into a coherent review of the NASA funded development of remote sensing digital image processing systems and raster-based GIS in the United States. A chronology of this synthesis is provided which highlights the major universities and research centers involved in the pioneering phase of GIS evaluation (Figure 10.1).

Most histories of geographic information systems provide only a passing reference or make no mention of how early developments in the field of image processing of remotely sensed data fit into the continuum of GIS development (Tomlinson and Petchenik 1988; Maguire, Goodchild, and Rhind 1991). Yet early developments in GIS are closely tied to the art and science of remote sensing. For example, Roger Tomlinson developed the initial idea for computer-based geographic information systems in 1960 while working for Spartan Air Services, a photogrammetric engineering firm that was in the process of bidding a forestry related survey in East Africa (Coppock and Rhind 1991). His immediate goal was to develop an efficient method of combining available map information to identify potential sites for plantations. This endeavor ultimately led to the development of the Canadian Geographic Information System in the early 1960s.

In 1958, the term "remote sensing" was introduced by Evelyn Pruitt, a geographer working with the U.S. Office of Navy Research, and defined as "the science of obtaining information about an object without being in direct physical contact with the object." The term was to be all encompassing and include optical, thermal, and microwave data collection. U.S. scientists were beginning to examine ways to process the digital remotely sensed imagery being acquired by more sophisticated remote sensing systems carried onboard aircraft and spacecraft. This work was primarily funded by NASA which was responsible for building the meteorological satellites for the National Oceanic and Atmospheric Administration (NOAA). The launch of the TIROS-1 meteorological satellite on April 1, 1960, was one of a number of indications that space technology would some day have a significant role to play in obtaining information to better measure, map, monitor, model, and manage Earth's finite resources.

Also in the early 1960s, a number of farsighted persons in universities and federal agencies began making plans for research programs that would apply the new remote sensing technology to useful applications. In 1961, the National Academy of Science National Research Council formed a committee on "Aerial Survey Methods in Agriculture," chaired by Professor J. Ralph Shay, head of Purdue's Department of Botany and Plant Pathology. In 1963, Dr. Shay's committee formulated a research program and received academy approval to proceed. The committee worked with NASA headquarters and the Bioscience and Manned Space Science divisions who were formulating remote sensing experiments. The committee also worked with the U.S. Army Electronics Command at the University of Michigan which was working on the development of military aircraft remote sensing equipment. The Army agreed to support the agriculturally oriented program to the extent of making available aircraft and sensing equipment operated by the Willow Run Laboratories of the University of Michigan's Institute of Science and Technology. In 1964, the U.S. Department of Agriculture's Economic Research Service

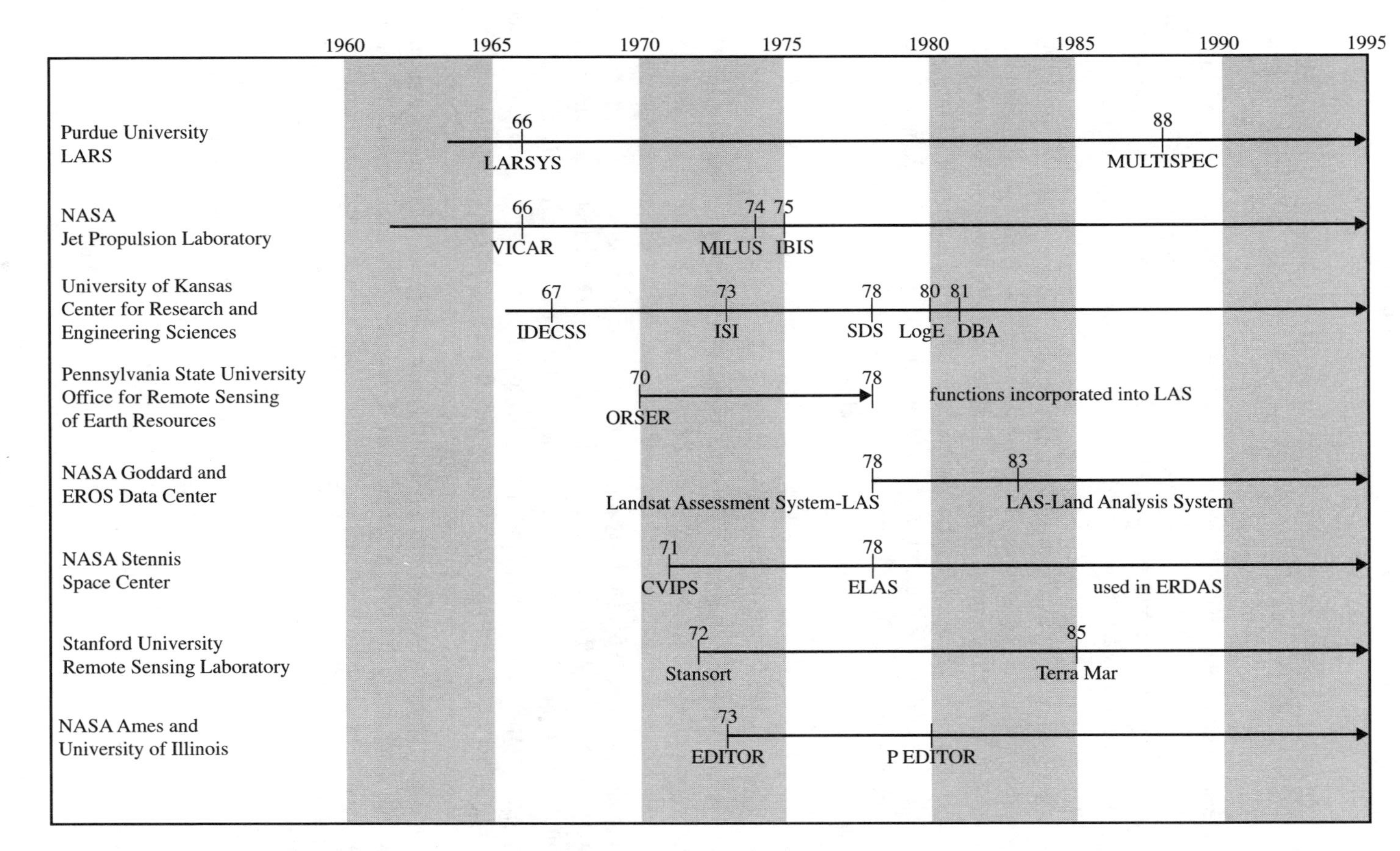

Figure 10.1 Chronology of the development of the major digital image processing systems sponsored by the National Aeronautics & Space Administration.

agreed to fund a small grant at Purdue to have data from the optical portion of the electromagnetic spectrum collected over Purdue agricultural sites by the Michigan equipment and analyzed at Purdue.

At the same time, significant efforts were also underway at the University of California Berkeley campus where Dr. Robert N. Colwell, Professor of Forestry, was leading a group in developing manual photo-interpretation techniques for forestry and agriculture. A group at the University of Kansas led by Dr. R. K. Moore was involved in the application of radar technology to agriculture.

As the program took form, the three universities working in the optical region (Purdue, Berkeley, and Michigan) began to work in a coordinated fashion. Although all three had some effort across the full breadth of the problem, each had its area of special concentration. Michigan personnel concentrated on designing, constructing, and operating the world's first airborne multispectral scanner. Berkeley scientists focused on perfecting manual and then digital data analysis procedures for forestry and agriculture applications. Purdue scientists focused on machine data-analysis methods applied to agricultural problems.

Purdue Agricultural Remote Sensing Lab

Dr. Shay was able to hire Dr. Roger Hoffer (Chapter by Hoffer) to begin work on the United States Department of Agriculture (USDA) Earth Resources Survey grant. Later, Professor Roger Holmes of the electrical engineering school began to work on the engineering aspects of the problem. In November of 1965, Dr. Shay presented a seminar to the electrical engineering faculty. Professor K. S. Fu and Professor D. A. Landgrebe began to participate. The Laboratory for Agricultural Remote Sensing (LARS) was formed in 1966 to take an interdisciplinary team approach to the creation and use of space-based technology for observing and managing agricultural resources (Hoffer 1967). The value of this approach is apparent from the fact that within a few short years, LARS became internationally recognized and was the largest research laboratory on the Purdue campus. In the late 1960s, LARS's interdisciplinary philosophy was unusual on a university campus and the only one of the three institutions that was aggressively approaching the problem in that fashion. This approach required a common physical location in which to conduct the research. There was no site available on the campus, so space was obtained in the Purdue Industrial Research Park.

During the first eight months, the LARS scientists focused on the development of a research plan for the next several years and the hiring of qualified personnel. By the end of the spring of 1966, three professional staff had been hired on a full-time basis, including R. B. MacDonald, who came to the program on temporary leave from IBM Inc. to serve as Technical Director, and T. L. Phillips. In addition, an electronics technician, a photographic technician, and 13 graduate student researchers were hired. MacDonald later headed the Large Area Crop Inventory Experiment (LACIE) and the Agricultural and Resource Inventory Surveys through Aerospace Remote Sensing (AGRISTARS) at the NASA Johnson Space Center.

From an engineering standpoint, the problem before the LARS scientists was to define:

- what physical quantities should be measured?
- what kind of instruments should be used?
- what sensitivity, resolution, precision should be used?
- what data analysis approach should be used?

It was known that remotely sensed data could generate valuable information rapidly. But doing so resulted in very large amounts of data to be analyzed. Thus, the question immediately arose: Couldn't emerging computer technology be useful in some way to accomplish the needed analysis rapidly and cost effectively? A study of the problem indicated that machine implemented pattern-recognition methods, which were rather new at that time, might be a feasible approach to machine analysis. However, applying such methods in a conventional image (i.e., picture) analysis mode would require much too fine a spatial resolution. To have spatial resolution fine enough that a machine algorithm would be able to identify plant species (e.g., by the shape of the leaf or plant) would unacceptably exacerbate the data-volume problem.

High spatial resolution remote sensor data are very expensive in several respects. First, data volume increases as the square of the resolution, e.g., 50-m resolution results in four times the amount of data as 100-m resolution. In addition, as resolution is increased the space platform stability and pointing accuracy must be increased, and so must the size (and thus the weight in orbit) of the sensor optics. Instead of relying on the spatial characteristics of the objects to be identified, LARS personnel decided to rely primarily upon the spectral response of the subject matter, thus creating a new type of spectroscopy. This concept was to be the central focus for LARS research for years to come. It also became the basis for the ERTS and LANDSAT series of Earth satellites and is the basis for a new generation of hyperspectral instruments that will be in orbit, it is hoped, before the end of this millennium.

At the outset, the feature extraction problem appeared to be a formidable one, given the detailed level of the information required by users and the high degree of variability which is present in Earth surface cover. It was clear that the work had to be soundly based upon a fundamental understanding of the spectral characteristics of Earth surface materials. Therefore, all research at LARS:

- had to be conducted in the laboratory, the field, and from aircraft before being ready to define space-based experiments;
- must define the correct instrument characteristics, information processing techniques, and Earth science understanding at each of these levels, and
- must have all of these efforts closely interrelated, so that what is learned at one level and in one discipline would immediately be available for use in the others.

The laboratory provided a research environment in which there was maximum control over experimental variables, thus facilitating rapid preliminary understanding of phenomena.

Next, the research was moved to the field where there was still a high degree of control of experimental variables while beginning to include more realistic characteristics present *in situ*. The aircraft environment added to the realistic nature, again at the expense of experimental control of observational variables, but began to allow for incorporation of the true statistical variation which occurred over large areas in nature. Then, satellite experimentation using suborbital remote sensing systems provided operational-like tests of satellite remote sensing concepts.

The instrument research effort was necessary because spectral measurements to the precision, detail, and volume needed had never been attempted outside of the laboratory. Thus, whole new remote sensing instrument concepts were developed. The same could be said for the handling and analysis of data such that new information processing concepts would also be needed. And clearly the best of efforts would be needed from agriculturists and other Earth scientists if the whole effort was to be successful.

It is difficult to overemphasize the importance of the disciplines working together and the contribution that rapid knowledge transfer between researchers of different disciplines made to image processing research progress. For many of the participants, perhaps the most stimulating aspect of LARS was to see how someone of a different discipline approached a given problem and what was learned from a given experiment. This experience clearly proved the value of the interdisciplinary team approach, and magnified the contrast in the participants' minds between true interdisciplinary research and research that was merely multidisciplinary. The name of the laboratory was changed to Laboratory for Applications of Remote Sensing when in 1969 the operation was funded by NASA to extend its work to include geology, hydrology, and geography in addition to agriculture.

The development of the LARSYS software system for analyzing multispectral image data began at LARS in the summer of 1966. Earlier work on the use of photographic and thermal infrared scanner imagery for discriminating between ground cover classes had begun at Purdue in 1964. Initial work was directed at agriculturally significant crops and soils; however, that was soon extended to forestry, geology, and geography. Though digital computers were not especially powerful at that time, they were expected to grow in power and thus rapidly become capable of handling the large quantities of data space sensors would produce. The initial algorithms were programmed on a campus IBM mainframe but were moved in 1966 to a new IBM 360 Model 44 at LARS. The approach was to center on the spectral aspects of such data and to use pattern recognition technology which was then just emerging.

Data first used in the development of LARSYS was from an 18-band scanner flown aboard a DC-3 aircraft of the University of Michigan's Willow Run Labs (later, during the Vietnam War to become the Environmental Research Institute of Michigan—ERIM) flown over sites near the campus during the 1966 growing season. The initial scanner system was actually four separate scanners providing one UV band, 12 bands in the 0.4-1.0 um range, four middle IR bands, and one 8-14 um band. Work was immediately initiated on algorithms for registering pixels from the four data sets into a single multispectral data set. This work resulted in one of the earliest image registration systems and was used and improved upon over a number of years.

The LARSYS software could process laboratory (DK-2) and field (interferometer and later field spectrometer) as well as airborne and associated ground truth data. However, its main utility was to analyze the multiband airborne data by assigning individual pixels to one of several preestablished classes, thus enabling the generation of thematic maps. The LARSYS-AA system (circa 1966) was used by Earth science researchers while being continually improved as associated engineering research produced improvements and additions to the algorithms. Within a short time, the system use had grown to the point that it was decided to split the separate functions into stand-alone systems. The aircraft data analysis system became known as simply LARSYS. It was used on a number of major remote sensing experiments, including the 1971 Corn Blight Watch Experiment, ERTS-1 (Landsat-1) investigations over the Wabash River Basin, the Central Plains States, and the Colorado Mountains, and in the generation of land use maps for each of the 192 counties in the U.S. portion of the Great Lakes Watershed.

Late in the 1960s, in addition to continuing the research, much emphasis began to be placed upon transferring the technology to others. In 1970, the system was moved to an IBM 360 Model 67 time-share system, making remote terminals to various sites possible. This was used as a primary technology transfer tool for more than a decade, and appropriate tutorial materials were assembled to be used with terminals remote to various sites. In addition to the remote terminal system, the LARSYS software was extensively documented and copies of it were given to all who requested it. NASA's Office of Technology Transfer (COSMIC) at the University of Georgia provided complete copies of LARSYS for the cost of media reproduction. Early adopters of LARSYS included Pennsylvania State University, Oregon State University, Texas A&M University, and NASA Johnson Space Center. A portion of the system was also given to the NASA Jet Propulsion Lab (JPL) for use in the VICAR system. Several commercial products are also descendants of the system.

In 1988, LARSYS was reestablished in a personal computer format as MultiSpec. The emphasis is now on the development of algorithms that can process both conventional multispectral data and the new hyperspectral data such as that produced by the Airborne Visual and InfraRed Imaging Spectrometer (AVIRIS) and future instruments (e.g., MODIS). Versions are available for 680x0 based machines and for Power PC based machines in native code. Updated versions incorporating new capabilities are issued about once per year. In November 1994, there were >400 registered holders of copies around the world. MultiSpec is provided at no cost. Information about the system is available from landgreb~ecn.purdue.edu or via the World Wide Web at the URL: http://dynamo.ecn.purdue.edu/~biehl/MultiSpec/.

VICAR-IBIS Image Based Information System

Digital image processing began at JPL for a specific purpose: to obtain the maximum information from images sent back by unmanned lunar and planetary exploration spacecraft (Castleman 1979). The software was the Video Information Communication and Retrieval system (VICAR) developed in 1966 by the NASA Jet Propulsion Laboratory Image Processing Lab under the direction of

Dr. Fred Billingsley, Stan Bressler, and Howard Frieden. VICAR was used to process, enhance, and display imagery from Ranger missions to the moon and Mariner data from Mars.

The Image Based Information System (IBIS) was developed in the mid 1970s in support of the NASA ERTS-1 program. At the time of the launch of ERTS-1, JPL was the only facility with a completely integrated image processing system. Dr. Billingsley, supervisor of the Earth Resources Application Group at JPL, hired Dr. Nevin Bryant in 1973 to augment the staff with vegetation and land use applications expertise. Billingsley and Bryant presented a concept paper at the spring 1974 convention of the American Society of Photogrammetry on a "Multiple Input Land Use System (MILUS)." MILUS was used to digitally overlay thematically classified ERTS information with digital census tract polygons to determine the amount and location of land use in each census tract. Jerry Clark prepared a binary mask of a census tract in Simi Valley, CA, registered it to a land use classification of ERTS imagery, subtracted the image, and prepared a histogram of the result. This first crude application of existing VICAR image processing routines convinced the JPL staff that a set of dedicated routines needed to be developed to have a viable approach. In the spring of 1974, Fred Billingsley hired computer scientist Dr. Al Zobrist to develop the set of software which came to be known as IBIS.

Al Zobrist developed 16 modular subroutines to extract land use statistics from ERTS imagery efficiently. The first example identified land use acreage in St. Tammany Parish, LA, and was presented at NASA Headquarters in March 1976. The integrated set of software permitted the conversion among raster, vector, and tabular data. The first refereed paper on the IBIS system was presented at the Purdue LARS Symposium that summer and later appeared in the *IEEE Transactions on Geoscience Electronics* (Bryant and Zobrist 1977). It became apparent that the IBIS technology was applicable to spatial modeling and the preparation of large area mosaics from multiple ERTS scenes. At that point, other NASA centers became interested in applying the capability and transferring the software to their own facilities. In 1977, after test cases for population expansion in Seattle and area pollution effects on the population in Portland were developed, NASA Ames adopted the system. In 1978, after test cases for urban expansion in Orlando, FL, and a mosaic of the state of Pennsylvania were prepared, NASA Goddard adopted the system. Applications were also prepared from 1977 to 1980 for the Kennedy Space Center (non-point source pollution runoff in the Occocquan River basin), Bureau of Land Management (range carrying capacity in the California Desert Conservation Area), the Department of Defense (ERTS image mosaics), and Department of Energy (solar photo voltaic potential inventory in Los Angeles, Miami, and Hartford). By 1981, IBIS had demonstrated a broad range of versatility for both remote sensing applications and the ability to perform what today is called raster GIS (Bryant and Zobrist 1981).

The IBIS software was delivered to NASA COSMIC for distribution. At last count, more than 300 copies of the software have been distributed. A review of the contribution made by IBIS to the remote sensing community and geographic information systems was prepared for the second edition of the *Manual of Remote Sensing* (Marble and Peuquet 1983). On October 31, 1994, the VICAR/IBIS software received honorable mention in the first NASA awards for software systems development.

IDECSS Software Development

In the mid 1960s, researchers at the University of Kansas Center for Research in Engineering Sciences (CRES) began the development of the Image Discrimination Enhancement Combination and Sampling System (IDECSS) to facilitate environmental analysis for the United State Army Engineering Topographic Laboratories at Fort Belvoir, VA. Researchers at CRES were working on ways to enhance the extraction of environmental information from optical multispectral photography. A great deal of research at CRES at the time was also directed at analysis of active microwave (radar) data. Much of this radar analysis was conducted for NASA by Drs. Richard Moore, Fawaz Ulaby, and David Simonett.

George Dalke, a graduate student in electrical engineering, was responsible for the development and implementation of IDECSS. In 1967, verification of the system's capabilities was initiated. This verification involved a subcontract to Dr. Robert N. Colwell of the School of Forestry at the University of California, Berkeley. Berkeley was the center established by NASA to advance the state-of-the-art of the interpretation of multispectral photography from aircraft and spacecraft. Multispectral photography was acquired for test sites in California and Kansas. IDECSS was used to video digitize the photo transparencies as input for processing and analysis. At this time, IDECSS had the capability to perform registration, enhancement, classification, video digitization, and simple GIS overlay functions. Work was conducted on relatively small geographic areas contained within single digitized aerial photographs. George Dalke received his Ph.D. at Kansas for developing the system. John Estes worked with the Dr. Robert Colwell and the Berkeley School of Forestry on the verification effort and subsequently received his Ph.D. at UCLA on the topic. Dr. Estes later founded the Geography Remote Sensing Unit at the University of California at Santa Barbara. Numerous colleagues came to Santa Barbara in the early 1970s because of Dr. Estes and Dr. David Simonett to work on digital image processing and early geographic information system concepts. Participating were Earl Hajic (retired), Jeff Dozier (at Santa Barbara), Alan Strahler (Boston University), John Jensen (University of South Carolina), and Larry Tinney (EG&G) (Jensen et al. 1979; Estes et al. 1983; Jensen 1983 1986).

The Kansas IDECSS contained sophisticated image processing capabilities for its time. Dalke founded Interpretation Systems Incorporated (ISI) in 1972. After this time, little further development of IDECSS capabilities took place at Kansas. ISI was later purchased by Spatial Data Systems of Goleta, CA. Spatial Data Systems was in turn purchased by Log Etronics, which was acquired by DBA, Inc. No further image processing or GIS related civil sector work was performed by these companies after about 1981.

Office for Remote Sensing of Earth Resources (ORSER)

The Office for Remote Sensing of Earth Resources (ORSER) began in 1970 as part of the Intercollege Research Program at the Pennsylvania State University. George J. McMurtry and Gary W. Peterson were ORSER's initial codirectors, with F. Yates Borden and Brian J. Turner as colleagues. Wayne L. Meyers joined the team in 1978. The purpose of the office was to develop a digital image processing system, ORSER, for analyzing airborne and satellite multispectral scanner remotely sensed data. The system was to process and interpret data for applications in natural resources, land use management, geology, hydrology, and environmental pollution. There was no precursor system from which the ORSER system was developed. It was one of the original software systems for analyzing remotely sensed digital image data with LARS and VICAR as contemporaries. ORSER was available through COSMIC for a time, but distribution was discontinued when its capabilities were substantially incorporated into the public domain Land Analysis System (LAS) at NASA Goddard Space Flight Center (GSFC).

Research experience with the ORSER system provided impetus to development of a prototype raster/vector GIS called SPIRAL at Pennsylvania State University. It functioned only on mainframes and became obsolete in an age of PCs and workstations. SPIRAL did, however, serve as progenitor for another unconventional approach to GIS called SPIN (Spatial Patterns in Networks) developed by Sidney L. Whitley at NASA's Earth Resources Laboratory at Slidell, LA.

Land Analysis System

The Land Analysis System (LAS) is a general purpose image processing system for manipulation of multispectral image data. It was developed by the Space Data and Computing Division (SDCD) at the GSFC to satisfy the image processing requirements of the Laboratory for Terrestrial Physics (LTP). The SDCD was responsible for system design, coding, integration, and documentation. The LTP was responsible for supplying functional requirements and for performing acceptance testing. The EROS Data Center (EDC) also participated in design reviews and provided programming support.

The LAS was originally developed as an engineering and scientific assessment tool for evaluation of data from the Landsat-4 Thematic Mapper (TM) sensor. It was used for one year after launch of Landsat-4 to perform the operational processing of TM data. As such, the system was initially designed to support the production requirements of one radiometrically and geometrically corrected TM image per day. A VAX 11/780 with a floating point array processor was used to provide the necessary computational support. It was proposed that, following the completion of TM ground processing in July 1983, the LAS could be expanded to an interactive general purpose image processing system to satisfy the requirements of the LTP Earth resource scientists.

Specification of requirements proved difficult at first because relatively few references were available on the subject of remote sensing software systems (Bracken 1983). Therefore, it was necessary to develop *ad hoc* criteria for the design and verification of LAS on the basis of experience with the strengths and weaknesses of the image processing systems then in use by LTP researchers. The first step was to identify a prioritized list of functional requirements from an analysis of the most frequently used image processing functions. The second step was to define further requirements regarding user interface, documentation, image display interface, programmer interface, maintenance, processing speed, and data management. Failure to satisfy these additional requirements can seriously compromise the utility of an image processing system even though it may otherwise provide all of the necessary functions.

In-house development provided the following advantages as compared to the option of acquiring a commercial system:

1. LTP users were able to exercise considerable influence in the end product by directly participating in requirements specification, design reviews, and system evaluation;
2. the system source code was made available to other government agencies at minimal cost; and
3. the system was easily maintained and/or modified internally to evolve with new or changing requirements.

However, it was not necessary to develop all of the system from scratch. An early version of the user interface, the Transportable Applications Executive (TAE), was available and numerous applications functions and support services were adapted from the TM ground processing system already in place. A further reduction in development effort was achieved by using commercial software for image display manipulation, geographic entry system, geographic information system, and mathematical and statistical processing.

LAS development was expedited by a close working relationship between the system developers in SDCD and the LTP users and by the efforts of an LTP "audit team" who formulated detailed requirements, assisted programmers in interpreting requirements, reviewed design specifications and documentation prior to coding, and performed testing and evaluation as part of the acceptance test. The baseline release of LAS with 240 functions and more than 200,000 lines of code was delivered by SDCD in July 1985. The LAS was first implemented on the SDCD's VAX system and then installed on the LTP VAX following the successful completion of the acceptance test in August 1985.

In 1984, EDC ported a subset of the LAS software to run under the UNIX operating system. The success of this port and the development of advanced UNIX workstations initiated another cooperative effort between GSFC and EDC in 1986 to modify LAS to be more transportable. LAS continues to evolve with additional capabilities and enhancements (Quirk 1987; Wharton 1988). It is still one of the primary image processing and GIS research tools for researchers at GSFC.

Earth Resources Laboratory Applications Software (ELAS)

ELAS was developed by personnel of the Earth Resources Laboratory at the NASA John C. Stennis Space Center (SSC) in 1978–1979 with continuing refinement to the present. Ronnie W. Pearson conceived the original idea for ELAS and was its principal architect. The detailed definition of the system was a cooperative effort of Ronnie Pearson and Sidney L. Whitley. When program development began, Pearson was assisted by Ray Seyfarth, Bobby Junkin, Marcellus

H. Graham, and Marie Kalcic. After the basic system was running, applications modules were added on a continuing basis by these civil servants along with a number of Lockheed Engineering Services contractors, including James Skipworth, Jammie Ramsay, Tiger Cheng, and Merle Beverly. ELAS came into being rather quickly because many of its capabilities were developed and refined in a predecessor system called CVIPS (Comtal-Varian Interactive Processing System), also developed by Stennis Space Center's Earth Resources Laboratory. During development in 1978–1979, Pearson proposed to call it ELAS for Earth Resources Laboratory Applications Software.

ELAS was created to support research and development remote sensing applications for NASA's Earth Resources programs. It was designed to accept and process data from space and airborne sensors, along with supporting correlative data from a variety of sources (maps, aerial photographs, field notes). The principal driver for the development of ELAS was the need for a system that could be easily transported to computer systems existing in state and local governments under NASA's Office of Space Sciences and Applications' Technology Transfer Program. ELAS was also designed to be easily transported to the "super minicomputers" that were widely available at that time. ELAS is a true GIS with all the classical functional elements of data acquisition, preprocessing, data management, manipulation and analysis, and product generation. ELAS can be used to process both raster and vector data, with preference being given to the raster format during manipulation and analysis.

ELAS's lineage can be traced to: 1) Aerojet General software for processing, display, and film writer input of data; 2) the Analog Computer System for processing multispectral scanner data at the Environmental Research Institute of Michigan (ERIM); 3) Purdue University's Laboratory for Application of Remote Sensing LARSYS-AA program, and; 4) some data registration ideas roughly based on capabilities found in the Center for Research in Engineering Sciences Image Discrimination Enhancement Combination and Sampling System (IDECSS) at the University of Kansas. These systems served as the basis for the Comtal Varian Interactive Processing System (CVIPS) that was implemented at the Earth Resources Laboratory in 1971. CVIPS evolved into ELAS in 1978. During the period of CVIPS development, many of the GIS capabilities which became central to ELAS were specified by the research scientists and managers of the NASA Earth Resources Laboratory at the Stennis Space Center. In the late 1970s in association with the NASA Regional Applications Program, the Earth Resources Laboratory began to have a need to port its image processing and GIS capabilities to other institutions and organizations. Researchers at the Earth Resources Laboratory determined that the time it would take to port CVIPS was such that a new, more integrated system was a more practical alternative. Thus ELAS was developed. The most direct external influence in the lineage of ELAS came from Purdue University's LARSYS-AA.

The NASA SSC Earth Resources Laboratory issued a grant to Nick Faust at the Georgia Institute of Technology to implement versions of CVIPS (ELAS's immediate predecessor) and ELAS on small computers. Dr. Faust became a principal in the ERDAS Corporation in Atlanta, GA, when it was formed. We understand that a significant part of the early ERDAS system was based on CVIPS and ELAS. The ATLAS system software developed by Delta Data Systems, Inc., was based on ELAS. DDS's president, Ferron Risinger, made significant inputs to the development of ELAS while he was employed by NASA SSC before forming DDS. Dr. Ray Seyfarth, who assisted Ronnie Pearson in the initial development of ELAS, developed a PC version of ELAS after leaving NASA employment. Because the complete source code and documentation of ELAS were distributed to NASA program participants and to purchasers of ELAS through COSMIC, it is not known how many other systems are based on ELAS. A great number of commercial users are currently using ELAS as distributed by NASA without modification. SSC personnel are currently implementing ELAS's capability for the geographical referencing of aircraft acquired scanner data into their commercially procured ERDAS package.

ELAS is still in use at the NASA Stennis Space Center by scientific researchers and by applications personnel working in support of the commercialization of remote sensing. Applications modules and refinements are still being added as needed but in smaller numbers than in years past. This is not surprising given the overall functionality of the current system. An ELAS User Group existed for several years and last met in 1991. ELAS remains the favored program by many users for the geographical referencing of aircraft acquired multispectral scanner data. It also has excellent functions for performing atmospheric correction of thermal data and for creating raster images from contoured physical data such as elevation, temperature, etc. ELAS is frequently used by NASA SSC personnel and others as one of a group of application packages such as GRASS, ARC-INFO, and ARCVIEW. ELAS software was inducted into the Space Hall of Fame in 1991.

Remote Sensing Laboratory and STANSORT Program

The Stanford University STANSORT program was initiated in August 1972 by Dr. Ron J. P. Lyon of the Department of Geology about one month after ERTS-1 was launched. Its purpose was to read and display ERTS-l computer compatible tapes received from NASA as a part of the "Evaluation of ERTS Cooperative Researchers Program" (Lyon et al. 1976). The program was initially used to evaluate ERTS data from the San Francisco Bay Area eastward to central Nevada (about 30 ERTS scenes) using both optical-photographic (photogeologic) processing and digital image processing.

From 1971 to 1973, Dr. Andy Green, a post-doctoral researcher from CSIRO Mineral Physics in Australia, evaluated ERTS-1 for possible Australian uses. This coincided with a year-long reciprocal visit of Dr. Lyon to CSIRO. Green developed TAPEREAD to read and process digital CCT's on the IBM 360 Mainframe. In 1973, Green returned to Australia and became head of the CSIRO Remote Sensing Unit. His Stanford post was filled by a second CSIRO post-doctoral researcher, Dr. Frank. R. Honey. Dr. Lyon came back to Stanford in June 1973.

Dr. Honey refined the batch programs into RIPPER, which could read CCTs, process the data, recognize lineaments, and perform a simple parallelepiped classification. RIPPER ran on a DEC 11/44 using the SAIL language (Stanford Artificial Intelligence language) and was interactive (Honey et al. 1974: Lyon et al. 1975). Dr. Alfredo Prelat, joined the group in 1974 (Prelat et al. 1978). He rejoined the group in 1976, after completing two years of research with the Norwegian Geological Survey, where he developed his statistically-based software programs. His Norwegian work on multivariate analysis set the stage for the GIS approach of the software with the integration of geological, geophysical, and LANDSAT digital datasets (Missillati et al. 1977), and the August 1977 evaluation of Landsat digital data for quaternary geological mapping of the Malaysian Peninsular (Prelat et al. 1978). Gary Ballew developed programs to display scaled ERTS data on a matrix printer (Ballew and Lyon 1977).

After Honey's departure in 1974, the programming tasks for STANSORT were carried out by John Prebus, Robert Maas, and Kai Lanz. Most of the programs were rewritten in FORTRAN for a DEC 11/34. Kai Lanz continued programming from 1975 to the present and was responsible for the source code that became the commercial TERRA MAR image processing software.

Development of EDITOR

The EDITOR (ERTS Data Interpreter and TENEX Operations Recorder) image processing software was developed in 1973 by Robert M. Ray III, Martin Ozga, Walter E. Donovan, John Thomas, and Marvin L. Graham, of the Center for Advanced Computation, University of Illinois at Urbana-Champaign. Its function was to serve as an interactive interface to the ILLIAC IV super computer located at NASA Ames Research Center. The ILLIAC and several TENEX computers were accessible over the ARPA Network via modem. The mode was distributed processing, making the ILLIAC available to multiple sites. Thus, EDITOR was a forerunner and a pioneer of many current trends such as distributed processing and network access via Internet.

The idea of using the ILLIAC for remote sensing is attributed to John DeNoyer, then chief of the USGS Earth Resource Observation Satellite (EROS) Program, and Hans Mark, director of NASA Ames Research Center. DeNoyer and Mark, who knew each other from U.C., Berkeley, decided remote sensing might be a good application for the new ILLIAC IV supercomputer installed at Ames in 1972. DeNoyer anticipated the promise of the MSS and digital processing of data from the scanner before the launch of ERTS-1 while most others were preoccupied either with the return beam vidicon (RBV) sensor or the promise of photo-interpreting the new "pictures" from space. A full ERTS scene was formidable for the computers of the time, and DeNoyer thought the parallel architecture of ILLIAC IV might be a real advantage given the nature of image processing operations. A project was begun with NASA, ARPA, and USGS funding to explore ILLIAC's potential for remote sensing.

Ray was the principal investigator, initially funded by NASA and ARPA grants, and collaborative support of the EROS Program of the U.S. Department of the Interior (USDI). Ray and his colleagues turned to Purdue University's Laboratory for Applications of Remote Sensing for the remote sensing techniques they were to implement. James R. Wray at the USGS, working in the Geographic Applications Program, had an ERTS-1 study called the Census Cities Project. ERTS-1 underflights were flown for a sample of U.S. cities using the NASA Ames' U2 aircraft contemporaneous with the 1970 census. Land use was to be mapped from the photos and later from ERTS images to detect change that would be correlated with population. One of the first acquisitions of ERTS was over San Francisco, a Census City test site. Richard Ellefsen, a USGS consultant and professor of geography at San Jose State University, was mapping land use from the U2 photos for Wray with colleague Duilio Peruzzi. Wray acquired a digital tape of the early San Francisco scene without a way to process it. With a "bird (or tape) in hand," San Francisco became DeNoyer's choice as the first data set to explore the potential of the ILLIAC. Wray's Census Cities project became a collaborator. Ellefsen was asked to journey to Purdue to explore the potential of making a land use map using a computer. Ellefsen and Philip Swain at Purdue tried out LARS software on the problem. Algorithms that seemed to work were given to Ray for implementation on the ILLIAC. Leonard Gaydos, a graduate student of Ellefsen's, was hired by Wray to help and to devote more time to the task.

Ellefsen and then Gaydos and student coworkers, Gail Thelin and Bill Newland, began to see that there was more to computer-aided analysis than batch processing on a big machine. Of perhaps greater importance was the interactive work of selecting and evaluating training sites and resulting multivariate statistics. Out of this necessity, EDITOR was born. It allowed "friendlier" use of LARSYS which was based on card decks and a mainframe computer. Early EDITOR users were able to execute new menu-driven commands using a terminal linked to BBN via modem. Gaydos created quite a stir in 1975 in Hawaii during a USGS workshop on land use mapping by pulling out a briefcase-size Texas Instruments terminal, connecting with Boston via phone, and analyzing ERTS data.

EDITOR's major sponsor became the USDA Statistical Reporting Service (SRS), now the National Agricultural Statistical Service. SRS used EDITOR and the ILLIAC for crop estimation in 1974. Their needs, endorsed by USGS, encouraged GIS-like appendages to the original system. A way was found to determine ERTS pixel coordinates by latitude and longitude based on the geometric correction transformation coefficients. The next step was a digitizing interface. Walt Donovan created software allowing for polygon creation (in vectors) using a pen digitizer. Polygons were attributed with an increasingly complex set of codes for USDA, then used to identify fields of pixels in the ERTS data. Both SRS and USGS used these tools in interactive training field selection.

Other GIS-like functions included a variety of plotting options, including a new film recorder. The digitizing option made it possible to overlay boundaries such as counties or enumeration districts to get acreage estimates or to stratify scenes by photomorphic regions for better control over training and classification.

In 1978, the EDITOR team was disbanded. It was too difficult trying to win support from a handful of sponsors each year to keep the grant going. Ray went to Research Triangle Park, and Marty Ozga got a job with SRS in Washington. Walt Donovan moved to Ames to support one of the EDITOR applications, the Digital Mapping of Irrigated Croplands project. SRS maintained a relationship with Ames to continue use of the ILLIAC (and later Cray) and EDITOR. Through this time (and beginning with the origins of the project), Buzz Slye, a NASA software engineer, was the major Ames contact. Buzz participated in the porting of EDITOR a few years later to PEDITOR (Portable EDITOR) for USDA. PEDITOR was used on the new Forward Technology computer (using the Stanford University Network architecture) and later SUN computer. EDITOR and PEDITOR continued to be the software of choice for the USGS group at Ames through the early 1980s and has continued as an integral component of USDA National Agriculture Statistics Service (NASS) programs to the present day. The latest activity was the development of a Computer Assisted Stratification and Sampling procedure using X Windows on an HP9000.

Conclusions

NASA played a significant role in the development of digital image processing systems from 1966 to the present. The availability of digital remotely sensed data in the 1960s spurred not only the development of digital image processing systems but also the initial development of raster-based GIS concepts and systems at a number of institutions. The rudimentary GIS functionality was principally related to the overlay of multiple bands of registered raster multispectral remote sensor data with political administration units. For example, Bryant and Zobrist at JPL overlaid and related census tract geography with remote sensing derived land cover information in the early 1970s. Others such as Roger Hoffer at Purdue registered digital elevation model data with remote sensor data to improve vegetation classification. More advanced raster-based GIS models evaluated registered elevation, slope, and aspect data with atmospherically corrected remote sensing radiance data to more carefully identify vegetation habitat. These applications were very successful and the raster-based GIS Boolean logic has been adopted widely (Congalton and Green 1992). As computer power and the demand for more geographic specificity increased, however, vector-based systems have become more widely used. The decision concerning which GIS system to use is application and situation dependent. Many of the raster-based digital image processing systems and GIS that are in public and commercial use today are based on NASA sponsored research. For example, continually improved versions of ERDAS, ELAS, VICAR-IBIS, LAS, TERRA MAR, and MULTISPEC are used widely today.

Information extracted from remotely sensed data will continue to be the primary method for updating many key GIS data layers. Therefore, remote sensing digital image processing and GIS represent linked, symbiotic technologies. While NASA has not continued the development of image processing capabilities as some of us believe they should, there are some NASA funded initiatives that will continue to impact both digital image processing and raster-based GIS. First, there is movement toward super computer implementation of digital image processing and raster-based overlay algorithms. Second, we see the integration of remotely sensed biophysical data into very large raster and vector spatial databases to solve global change issues. These developments will continue to have an impact on GIS and remote sensing digital image processing systems.

Bibliography

Ballew, G. I., and R. J. P. Lyon. 1977. "The Display of Landsat Data at Large Scales by Matrix Printer." *Photogrammetric Engineering & Remote Sensing,* 43 (9): 1147–1150.

Bracken, P. A. 1983. "Remote Sensing Software Systems." Chapter 19 in D. S. Simonett and F. T. Ulaby, eds., *Manual of Remote Sensing,* Vol. 1, 2nd ed. Falls Church, VA: American Society for Photogrammetry and Remote Sensing.

Bryant, N. A., and A. L. Zobrist. 1977. "IBIS: A Geographic Information System Based on Digital Image Processing and Image Raster Datatype." *IEEE Transactions on Geoscience Electronics,* GE-15 (3): 52–159.

Bryant, N. A., and A. L. Zobrist. 1981. "Some Technical Considerations on the Evolution of the IBIS System." *Proceedings, Pecora VII Symposium,* 465–75.

Castleman, K. R. 1979. *Digital Image Processing,* Englewood Cliffs, N.J.: Prentice Hall, 429.

Congalton, R. G., and K. Green. 1992. "The ABCs of GIS." *Journal of Forestry,* 90 (11): 13–20.

Coppock, J. T., and D. W. Rhind. 1991. "The History of GIS." In D. J. Maguire, M. F. Goodchild, and D. W. Rhind eds., *Geographical Information Systems,* Vol. 1. New York: John Wiley & Sons, 21–43.

Estes, J. E., E. J. Hajic, and L. Tinney. 1983. "Fundamentals of Image Analysis: Visible and Thermal Infrared Data." Chapter 24 in R. N. Colwell, ed., *Manual of Remote Sensing.* Falls Church, VA: American Society for Photogrammetry and Remote Sensing, 987–1125.

Hoffer, R. M. 1967. *Interpretation of Remote Multispectral Imagery of Agricultural Crops.* LARS Annual Report No. 1. West Lafayette, IN: Purdue University Agricultural Experiment Station.

Honey, F. R., A. E. Prelat, and R. J. P. Lyon. 1974. "STANSORT: Stanford Remote Sensing Laboratory Pattern Recognition and Classification System." *Proceedings, Ninth International Symposium on Remote Sensing of Environment.* Ann Arbor, MI: Environmental Research Institute of Michigan, 897–905.

Jensen, J. R. 1983. "Educational Image Processing: An Overview." *Photogrammetric Engineering & Remote Sensing,* 49 (8): 1151–1157.

Jensen, J. R. 1986. *Introductory Digital Image Processing: A Remote Sensing Perspective.* Englewood Cliffs, N.J.: Prentice Hall, Inc. 279.

Jensen, J. R., F. A. Ennerson, and E. J. Hajic. 1979. "An Interactive Image Processing System for Remote Sensing Education." *Photogrammetric Engineering & Remote Sensing,* 45 (11): 1519–527.

Lyon, R. J. P., F. R. Honey, and G. I. Ballew. 1975. "A Comparison of Observed and Model-Predicted Atmospheric Perturbations on Target Radiances Measured by ERTS-l: Part I—Observed Data and Analysis.: *Proceedings, IEEE Conference on Decision & Control,* 244–248.

Lyon, R. J. P. et al. 1976. "Evaluation of ERTS: Multispectral Signatures in Relation to Ground Control Signatures Using a Nested-Sampling Approach." Final Report, NASA Contract NAS-21884, 288.

Marble, D. F., and D. J. Peuquet. 1983. "Geographic Information Systems and Remote Sensing." Chapter 22 in R. N. Colwell, ed., *Manual of Remote Sensing,* 2nd ed. Falls Church, VA: American Society for Photogrammetry and Remote Sensing.

Maguire, D. J., M. F. Goodchild, and D. W. Rhind, eds. 1991. *Geographical Information Systems.* New York: John Wiley & Sons.

Missallati, A., A. E. Prelat, and R. J. P. Lyon. 1977. "Simultaneous Use of Geological, Geophysical, and Landsat Digital Data in Uranium Exploration." *Remote Sensing of Environment,* 8: 189–210.

Prelat, A. E., R. J. P. Lyon, and E. A. Van de Meene. 1978. "Landsat—Digital Data as a Tool in Quaternary Geological Mapping in the Coastal Plain of the Malaysian Peninsula." *Proceedings, 29th*

International Symposium on Remote Sensing of Environment. Ann Arbor, MI: Environmental Research Institute of Michigan, 1985–1992.

Quirk, B. K., and L. R. Oleson. 1987. "Overview of the Land Analysis System (LAS)." *Proceedings, Pecora XI Symposium.* Falls Church, VA: American Society for Photogrammetry and Remote Sensing, 133–148.

Tomlinson, R. F., and B. B. Petchenik, eds. 1988. "Reflections on the Revolution: the Transition from Analog to Digital Representation of Space, 1958–1988." *The American Cartographer,* 15 (3): 243–334.

Wharton, S. W. et al. 1988. "The Land Analysis System (LAS) for Multispectral Image Processing." *IEEE Transactions on Geoscience and Remote Sensing,* 26 (5).

CHAPTER 11

GIS Development in the Department of Interior

David D. Greenlee and Stephen C. Guptill

This story is told from the standpoint of two people in the United States Geological Survey (USGS). We have tried to tell the story of Geographic Information Systems developments in our agency and others in the Department of the Interior (DOI) through the time period 1970 through 1985. We have had some assistance from our colleagues, but we do not pretend to present a complete picture. We have tried to provide a few more pieces to this interesting puzzle. We encourage others who read our account to help us develop the story and to contribute to the historical record of GIS developments in the Department of the Interior.

Introduction

Scientists at the United States Geological Survey (USGS) were among the early pioneers in the application of computer technology in the handling of spatial data. The applications were varied

David D. Greenlee is a principal scientist at the USGS EROS Data Center, Sioux Falls, SD. He began his GIS career at ESRI in Redlands, CA, during the early 1970s. For more than two decades, his research has involved advance applications on the integration of GIS with remote sensing.

Stephen C. Guptill is a scientific advisor for Geography and Spatial Data Systems, National Mapping Division, USGS. He was an original participant in the development of GIRAS. Research studies concentrate on conceptual modelling of geographical and cartographical features, scale-independent databases, spatial accuracy, and the design of advanced GIS.

Contributing Authors: Charles J. Robinove, Maurice O. Nyquist, and Charles M. Trautwein

Authors' Addresses: David Greenlee, EROS Data Center, Science and Application Branch, Sioux Falls, SD 57198. Stephen C. Guptill, USGS National Mapping Division, Office of Research, 519 National Center, Reston, VA 22092.

and widespread. A report done for USGS by the International Geographical Union Commission on Geographical Data Processing characterized the activities as follows:

> The USGS is directly concerned with gathering and handling information in the fields of geology, geography, topography, and water resources. Many of the data that the survey gathers are "spatial," in that they record some attribute of specific places; they are individually or collectively attached to location. Spatial data handling techniques, then, allow the locational information of the data to be manipulated in concert with their other values.
>
> Before 1960, the only form of storage for spatial data was graphic, and spatial data handling techniques were provided manually by well-established methods of statistical analysis and cartography. Manually in this sense applies to tasks accomplished by human skill, aided by drafting and photographic equipment but not by electronic data processing equipment. Since 1960, however, computers that can store more than one million characters have been developed, and their use for information handling has rapidly increased. In particular, the door has been opened to nongraphic, digital storage and handling of spatial data.
>
> Numerous independent activities currently in progress in the USGS are moving toward the design and implementation of sets of spatial data in digital form. These projects vary in size from small, even one-person, efforts (which are not for that reason insignificant) to ones that are national in scope and that will make increasingly large demands on the total resources of USGS. Taken as a whole, the activities involve the development of new skills and new technology within USGS; they will create a set of new products for the survey, and they will be of concern to new groups of users. (IGU 1976)

At the Geological Survey, computerized spatial data handling activities began in the 1960s. Experiments in automated cartography included the drafting of projection lines, grid ticks, and control points in various map projection coordinate systems (Thompson 1969). In addition, work was begun on computer programs for photogrammetric, geodetic, and field surveying applications. Although dramatic changes have occurred in these three fields, their evolution will not be chronicled here, as it took a path fairly independent from the path taken by GIS. An activity that will be touched on is automated stereo compilation. A major design goal of this process was to produce an orthophotograph (an aerial photograph from which the effects of relief displacement have been removed). A significant by-product was the production of elevation data that could be fashioned into a digital terrain (elevation) model.

Automated cartography activities quickly branched into two components. One was concerned with automating the production of traditional topographic base map graphics. The second was directed toward the use of base map information combined with statistical or thematic data to automate the production of atlases. These different approaches identified the theme of an early debate in the field. Was automation of the cartographic process the end goal, with the digital data merely an intermediate by-product, to be used to produce the "true" product, an analog graphic map? Or was the digital data a valuable product, in and of itself, to be used not only to produce analog maps but also as information to be combined with other data sets, manipulated, and then used to produce other sets of products? This debate continues, in various and more subtle ways, even to today. However, in the early days of the automated cartography, it was primarily the community of "atlas builders" that articulated the position of the nascent GIS field, advocating digital spatial data as an informational product to be used in a host of subsequent steps.

The people involved in these endeavors made up a small and cosmopolitan group. This included people such as Dean Edson, Roy Mullen, Hugh Loving, Morris Thompson of the Topographic Division, and Arch Gerlach, the chief geographer. Their interests spanned the spectrum of digital activities of the time. There was also a good deal of interaction with experts from other nations, such as David Bickmore of the Experimental Cartography Unit in London, England. Bickmore and Edson organized one of the first international meetings on the topic, the Symposium on Map and Chart Digitizing, held at the National Science Foundation, Washington, D.C., in 1969. Some of Bickmore's comments still speak to us with great clarity:

> . . . there are many people liable to be involved in the new cartography who have not previously been concerned with the subject before. One has, as a cartographer, been well aware of the dichotomy between surveyors and geographers—the geographers perhaps not so good at the discipline of fixing a position accurately but the surveyors not so clear about features they were trying to fix accurately. Both sides, it seems to me, have a lot to contribute.
>
> Beyond that, it seems to me that cartography has expanded with the recent incursion into it of mathematicians, of physicists, of computer programmers, and even of perception psychologists. The impact of automated cartography seems to be sending out ripples that affect a great number of disciplines. (USGS 1970)

Bickmore had been working in the field of digital cartography and spatial analysis beginning in the early 1960s after he directed the production of the Oxford Atlas of Britain in 1959. In 1967, after six years of modest research funding from the Clarendon Press, the Experimental Cartography Unit was established in the Royal College of Art, London. This location was not as odd as it may seem.

> The involvement of the Royal College of Art, which at first sight seems a strange place for the development of a new branch of computer technology, came about because of Bickmore's insistence that the new-style map-making should not be hampered by the graphic conventions that had grown out of the methods of cartographic draughtsmen. If automatic cartography was possible, all manner of new graphical techniques might be used to convey information meaningfully, and the Royal College of Art, with its strong Graphic Design Department and general interest in design, could provide an environment in which creative ideas would develop. (Margerison 1976)

Bickmore saw the value of the digital data, as noted by Margerison (1976):

> Bickmore's original idea had been to store data on a magnetic tape and to build up a library of such tapes, called a "database.". . . the ultimate store of information in the computer-aided system is the database . . . it must be arranged in the most general possible way so that its structure did not limit its usefulness. The concept of generalised data is difficult for the manual cartographer to understand, since he is always working towards a particular completed map with a series of specified graphic conventions. But the database must contain nothing which restricts the data it contains to the convention of a particular form of map.

Bickmore's ideas had a significant impact on developments at the Geological Survey. In addition to his work with Edson, Bickmore also worked with Arch Gerlach on the preparation of the 1970 *National Atlas of the United States*. Jim Anderson, Gerlach's successor as chief geographer, set in place plans for the 1980 edition of the *National Atlas* to be produced with computer assisted methods, but funds for another edition of the *National Atlas* were not secured.

The early 1970s was a critical time from a technical standpoint in testing applications of digital technology; developing software to collect, edit, and manipulate digital spatial data; and designing data structures and databases. It was important from an institutional point of view in establishing programs to collect and utilize digital spatial data; creating mechanisms to deal with coordination, duplication of effort, and standards; and initiating the change from traditional methods to the methods of the information age.

Geography Program: Land Use/Land Cover Database

While the engineers of the Topographic Division were concentrating on automating the cartographic production process, the geographers of the Geography Program of the Geological Survey were approaching the automation process from a different perspective.

The major thrust of the Geography Program was to compile a consistent set of land use and land cover information for the nation from remotely sensed imagery (usually high altitude color infrared photography) (Anderson et al. 1976). A set of polygonal maps resulted from the initial compilation. In addition to the land use map, a set of associated maps was prepared. These maps included hydrologic units, political boundaries, census tracts, federal lands, and state lands (when information was provided by a state cooperator). The desired end product was to be a set of tabulations showing the number of hectares of each type of land use within a county or a census tract or a hydrologic unit. A GIS was needed to prepare these tabulations, so work was begun on a system that could perform the necessary full range of data collection, editing, and analysis functions.

Virtually all of the systems of this era were of a "roll your own" type, that is, software that was developed (either by in-house staff or via a contract) to fill a specific function. This was necessary for several reasons. First, full-function, commercial GIS systems simply did not exist. For example, the Geography Program produced land use/land cover maps that might have 10,000 polygons and needed to overlay a census tract map that could have 1,000 polygons. Maps of such complexity challenged the data processing capabilities of the time. Secondly, the hardware environment of each agency was different. Most processing was done on mainframe computers (usually IBM or CDC), with limitations on disk space and main memory (USGS users could get a maximum of about 720K for a given job). Most processing was done in batch mode, although interactive access to the mainframe was available through time share. In order to function properly, software had to be somewhat customized to the operating environment within the agency. In effect, the arrival of commercial GIS capability needed to await the development of standardized computing environments that were offered by minicomputers like the VAX 11/780 and PRIME machines that were not available until the early 1980s.

The GIS branch of the Geography Program was led by William B. Mitchell who worked to push the envelope of the technology of the day and convinced senior USGS management of the potential of GIS technology. The staff of the branch (including Eric Anderson, Robin Fegeas, Steve Guptill, and Cheryl Hallam) created a GIS software package named GIRAS (Geographic Information Retrieval and Analysis System) (Mitchell et al. 1977) that was on the leading edge with respect to data editing routines such as topological structuring and attribute editing, computer generated color map production, and the production of cross tabulated statistics resulting from the overlay of six data sets.

As newer GIS technology has provided users with the tools to process the land use/land cover data products, the digital products have become more and more popular. As part of the USGS's attempts to utilize the Internet to deliver spatial data, data products from this Land Use Data Analysis (LUDA) program have been staged for public access via Internet (http://www.usgs.gov). LUDA provides a baseline of complete coverage land use information that has been used to calibrate remotely sensed land cover characterizations and is being increasingly used to provide "historic" (mid 1970s) data by which to analyze and measure temporal changes.

Collaboration with International Geographical Union

Early adoption of GIS technology in the USGS was stimulated by a series of studies and seminars carried out cooperatively by the International Geographical Union (IGU) Commission on Geographical Data Sensing and Processing and USGS personnel. This series of events, occurring in 1975 and 1976, was the brain child of Mitchell, Rupe Southard, and Roger Tomlinson, at that time chairman of the IGU Commission. The initial activity focussed on preparing a report that described major spatial data handling activities ongoing at USGS (more than 50 in 1975). Major programs included: the land use mapping and data project, the national coal resources data system, the national earthquake information service, computer composite mapping, and the national water data storage and retrieval system.

In parallel, the IGU inventoried spatial data handling software developed outside USGS and found 258 software modules and documented:

> a large amount of duplication in the field. . . . very little software existed for handling large volumes of spatial data, and that no one had developed any universal spatial data handling program. (IGU 2nd Report)

A second phase of activity included a series of 16 one-day seminars on various topics of spatial data handling. The audience included senior USGS management staff. Their exposure to technology in its early stages of development had a positive influence in the later adoption and widespread use of GIS throughout USGS.

A Remote Sensing Connection: The EROS Data Center

In the early 1970s, while activities in Reston, VA (USGS Headquarters), centered on automated cartography and an essential national mapping program, a new facility was being built in Sioux

Falls, SD, to support the first Earth Resources Technology Satellite (ERTS). This satellite, launched in 1972 and later renamed Landsat, was the first of several satellites that have provided continuous imaging of Earth resources. The facility in Sioux Falls was called the EROS Data Center (EROS stands for Earth Resources Observation Systems). The EROS staff included an interdisciplinary group of applications scientists with training in diverse fields such as forestry, geology, geography, range science, and land planning. Most were trained in the methods of remote sensing, including image interpretation and digital image processing. Some had specialized training in spatial analysis (i.e., GIS) and other related specialties such as computer programming, signal processing, forest biometrics, and statistics. As they learned to interpret the first satellite images and teach others to identify surface cover themes, it was an exciting time of opportunity and discovery.

One of the fathers of the EROS Program, geologist W. A.. (Bill) Fischer, chided students of EROS's International Training Programs to "think big" as they learned to interpret the first satellite images. This was the first time we had been able to see Earth in this fundamentally new and synoptic way. Bill's enthusiasm was born of many years of planning and negotiating with NASA to define satellite and sensor configurations for analyzing Earth resources. As he addressed the international scientists, he would point out the appearance of a huge circular ring covering about a third of an image mosaic of the U.S., listing sites of alteration and mineralization along the perimeter. Unafraid to take a risk, Bill would prompt the class with questions like, "Is this an astro-blem?" "What is it telling us?" With a twinkle in his eye, Bill predicted that these images would be able to answer questions that we weren't yet smart enough to ask. In the 25 years that have passed, we have learned some of the questions and have gotten some of the answers. In addition, we have managed to collect imagery that documents changes in Earth over that time—new questions and answers to provide us with new challenges. Fischer's career is memorialized in an address given by Gary North at the Annual Convention of the American Society of Photogrammetry (North 1995).

It wasn't long before the first digital satellite data were registered with other ancillary or collateral data, and when that happened, remote sensing met GIS. Geographic registration of remote sensing and GIS data is contingent on precise reference to the Earth, or georeferencing. Prior to satellite remote sensing, our understanding of the geometry of remotely sensed data was developed from experience with aircraft scanners. Aircraft are affected much more by flight perturbations and terrain distortions caused by the steep "look angles." From the Landsat system, data were acquired from a much more stable platform. Images acquired above the atmosphere and with nearly vertical look angles yielded surprisingly superior geometry. Evidence that the geometry was better than expected can be found in the original Landsat *User's Guide*. In the handbook, a rigorous "precision correction" processing capability is described for the ground processing of the data. It would have required extensive ground control point selection, an expensive and time consuming task. Almost immediately, it was recognized that a simple systematic correction gave acceptable results and that the additional cost of selecting a dense set of ground control points was not justified. Precision correction was almost immediately abandoned.

In those early days, the term "GIS" was seldom used by the remote sensing scientists, in favor of the term *spatial analysis* or more specific techniques such as *overlay analysis, composite mapping*, or *suitability modeling*. Readers should take this into account when performing a literature review for publications of this era. Even 25 years later, when asked for a definition of GIS, some remote sensing scientists will just smile and attempt to answer in general terms or otherwise evade the question.

In 1977, NASA funded a workshop to examine GIS Impacts on Space Image Formats at Santa Barbara, CA (Simonett et al. 1978). This conference was organized and conducted by such visionaries as Dave Simonett, Jack Estes and Waldo Tobler of the University of California at Santa Barbara, and Fred Billingsley of the Jet Propulsion Laboratory. Note that this was long before NSF funded the establishment of the National Center for Geographic Information and Analysis (NCGIA) at Santa Barbara, Buffalo, and Orono. Viewed in retrospect, this conference did much more than just examine "GIS impacts on space image formats." For many who had the privilege of attending, it served to make the integration of GIS techniques with the tools of image processing systems all the more purposeful. While there were few great technical breakthroughs from this gathering, it helped to fortify the ties and build consensus between the GIS and remote sensing communities.

Building GIS techniques into digital image processing systems was not really a big leap but rather a series of small steps. Digital image processing required fast and powerful computer systems with robust software, so it was relatively simple to augment these systems with GIS tools. In 1976, a software package for calculating topographic derivatives was procured from a small consulting company called the Environmental Systems Research Institute (ESRI). Most of these techniques found their way into image processing systems over the next few years. With map-based information registered to remotely sensed data, techniques such as *stratification* and *post-classification refinement* (Hutchinson 1978) were made possible. EROS's systems were also augmented to provide overlay analysis and allocation models, minimum distance calculation, surface interpolation, and proximal zones or Thiessen polygons (Greenlee 1981).

In the early 1980s, Sue Jenson developed a method for "filling depressions" in digital elevation models (Jenson 1985). By identifying and filling these self-draining (noncontributing) areas, this algorithm provided a breakthrough for routing the drainage from surface water flow and for the delineation of drainage basins or watersheds. By the early 1990s, these techniques were combined with the surface interpolation contributions of Michael Hutchinson, Australian National University (ANUDEM), to be included in the hydrologic analysis capabilities of ARC/INFO (Hutchinson 1989).

The raster domain has generally provided an acceptable data structure for studies that don't require high quality cartographic output. Many systems used the raster construct exclusively. One of the most notable of these all-raster systems was the Image Based Information System (IBIS) that was developed at the Jet Propulsion Laboratory (Bryant and Zobrist 1981).

Vector Based GIS—The Need for Robust Polygon Overlay

When the raster structure proved to be less than adequate, a raster-to-vector conversion technique developed by Dave Nichols (Nichols 1981) made it possible to generate vector maps and perform spatial analysis in the vector domain. Using polygonal regions with unique labels, a method of postclassification processing was developed by Gene Fosnight that has become a useful tool for "nominal filtering" (Fosnight 1988). Polygonal areas that were deemed small and/or insignificant could be eliminated based on rules established for each class.

The conversion to vector also allowed for more efficient calculation of "corridors" (i.e., proximal zones) without the need to iteratively "grow" them in the raster domain. This technique was demonstrated on a "pre-PC" type microprocessor based Remote Image Processing System (RIPS) (Greenlee and Wagner 1982).

It should be noted that vector data structures were used sparingly at EROS until 1984 when ARC/INFO was first installed on a VAX 11/780. Unable to muster the substantial resources and talent to develop a robust vector processing system, EROS had been waiting for a reliable polygon overlay and topological structuring function. Jan Van Roessel, a prominent GIS researcher who worked at EROS during the 1980s, observed that the same three structuring/overlay algorithms seemed to keep showing up in new GISs and that all three had some fundamental problems (Van Roessel and Fosnight 1985). These three overlay processors were: 1) United States Forest Service's MIADS, 2) Oak Ridge National Laboratory's GOVERLAY, and 3) Environmental Systems Research Institute's PIOS.

When EROS staff attended an ESRI User Conference in 1981, vector processing was performed by a Polygon Information Overlay System (PIOS). The conference was memorable for its small size (nine users plus ESRI staff) and the collegial bonds formed between a group of dedicated users who daily faced PIOS's quirks and idiosyncrasies. As the first ARC/INFO prototypes were used by the application project staff at ESRI, they began to relate impressive success stories to the 1983 user conference. Later that year, the user community got the chance to try the software on PRIME computers and soon thereafter on the VAX.

By 1984, users had composed their own success stories. Over the next few years, the user conference saw a geometric increase in size, and the tone was transformed from a support group of die-hard "techies" to a buzzing convention of energetic zealots. The contrast is difficult to overstate. The "overnight" success of ARC/INFO can be attributed to many factors, but among them must be the robust functioning of the overlay and topological structuring software. That is to say, the system worked, and it worked well. The history of ARC/INFO, and the role of Scott Morehouse and others in the development of a robust overlay and topological structuring system, is an important part of the history of GIS. It was not until this time that the USGS and other DOI agencies began to move from test and evaluation to fully integrate vector GIS tools into applications projects.

In 1986, EROS evaluated several GIS software packages using small but tricky test cases to identify potential flaws in topological structuring, polygon overlay, and attribute handling (Greenlee et al. 1986). This study was conducted for internal planning purposes and was not published, but many copies were made based on request. While we researchers at EROS had developed concern about the effects of dysfunctional software on otherwise high quality cartographic data and we wanted to be helpful to others who shared this concern, we had no mandate to provide system certification or provide consumer testing. In a later attempt properly to assist users in their own testing, a procedure for GIS test and evaluation was published and offered to the public (Guptill 1988).

USGS as a GIS Data Supplier

The USGS has been guided by two important policies that have had far-reaching effects on the development and acceptance of GIS technology in the United States.

1. "Free flow of information from the government to its citizens" (OMB 1985) was called for with no copy restrictions. The reader is referred to Nancy Tosta's review of this guiding principal (Tosta 1992). This policy is elaborated in Office of Management and Budget Circular A-130, published in 1985 (and revised in 1993), and builds on the tradition of open information flow reflected in the Freedom of Information Act. The result of this guiding principal is that most agencies have promoted data sharing to avoid redundancy and have developed policies that provide spatial data for the cost of reproduction.

GIS database builders in the United States have received a unique benefit in getting access to inexpensive and readily available data. As most users have learned, the establishment of spatial databases is one of the most expensive and time-consuming aspects of GIS implementation. In contrast, many other countries have established substantial user fees (Tosta 1993) that have the effect of constraining data exchange and impeding the development of GIS technology.

2. An "Open Skies Policy" has guided the remote sensing industry and provided us with unrestricted access to remotely sensed data. This policy was proposed in the '50s by the Eisenhower administration but was never officially established as policy. Nonetheless, it has been stated repeatedly and argued for over such a long time that it has become the basis for free availability of data from Earth orbiting satellites. Open skies may have even set the stage for data sharing and improvements to GIS databases in the United States. Certainly, it has helped to provide us with a globally consistent set of remotely sensed images and a base for measuring future changes.

Digital Elevation Models

The availability of low-cost Digital Elevation Models (DEM) created from automated methods is a cornerstone to the development of GIS technology. Prior to this, only the most dedicated users were able to digitize the necessary elevation data to calculate topographic slope, aspect, and perform cut and fill analyses. The most typical method for creating the data involved placing a transparent grid on the map and then estimating the elevation cell by cell by manually interpolating between the mapped contours. Needless to say, this was a tedious and expensive process. On the other hand, elevation data could be used to derive several useful parameters for environmental planning and analysis of engineering constraints. By any measure, the user community has made good use of the low-cost DEMs provided by the USGS.

Some of the earliest publicly available DEMs were called Digital Terrain Tapes (DTTs) and were provided by the Defense Mapping Agency. These data have evolved to become the Level-1 Digital Terrain Elevation Data (DTED) that DMA now uses and the 3 arc-second DEMs that the USGS distributes. The first datasets were a by-product from generating raised relief maps from the 1:250,000 scale topographic maps. Contours were digitized and converted to south-to-north profiles every .01 inch by planar interpolation. This formed a grid that was projected to the Universal Transverse Mercator projection with elevation postings every 208.33 feet. Later, USGS and DMA worked together to provide these data reprojected to geographics with a cell size of 3 arc-seconds, or about 90 meters square. Gradually, the planar data have been replaced by "photogrammetric" data that are improved.

Larger scale DEMs from 1:24,000 scale mapping became available from the USGS beginning in 1976. In the early days of automation, many activities at the USGS were devoted to the preparation of digital elevation models and orthophotographs. These two products are often linked, in that an elevation model is needed to create the orthophotograph (whether analog or digital). Systems to collect this data from stereopairs of photographs or from digitized contour maps were developed. One system, the Gestalt PhotoMapper (GPM), was in operation from the mid 1970s to the early 1990s. This device generated more than 10,000 of the 7 1/2 minute digital elevation models. The system was originally acquired to produce orthophotographs, and the elevation model was first considered a throwaway by-product. Fortunately, someone at USGS had the foresight to convert these raw GPM data into more useable DEM.

Perhaps the most notable characteristic of the early GPM-derived DEM is the tendency to show steps or "cliffs" in water bodies, large agricultural fields, and other areas of low relief and featureless terrain. With few features to provide numerical correlation, the system would wander a bit and then jump back into registration, causing the appearance of a cliff. This lack of correlation did not matter much in the generation of the orthophotoquad as the image content was about the same, even though elevation and georeference might be in error. Over the years, there were many incremental improvements made to the DEMs (e.g., water body edits and cell-by-cell edits) and also some new and improved methods for producing DEMs.

With more than 55,000 of the 7 1/2 minute quadrangles to map and revise, DEMs had to be created over a considerable time. The 30-meter data have been added to the USGS's National Digital Cartographic Database at the rate of few thousand per year. An elaborate cost sharing and prioritization procedure is now formalized in an Office of Management and Budget Circular A16 (OMB 1990). By 1997, about 90 percent of the conterminous United States has been completed by one method or another.

Digital Line Graphs (DLG)

Digital Line Graphs were extracted from the 7 1/2 minute map series, and most were created by semiautomated digitizing of the maps to generate line graphs of the mapped hypsography, hydrography, transportation, manmade features, and boundaries. These files were designed to create a digital base of the mapped features. GIS users observed that there are sometimes slight offsets between the cartographic location of a feature and the true geographic location. Often cited examples include the separation of contours in high slope areas to avoid intersection of contours (and confusion about the line continuity) or the case of a road and railroad that are essentially coincident at a given mapping scale but have been purposely separated for clarity in mapping and map reading. The capture of the cartographic locations versus their geographic locations is an issue that has caused some lengthy and sometime heated debates among the USGS staff of cartographers and geographers. In the end, it is agreed that it would be best to have both locations (or the correct geographic location and the necessary cartographic offsets), but the cost of collecting this information has historically been deemed prohibitive.

Over time, evolutionary changes have been made to the procedures for creating DLGs, especially after the USGS began to use its own DLGs in a program for Digital Map Revision and Product Generation. Seemingly small but collectively significant changes gave rise to the need for a rules database and more explicit standards for the digital mapping process. The simple DLGs are based on graph theory and have some marked limitations, especially in the way attributes are accommodated. Recent work at the USGS has led to a reconsideration of feature attribute processing, how features change through time (e.g., temporal GIS, a need for permanent feature IDs), and how features relate to each other (e.g., topology checks, geographic vs. cartographic locations, vertical and horizontal integration, and consistency checking).

Public Domain Software

The USGS's development of the General Cartographic Transformation Package (GCTP) is a technical advancement worthy of note (USGS 1982). GCTP evolved over at least two decades, and the USGS nurtured and reengineered it to become a de facto standard. Quietly but surely, the subroutine library has found its way into most major GIS packages. In a time when there is much discussion about data formats, spatial data transfer, and metadata standards, projection conversions are not usually problematic. Aside from some difficulties caused by differing datums and ellipsoids, projection conversions are relatively straightforward. This may be due to the competence and foresight of a few people, most notably John Snyder, who saw to it that the projections were properly described, and software was carefully crafted, then tested, refined, and freely provided to the public domain. Another factor that may have contributed to such de facto standards is the inherent complexity and formalized mathematics necessary to do the job correctly (i.e., few people would want to take this task on if there is an alternative available).

Over the years, the USGS has also offered public-domain software to read and display DLG, DEM, and remotely sensed imagery (e.g., Land Analysis System-LAS, X-Image Display-XID, Remote Image Processing System-RIPS). In general, this software was developed for internal use and was made available to assist users on an as-is basis. The offering of public-domain software has been problematic because it typically can only be made available without providing adequate support. Support would require supplemental funding and an expanded mission. Neither is likely, because these services can be provided by the private sector, and competition with the private sector is not desirable. With such a circle of logic, users continue to develop unrealistic expectations for software support, and this has caused the USGS to include disclaimers and to become increasingly conservative in distributing software. As users, history should teach us that when it comes to public-domain software, there is no free lunch.

The Mid 1980s: GIS Blossoms in the USGS

In the early to mid 1980s, GIS development became visible in almost every corner of the USGS. While the National Mapping Division was developing vector data, the Water Resources and Geologic Divisions were digitizing maps of geology, hydrology, and all sorts of ancillary data. The users of the data often didn't talk to the technique developers, and many of the working scientists didn't let their management know that they were working with digital data. At this early stage, management lacked foresight and thought that much of GIS was a waste of time and effort.

In 1983, the "planets became aligned," and the Water Resources Division (WRD) decided to demonstrate GIS for national water resources to the Secretary of the Interior. Six months were available to do the job from scratch. A team under the direction of C. J.. Robinove was formed with the Wisconsin District office of WRD and the EROS Data Center. The Fox-Wolf River basin in Wisconsin was selected for the demonstration project. Interestingly, the only software available to do such an operational job was the Remote Image Processing System (RIPS) developed at EROS for analysis of Landsat images. The relevant maps were digitized and converted to raster images. A set of map related questions was developed, and the questions were answered with the raster software. The project was ready in time for the demonstration.

The Fox-Wolf demonstration project spurred the adoption of GIS as a tool in the Water Resources Division. Within a year, WRD was ready with a major procurement of commercially available software for the computers in their Distributed Information System network. A few years later, a second round of procurements for Unix-based systems was let for Distributed Information System 2 (DIS2) and for GIS software (GIS2). This procurement process provided cost effective and standardized systems not only for WRD but also for several other DOI agencies that desired GIS and had collaborated on the procurement. The success of this procurement may be tracked to an excellent and thorough specification, phased upgrades that allowed for incremental advancements in hardware and software, and a large set of optional quantities that allowed many DOI agencies to exercise options and purchase from the contract.

Using a distributed database approach, the Water Resources Division pioneered a Wide Area Network that began as networked PRIME computers (PRIME-Net), evolved to WRD-Net, then GeoNet, then to DOI-Net, and has now become an important part of the Internet. USGS's early acceptance of the Internet and development of innovative data delivery systems have WRD to thank for helping to move us into a distributed environment.

In 1979–1981, research at EROS Data Center was expanded to investigate the utility of "geoscience data bases" for mineral resources exploration. A study area in Alaska in the Nabesna quadrangle provided a place to assemble output from the Alaskan Mineral Resources Appraisal Program (AMRAP) together with Landsat MSS (multispectral sensor) data and derivatives. The Nabesna database and GIS tools were demonstrated to Geologic Division (GD) staff via a series of interactive workshops and formal presentations. These demonstrations eventually resulted in a collaboration between GD and EROS staff and a further refinement of GIS applications for mineral resources investigations (Trautwein et al. 1982; Trautwein 1983).

In 1982–1983, EROS and GD staff evaluated raster based GIS techniques applied to a mineral resource assessment of the Rolla Quadrangle, MO. This study compared the results obtained from a GIS-based approach with those obtained from the conventional approach of overlaying geologic, geophysical, geochemical, and other hardcopy maps and manually delineating favorable environments for specific types of mineral deposits. In addition to duplicating the results of the manual assessment, the GIS approach also provided additional detail and refinements in the mineral resource model (Pratt et al. 1983). In 1984-1989, EROS and GD staff continued to collaborate in several mineral resource investigations that were being carried out as part of the then operational Conterminous U.S. Mineral Appraisal Program (CUSMAP). Results of this work included the realization that handling of tabular, raster, and vector data required specialized tools and enhanced GIS interfaces in order to process efficiently the diverse types of geoscience data used in mineral resource studies. Study areas included Dillon and Butte quadrangles in Montana and Tonopah quadrangle in Nevada (Pearson et al. 1986; Dwyer et al. 1987; and Trautwein et al. 1988).

In 1985, the National Mapping Division began funding GIS Laboratories in Reston, Denver, and Menlo Park. In 1986 to 1990, Dallas Peck, then Director of USGS, established a special $2M fund for innovative GIS projects, to be awarded annually to projects on a competitive basis. A principal criterion for funding was cooperation between the USGS's Geologic, Water, and Mapping Divisions. The process was referred to internally as "the sweepstakes," and some of the funding was used to purchase hardware and software that eventually ended up in the newly established GIS labs.

One of these sweepstakes projects launched EROS and the Geologic Division into yet another evaluation of GIS. This involved development of a GIS for deciphering and reconstructing geologic terranes of the Proterozoic-aged Belt Basin of the Northwestern U.S. and adjacent areas in Canada. By 1987, the Geologic Division's Office of Mineral Resources decided to require all new CUSMAP projects to use GIS techniques. This decision was an important administrative step in the integration of GIS technology within the division's programs. Also in

1987, Geologic Division conducted some interesting experiments allowing comparison of vector with raster GIS analysis, a study known internally as the "bake-off."

United States Fish and Wildlife Service

In 1976, the Western Energy and Land Use Team (WELUT) of the U.S. Fish and Wildlife Service (FWS) in Ft. Collins, CO, began the development of the Map Overlay and Statistical System (MOSS). A digitizing system called the Wetland Analytical Mapping System (WAMS) provided the input mechanism. Over the next ten years, while FWS used MOSS for mapping and GIS applications, the development and maintenance gradually moved to the Bureau of Land Management (BLM) and more specifically to Autometric, who maintained the system under contract. WELUT was later renamed the National Ecology Research Center (NERC).

Many of the GIS applications involved baseline inventory of vegetation and habitat characteristics for studies on land managed by the FWS. As a data provider, the FWS is perhaps best known for the National Wetland Inventory, an extensive database of wetland resources collected to 7 1/2 minute quadrangles. GIS applications in FWS include habitat suitability determinations, national wetland inventory mapping, refuge mapping and master planning, and land records information management.

The FWS staff were regular and consistent members of the Denver Subcommittee of the Interior Digital Cartographic Coordinating Committee (IDCCC). The IDCCC later became known as the Interior Geographic Data Committee, closely tied to the Federal Geographic Data Committee. FWS staffers Charlie Gish, Barb White, and Duane Asherin established coordination with other agencies as a part of their jobs in the early 1980s.

Bureau of Land Management

The Bureau of Land Management (BLM) has the responsibility to compile, map, and inventory environmental resources. BLM began using GIS technology in the early 1980s, almost as soon as FWS, and quickly recognized the power of GIS. Early applications included collection of Public Land Survey data, soil, range site, range allotment, ownership, and so forth, to support soil mapping, land use planning, range assessment, oil and gas well field mapping, habitat assessment, and fire mapping.

In early 1985, BLM and FWS teamed up to develop a joint procurement of computer hardware and a conversion of MOSS software. At about the same time, BLM established the concept of a Common Technology Package for DOI and submitted this to the Information Resources Management Review Council (IRMRC). This led to the purchase of PRIME computers and the porting of MOSS to the PRIME. According to a 1985 DOI GIS Implementation Planning Report, BLM's plans were to establish GIS capabilities in state, district, and resource area offices (219 total). Faced with such an ambitious undertaking, it may be understandable why the continued development of the public-domain MOSS package was chosen over licensing a commercially available system. The USGS objected to the standardization approach, because it was felt that GIS technology was undergoing a time of "rapid evolution" and that standardizing on one system might hold us back. The plan for a standard GIS for DOI didn't really materialize, but the procurement specifications did lead to GIS hardware purchases by BLM and FWS, and several systems were purchased by other DOI cooperators.

Bureau of Indian Affairs

In 1985, the Bureau of Indian Affairs (BIA) developed a plan for the Indian Integrated Resource Inventory Program (IIRIP), a "systems approach" to resource management that utilized GIS technology. As this concept evolved and BIA staff worked with individual tribes, there have been many adjustment and refinements in the decade since.

The BIA's Geographic Data Service Center (GDSC) developed a "client focus" that has helped the tribes to build databases and conduct applications on tribal lands. The GDSC spends considerable effort in training, client services, computer systems support, and applications development.

The BIA is a charter member of what is now known as the Denver Subcommittee of the Federal Geographic Data Committee. The BIA's GDSC is in the unique situation of not being able to offer datasets that are held proprietary to the tribes. Even though BIA cannot offer many of the datasets collected for the tribes, their use of GIS technology and development of training materials is long standing and exemplary.

National Park Service

The National Park Service (NPS) was involved in several project oriented GIS endeavors in the mid to late 1970s, the most notable of which was the massive effort for the Yosemite Master Plan. Concurrently, Maury Nyquist at the NPS Denver Service Center (DSC) was engaged in a testing program with NASA-Earth Resources Laboratory, EPA Environmental Monitoring Systems Laboratory, and USGS EROS Data Center to evaluate the utility of remote sensing and GIS technologies for NPS planning and resources management applications.

In the early 1980s, an integrated approach to remote sensing and GIS became operational at the DSC on an in-house 16-bit minicomputer via modem connection to a mainframe computer at the Bureau of Reclamation. The integrated activities were aligned along basic functional lines, with Harvey Fleet leading vector based activity and Maury Nyquist leading raster based and remote sensing activities. These activities were still based on individual project funding and lacked the continuity necessary to have enduring value and NPS-wide impact.

By 1985, the potential of remote sensing and GIS for NPS use and impacts of project funding on successful implementation was starting to be recognized at higher NPS management levels. Consequently, the DSC function was reorganized as a Washington Office (WASO) GIS Division with base funding to supplement project activities. The creation of the WASO GIS Division not only spawned much greater remote sensing and GIS activity within the NPS but also facilitated the development and implementation of geospatial data and technologies policy and guidelines. Probably the most significant policies at this time were related to hardware/software independence and the investment of very limited resources into data. Hence, there was a reliance on public domain software and "open systems" architecture.

The later half of the 1980s led to the codevelopment and cosupport of a suite of public domain software, based primarily on GRASS (Geographical Resources Analysis Support System) (Chapter by Goran), SAGIS (System Applications Group Information System), and ELAS (Earth Resources Laboratory Applications System) for UNIX workstations or enhanced PC based very inexpensive software for DOS-based systems. This complimentary combination of raster and vector based GIS and image processing software with straightforward data interchange capability greatly facilitated the NPS goals (Nyquist 1987) of: an integrated remote sensing and GIS approach to geospatial data analysis; an "open" and comparatively inexpensive and effective hardware/software environment; the preponderance of financial resource allocation going to data acquisition and the movement of technology to the field for day-to-day applications.

With the advent of more powerful and less costly computers, increased availability of digital data and a broad-based regard for the utility of GIS in the NPS, GIS operations in the field exploded from fewer than six installations and about a dozen partial databases in the mid 1980s to around 75 installations and more than 100 robust databases by the end of the decade. Field GIS installation ranged from a partial FTE using PC-based equipment to full-fledged laboratories with multiple dedicated FTEs and multiple, networked UNIX workstations and peripherals.

Much more of the routine database development and applications work was being performed in the field at this time, which allowed the WASO GIS Division to focus more of its efforts on programmatic, technical support and research issues. Some of the more salient developments included more resources being directed to training and program coordination. For example:

1. A nominal database, with contained the basic set of thematic data (e.g., vegetation, soils, geology) for multiple applications was envisioned for parts, and its development was linked and funded through the NPS Inventory and Monitoring Program.
2. The NPS GIS Sourcebook was developed to give guidance on policy, data, hardware/software, applications, and related technology issues, and the Interior Geographic Data Committee (IGDC) has recommended that this document be updated and expanded for department-wide use.
3. Programs coordination was strengthened by instituting Regional GIS coordinators and a National Park Service Geographic Data Committee that paralleled the FGDC and was comprised of decision-makers from the basic administrative and functional areas of the NPS—Science, Natural Resources Management, Operations, Planning, Cultural Resources, WASO, Regions and Support Centers.
4. GIS training was developed for management as well as practitioners and biennial NPS-wide GIS conferences (the last of which had more than 300 attendees) were held to bring together all NPS staff and management engaged or interested in GIS.

In summary, GIS in the NPS during the decade of the 1980s went from a few project funded activities at a few sites to a fully institutionalized and effective program at scores of sites. The early 1990s saw the NPS developing funding initiatives to add additional regional support centers and more personnel.

Bibliography

Anderson, J. R., E. E. Hardy, J. T. Roach, and R. E. Witmer. 1976. "A Land Use and Land Cover Classification System for Use with Remote Sensor Data." *U.S. Geological Survey Professional Paper,* 964: 28.

Bryant, N. A., and A. L. Zobrist. 1981. "IBIS: A Geographic Information System Based on Digital Imager Processing and Image Raster Datatype." *IEEE Transactions on Geoscience Electronics*, GE-15 (3): 52–159.

Dwyer, J. L., S. H. Moll, C. M. Trautwein, J. E. Elliott, R. C. Pearson, W. P. Pratt, and J. T. Nash. 1987. "Geographic Information System Requirements in Mineral-Resource Assessment—Lessons Learned through Cooperative Research." U.S. Geological Survey Open-File Report 87-314. Third Annual V. E. McKelvey Forum on Mineral and Energy Resources, Denver, CO, 9–10.

Fosnight, E. A. 1988. "Applications of Spatial Postclassification Models." *Proceedings, 21st International Symposium on Remote Sensing of Environment,* Ann Arbor, MI: Environmental Research Institute of Michigan, 469–485.

Greenlee, D. D. 1981. "Application of Spatial Analysis Techniques to Remotely Sensed Images and Ancillary Geocoded Data." In P. A. Moore, ed., *Computer Mapping of Natural Resources and the Environment, Plus Satellite Derived Data Applications,* vol. 15. Cambridge, MA: Harvard Graduate School of Design, Laboratory for Computer Graphics and Spatial Analysis, 111–120.

Greenlee, D. D., and H. L. Wagner. 1982. "An Evaluation of a Microprocessor-Based Remote Image Processing System for Analysis and Display of Cartographic Data." *Proceedings, International Symposium on Computer Assisted Cartography (AUTO-CARTO V).* Falls Church, VA: ASPRS/ACSM: 357-365.

Greenlee, D. D., J. W. Van Roessel, and M. E. Wehde. 1986. "An Evaluation of Vector Based Geographic Information Systems at the EROS Data Center." Unpublished internal working paper.

Guptill, S. C., ed. 1988. "A Process for Evaluating Geographic Information Systems, Federal Interagency Coordinating Committee on Digital Cartography." Technology Exchange Working Group, Technical Report 1. U.S. Geological Survey: Open File Report 88-105: 142.

Hutchinson, C. F. 1978. "Techniques for Combining Landsat and Ancillary Data for Digital Classification Improvement." *Photogrammetric Engineering and Remote Sensing*, 48: 123–130.

Hutchinson, M. F. 1989. "A New Method for Gridding Elevation and Stream Line Data with Automatic Removal of Pits." *Journal of Hydrology*, 106: 211–232.

International Geographical Union Commission on Geographical Data Sensing and Processing. 1976. *Second Interim Report on Digital Spatial Data Handling in the U.S. Geological Survey.* Unpublished report.

Jenson, S. K. 1985. "Automated Derivation of Hydrologic Basin Characteristics from Digital Elevation Model Data." *International Symposium on Computer-Assisted Cartography (AUTO-CARTO 7).* Falls Church, VA: American Society of Photogrammetry and American Congress on Surveying and Mapping, 301–310.

Margerison, T. A. 1976. *Computers and the Renaissance of Cartography.* London, U.K.: National Environment Research Council, Experimental Cartography Unit, 20.

Mitchell, W. B., S. C. Guptill, E. A. Anderson, R. G. Fegeas, and C. A. Hallam. 1977. "GIRAS—A Geographic Information Retrieval and Analysis System for Handling Land Use and Land Cover Data." Professional Paper 1059. Reston, VA: U.S. Geological Survey.

Nichols, D. A. 1981. "Conversion of Raster Coded Images to Polygonal Data Structures." *Proceedings of Pecora VII Symposium.*

North, G. W. 1995. "Memorial Address: William A. Fischer, Photogrammetric Engineering and Remote Sensing." *American Society of Photogrammetry,* 61 (7): 857–862.

Office of Management and Budget (OMB). 1985. *Management of Federal Information Resources.* Circular A-130. Washington, D.C.: Office of Management and Budget.

Office of Management and Budget (OMB). 1990. *Coordination of Surveying, Mapping and Related Spatial Data Activities.* Circular A-16. Washington, D.C.: Office of Management and Budget.

Pearson, R. C., J. S. Loen, C. M. Trautwein, S. K. Jenson, and S. H. Moll. 1986. "An Empirical Resource-Potential Model for Vein and Replacement Deposits, Dillon 1 Degree by 2 Degree Quadrangle, MT." *Abstract of the Geological Society of America,* 18 (5): 402.

Pratt, W. P., K. M. Walker, S. K. Jenson, J. R. Francica, D. A. Hastings, and C. M. Trautwein. 1983. "Mineral-Resource Appraisal of the Rolla 1×2 Degree Quadrangle, MO—Manual vs. Digital Synthesis." *International Conference on Mississippi Valley Type Lead-Zinc Deposits.* Rolla, MO: University of Missouri, 584–595.

Simonett, D. S., T. R. Smith, W. Tobler, D. G. Marks, J. E. Frew, and J. C. Dozier, eds. 1978. "Geobase Information System Impacts on Space Image Formats." Report of a Workshop at La Casa De Maria, Santa Barbara, CA, NASA Contract NASW-3118. Unpublished report.

Thompson, M. M. 1969. "Automation in Topographic Mapping." *Proceedings, 1969 ASP-ACSM Fall Convention.* Bethesda, MD: American Congress on Surveying and Mapping, 316–326.

Tosta, N. 1992. "Who's Got the Data?" *Geo Info Systems,* 2 (8): 24–27.

Tosta, N. 1993. "The Data Wars: Part I and II." *Geo Info Systems,* 3 (1–2): 25–27, 22–26.

Trautwein, C. M., D. D. Greenlee, and D. G. Orr. 1982. "Digital Data Base Application to Porphyry Copper Mineralization in Alaska—A Case Study Summary." U.S. Geological Survey Open-File Report 82-801: 14.

Trautwein, C. M. 1983. "Combined Data Sets." In R. N. Colwell, ed., *Manual of Remote Sensing,* vol. 2, 2nd ed. Falls Church, VA: American Society for Photogrammetry and Remote Sensing, 1757–1762.

Trautwein, C. M., R. C. Pearson, and J. E. Elliott. 1988. "GIS Applications to Conterminous United States Mineral Assessment Program Investigations." *GIS: Integrating Technology and Geoscience Applications Symposium Proceedings.* Washington, D.C.: National Academy of Sciences/National Research Council, 20–21.

U.S. Geological Survey (USGS). 1970. "Computer Contribution Number 5." *Proceedings of the Symposium on Map and Chart Digitizing.* Washington D.C.: U.S. Geological Survey, 81.

U.S. Geological Survey (USGS). 1982. *Software Documentation for GCTP General Cartographic Transformation Package: National Mapping Program Technical Instructions.* Washington D.C.: U.S. Geological Survey National Mapping Division, 81.

Van Roessel, J. W., and E. A. Fosnight. 1985. "A Relational Approach to Vector Data Structure Conversion." *International Symposium on Computer-Assisted Cartography (AUTO-CARTO 7).* Falls Church, VA: American Society of Photogrammetry and American Congress on Surveying and Mapping, 541–551.

CHAPTER 12

GIS Technology Takes Root in the Department of Defense

William Goran

Introduction

This chapter focuses on the processes and events relating to GIS technology examination, exploration, and adoption by environmental, engineering, and facility management organizations within the Department of Defense (DOD). The time frame covered is the two decades between 1975 and 1995 (Figure 12.1).

The title of this chapter implies GIS for all aspects of DOD, but no mention is made of the many GIS applications that have been developed in direct support of military operations, such as military engineering and terrain analysis, battlefield and theater of war operations, and train-

William Goran is a geographer and soil scientist (with master's degrees in both fields) but generally regards himself as an environmental researcher focused on landscape management issues. His current responsibility is the development of the LMS program, with goals of development watershed/ecosystem based tools for modeling the complex landscape dynamics that are affected by and affect both military training and testing and civil works (navigation, wetland management) operations. The focus of this program is to better model these landscape dynamics and incorporate these improved modes into decision support processes to ultimately improve our understanding of the consequence tomorrow of today's decisions.

Acknowledgments: The author would like to thank several individuals who provided valuable data, critical review, and comments during the development of this chapter. These include Andy Bruzewicz, Kurt Buehler, Penny Capps, Brian Cullis, Kelly Dilks, Jim Farley, Randy Hanna, M. K. Miles, and Jim Westervelt. Thanks also to Dana Finney for several editorial reviews and to Teresa Aden for assistance with preparation of figures.

Author's Address: William Goran, US Army CERL, Environmental Laboratory, P.O. Box 9005, Champaign, IL 61826.

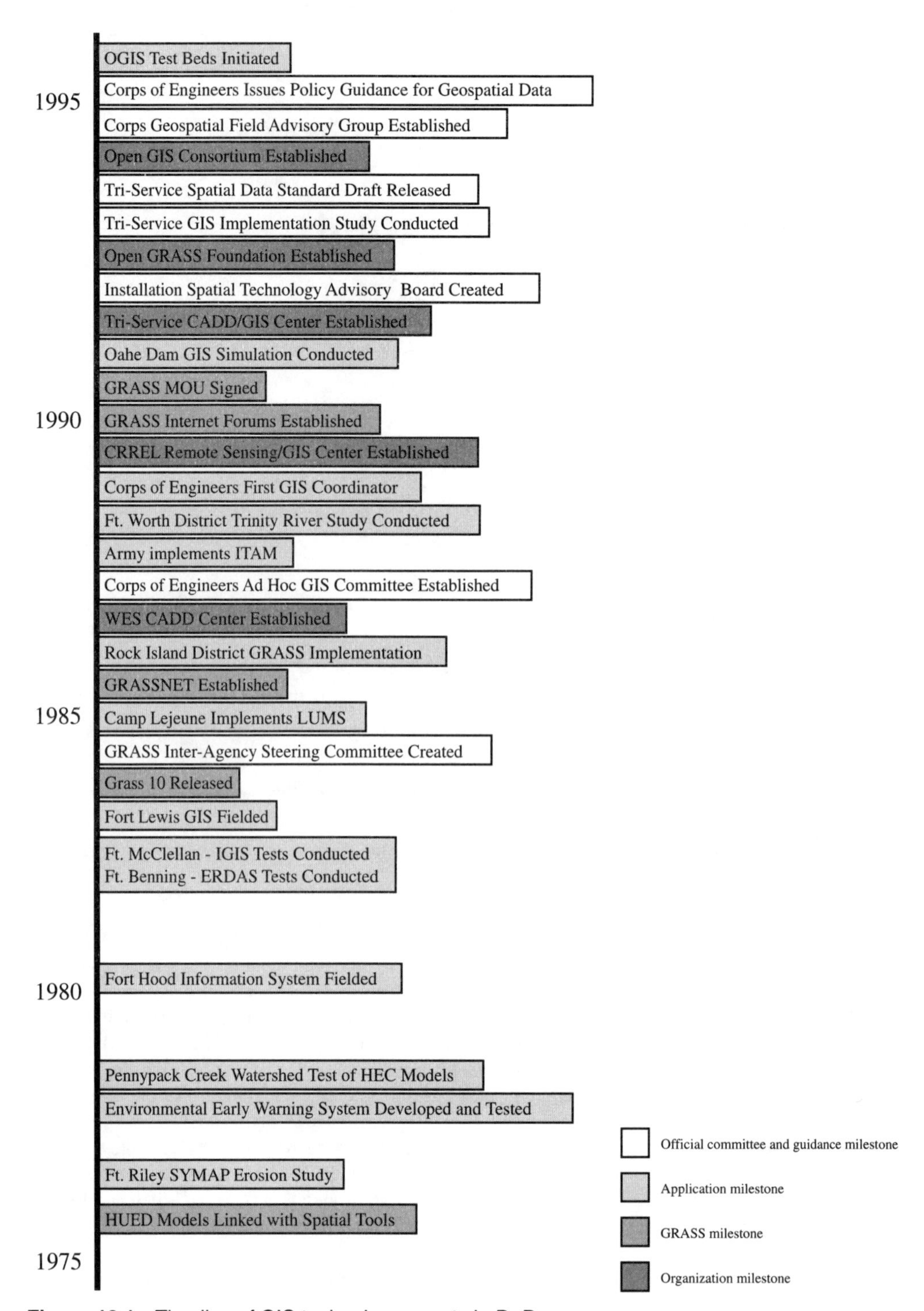

Figure 12.1 Timeline of GIS technology events in DoD.

ing simulations. Instead, the targets for GIS implementation, within the context of this chapter, are the military installations, primarily those installations with large land holdings and with troop training and/or weapons testing missions. The U.S. Army Corps of Engineers districts and divisions are also considered as targets of technology adoption, in relation to their support to military installations and their civil (navigation, regulations of wetlands, reservoir management, emergency operations) activities.

At the beginning of this time period (1975), GIS was essentially an unknown technology within the DOD installation management and civil works communities. The potential of GIS technology was known by only a few graduate students and junior staff at the laboratories and a few imaginative planners and resource managers at installations and districts, searching for a better way to integrate and analyze diverse data about the physical land and seascapes. These groups teamed up to become champions of this promising technology. Managers naturally needed to be convinced that the promise of the technology would be worth the significant investment.

In the years following 1975, the cost for GIS implementation often did exceed short term returns, but demand for the technology continued to build. Why? No doubt GIS technology rode the computer revolution wave, benefiting from remarkable computational advances and cost reductions, as well as a series of generational improvements in data input and output devices. Also, the increasing availability of digital spatial data continues to improve the organizing, understanding, and presenting of data. Two decades later, GIS technology is fast becoming an everyday feature of doing business in DOD, plus the formidable task and cost of building a GIS database. But an important factor was and is the role of technology champions—those who fought for the resources, marketed successes, and pointed budget conscious managers, hard-to-please environmental regulators, anxious military trainers, and even traditional foresters and wildlife managers toward the promise of geographically organizing, understanding, and presenting data.

Installations Environmental Requirement for GIS

In the late 1970s and early 1980s, military installations were struggling with the complex issues associated with the requirements of the National Environmental Policy Act (NEPA) for an analysis of alternatives and potential impacts of proposed actions. Traditionally, troops had cut trees and built roads, ranges, or drop zones as necessary to meet their military use requirements. But compliance with NEPA meant that the potential impacts of any of these actions needed to be evaluated and alternatives considered before making changes to the physical landscape. Such evaluations require data on soil erodibility, species habitat, the potential or known location of cultural sites, topography and vegetative cover, analysis of line of sight (for target placement and visibility), road distance, and the relative costs of alternative routes. Flipping through paper maps and reports from contractors to try and integrate all these factors took on nightmarish dimensions, and the environmental managers were crying out for help.

In the Army, many environmental managers had already beaten a path to the doors of the U.S. Army Corps of Engineers Construction Engineering Research Laboratory (CERL, now Laboratories) for assistance with environmental databases and computer systems. CERL was established in the late 1960s, adjacent to the campus of the University of Illinois Urbana–Champaign to provide research and technical assistance to support the Corps of Engineers construction mission. After Congress passed NEPA, CERL was also assigned an environmental mission. Researchers at the lab had developed a suite of computerized environmental databases and decision support systems, called the Environmental Technical Information System (Webster 1975) and were now exploring the potential of automated map and map analysis capabilities to help evaluate the impacts of training activities on military lands.

In the late 1970s, there were only a few off-the-shelf options in the emerging GIS marketplace, and these options were generally mainframe or minicomputer based and quite expensive to acquire and operate. In 1980, CERL researchers carried out a market survey to analyze all the GIS marketplace options, with the intention of developing selection criteria for an eventual purchase. They also began testing selected GIS capabilities, to better understand installation requirements and to assist with some specific NEPA assessments. Harvard's SYMAP (Synagraphic Mapping System) software was tested in an erodibility study on a portion of Fort Riley, KS, by one group of researchers, while another group designed and contracted the development of a spatial analysis component for the Environmental Early Warning System (EEWS) (Lozar 1986). EEWS was a CERL product intended to identify environmental and facility impacts of military stationing and subsequent training actions. The spatial component was tested by evaluating a proposed increase in training activities on woodpecker habitat at the EEWS test database site, Fort Polk, LA. But these efforts were pilots. The results never left the lab. Such applications represented typical laboratory exploration of a new technology opportunity.

Fort Hood, TX, was the first installation sponsor of these lab explorations, and the first fielded pilot was named the Fort Hood Information System (FHIS). Running under the UNIX operating system on a Digital Equipment Corporation's VAX minicomputer located at the University of Illinois, the system was designed as a simple fixed raster matrix with a suite of data layers including satellite imagery. FHIS was successfully used by USACERL staff for an environmental impact assessment of a proposed training lands upgrade and an initial evaluation of training land carrying capacity (Goran 1983). Local use at Fort Hood, however, was quite difficult, as it required uninterrupted direct phone line connection to the Illinois based VAX computer.

As a pilot, FHIS was very useful. Among the lessons learned were (1) the system must be run on a local host, (2) the system should be designed with maximum flexibility and modularity to allow continuous enhancements to both the software capabilities and the database, and (3) the system should take advantage of a multitasking (so longer tasks can be relegated to background) and have a multiuser operating system (like UNIX) that is hardware independent, to allow ease of porting. Ease of porting was a particularly important issue, so that the software was not bound up within a particular hardware environment.

CERL staff also learned that the installation users were eager for the technology, despite troubles with access and the limited functionality of FHIS. Without a system to integrate and analyze the complex spatial relations between diverse landscape elements (and data sources), installations were essentially unable to perform NEPA assessments adequately or to assess the condition and capability of their training lands.

Starting in 1983, a double pilot was attempted. An early personal computer hosted system from the Atlanta-based vendor Earth Resources Data Analysis Systems, Inc. (ERDAS) was placed at Fort Benning, GA, and a second generation of FHIS (this one called the Installation Geographic Information System or IGIS) went to Fort McClellan, AL, running on one of the first computers (126) manufactured by Sun Microsystems of Mountain View, CA. After a year of testing, the sponsoring organization, the U.S. Army Training and Doctrine Command (TRADOC), hired a geography professor (Dr. Thomas Baucom of Jacksonville State University in Alabama) to provide an independent evaluation of both installation GIS. The GIS were found to be short of meeting reasonable cartographic requirements (both were raster systems, with limited vector and cartographic referencing functions). Despite the significant limitations of these two systems, staffs at the installations were busily turning out analyses and adding new data sets to those initially delivered as part of their systems.

At Fort McClellan, the system was designed to support all elements of the environmental and natural resources program, and applications included siting a training course for the then new Humvee combat vehicles, evaluating the interrelationships between forestry and wildlife management plans, conducting a safety fan analysis of various tank firing positions (for which Jim Westervelt of CERL wrote some additional specialized software), and reevaluating a recently completed landfill siting study. At Fort Benning, the ERDAS PC system found a warm home and willing hands in the forestry branch, where it was used to support the installation's forestry planning, harvesting, planting, and protection projects.

The demands for similar systems at other installations kept growing. CERL researchers, working with funds from the installations, continued to develop pilot capabilities. Fort Lewis near Tacoma, WA, received the next new and improved pilot system, this time running on a MASSCOMP (later acquired by Concurrent Computer Corporation) graphic workstation. CERL researchers were determined to test hardware independence by developing and fielding on both Sun and MASSCOMP.

As GIS development work continued at CERL, another guiding principle emerged. Military environmental managers were concerned with tracking changes in vegetation and soil conditions over time, to match the military and other land uses to the current conditions and overall capacity of the resource base. Tracking changing conditions requires access to and analysis of new data streams such as aerial photographs and satellite imagery. Thus, another system requirement was the dynamic incorporation of remotely sensed imagery and the inclusion of software tools to read, manipulate, and analyze these images. At the time, GIS and image processing functions were performed in separate software environments. CERL defined a requirement for seamlessly interweaving the two technologies.

Emergence of GRASS

In autumn 1984, a modest direct research program was initiated at CERL, and the pilot capabilities were standardized and released (to Forts Hood, Lewis, and McClellan, Central Washington University, and University of California at Berkeley) in March 1985 as the Geographic Resources Analysis Support System or GRASS, version 1.0. In autumn 1985, the first user group meeting was held at Fort Hood. Attendants included representatives of the U.S. Department of Agriculture (USDA) Soil Conservation Service (SCS), which had developed a similar set of system criteria and was evaluating public domain options, and the American Farmland Trust, which was interested in pursuing methods to encourage the use of spatial technologies for land use planning (and farmland preservation) at the local level nationwide.

In the following years, the SCS established a formal pilot project and eventually adopted GRASS as a standard software tool for soils mapping, watershed, and farm planning (Rohaley 1990). The American Farmland Trust's GIS interest eventually resulted in the establishment of the National Centers for Resource Innovations (NCRI), with a mission funded through USDA, of training and assistance to counties and other local governments interested in implementing spatial technologies (Maizel 1990).

After 1985, CERL pulled together a series of new software releases, and GRASS grew from a raster-only system to more fully functional GIS, with separate data structures for vector, raster, point, and imagery data. Funding remained low, however, patched together from a small project in the Army's environmental quality research program, the Cartography and GIS Division of the SCS, and, in some years, the National Park Service. A coordinating body was established to accommodate the interest of the different agencies involved with GRASS, and this body soon took upon itself the primary role of shepherding GRASS software development and interacting with the emerging user community (Goran 1991).

Within the Army, the implementation of GIS for land management and environmental planning continued at a slow and measured pace, one installation at a time, until the late 1980s, when the Integrated Training Area Management (ITAM) program was adopted across the Army as a set of tools for monitoring and managing training and testing lands (Severinghaus 1991). ITAM was also a CERL product, and GRASS was tucked into the ITAM package which was being rapidly fielded to every Army installation with a significant land impacting mission. Soon, wherever there were tanks and trucks disturbing the soil, there also was GRASS to help select monitoring sites, to present information on soil erosion and vegetation condition status, and to interface training schedules with land restoration and protection projects.

Besides these ITAM applications, installation personnel began using GRASS to help: analyze proposed land use changes and acquisitions; model and predict the location of archaeological sites; suggest site facilities such as radar towers, vehicle training courses, pipelines, and sanitary landfills; target surveys for and develop habitat protection plans for threatened or endangered species; develop joint land use and noise assessment plans with neighboring communities; aid in forestry and wildlife management plans and practices; develop spill protection and corrosion hazard maps and plans for fuel storage tanks; and model, predict, and evaluate risks associated with contaminated ground water at restoration sites.

Data gathered to support these applications varied by installation, but CERL staff developed a standard suite of themes for delivery with ITAM. This suite included satellite imagery, a soils map, a training or testing unit boundary, and a road map. These themes were used in a model for siting field sampling locations that represented different landscape and land use units. Often, digital elevation models were also included with this suite of data themes.

The systems were fielded in a turnkey manner to the installation from CERL in a package that included (1) a workstation fully loaded with (2) software, on-line training manuals, and help systems, (3) the initial data sets, (4) and all the required peripherals. Training was provided on system management, data maintenance and updates, and software use both before and after the system was delivered. Then, a few months after delivery of the system, the primary software users on the installation staff would travel to the nearest university offering GRASS training workshops (over the years, this list included Central Washington State, Colorado State, University of Illinois, University of Arkansas, Rutgers, University of Cincinnati, Purdue, University of California at Berkeley, and George Mason) and receive one or two weeks of formal instruction.

CERL also established relationships with some Corps of Engineers Districts to provide this same set of turnkey installation support services. To help both installations and districts understand implementation planning and operational support issues for GIS, CERL researchers developed an implementation planning guide (Martin 1989), a cost benefit analysis guide (Sliwinski 1989), and a database development guide (Ruiz 1990).

Despite the CERL turnkey approach and the post implementation support services, the Army installations had mixed success with GRASS implementation. At many sites, operation and maintenance of UNIX workstations proved daunting. Installations were able to find funds for acquiring equipment and creating data files, but they were often unable to dedicate the appropriate staffing to become technically proficient. In several cases where lower grade staff did become technically proficient, their newly found GIS skills opened doors to higher paying jobs elsewhere, and an installation was again without a trained operator, having acquired expensive equipment and complicated software that no one else was capable of using.

Role of Technology Champions

At installations where the technology was successfully being implemented, the critical difference was often the impact of champions of GIS technology. Several factors influenced the impact of these champions: position, technical skills, duration in assignment on-post, personal influence, and the ability to conceive and achieve useful technology applications.

At Fort Hood, the chief of the Environmental Division, Emmet Gray, combined the roles of champion and guru in his innovative approaches to integrating GIS technology into standard business practices. He developed a system, called the Maneuver Area Damage Assessment Model (MADAM) that provided Fort Hood trainers with land use scheduling options that minimized environmental damage and institutionalized the use of the technology. As a technology guru, he also made extensive use of the computing capabilities of the GIS hardware platforms.

The chief of the Fort Lewis Environmental and Natural Resources Division, Randy Hanna, proved another important champion. Hanna assembled a dedicated GIS staff that developed applications to support all aspects of his division. With staffing and technology support from CERL and Battelle Pacific Northwest Laboratories of Hanford, WA, Hanna's GIS operation eventually developed into the Northwest GIS Center, providing GIS support services to a string of military installations along the west coast and occasionally to other organizations in Washington state. Among the many successful GIS applications of Fort Lewis was an impressive NEPA assessment completed for a proposed land acquisition at Yakima Firing Center between the Yakima and Columbia rivers in central Washington state. In another application, Fort Lewis GIS professionals provided line of sight assessments for a radar tower that greatly reduced the projected requirement for tree clearance. Over the years, Hanna's GIS staff shifted most of their applications from GRASS to ESRI products, primarily because of their requirement for stronger database interface and cartographic output capabilities than those provided with GRASS.

Champions were important at many other installations, but often these champions moved on to other assignments and their impact proved ephemeral. Some, however, moved on to new responsibilities within the department, where they continued as proponents of the technology. Ray Clark, for example, moved from his position as chief of the Environmental and Natural Resources Division at Fort McClellan, where his staff performed an early test of the technology, to the Army Environmental Office at headquarters in the Pentagon, where he promoted the use of GIS to improve the quality of NEPA assessments throughout the Army. John Brent was a strong GIS proponent in his position as chief of the Environmental Office at Hohenfels Training Center in Bavaria, Germany, and brought that same technology enthusiasm with him when he transferred to a similar position at Fort Benning, GA, in the mid 1990s. Fort Benning had been an early GIS test site, but the system was neglected after the loss of skilled operators and interested managers.

The Army leadership at various levels tended to view GIS as simply a tool to help installation personnel perform their current duties rather than as a technology that transforms information flows and processes and therefore requires extensive planning, coordination, and leadership for successful implementation. GRASS was fielded primarily within the environmental, natural, and cultural resources offices on military installations, essentially all within one stovepipe. But the capability of the technology and the specific data sets used within this stovepipe clearly had value in many other installation management areas, such as master planning, safety and security, engineering services, training planning and scheduling, emergency vehicle routing and emergency planning, and military exercises being conducted on the installation training lands. However, while there was some discussion of the technology across these mission areas, separate GIS and computer-aided design and drafting (CADD) solutions were developed and fielded in discipline specific management chains, or stovepipes.

Corps of Engineers Technology Centers

While GIS was being fielded to environmental and natural resources staff at Army installations, many of the master planners at these same installations were acquiring and expanding their CADD systems to include an installation-wide mapping base, with data on utilities, terrain contours, built facilities, and sometimes also flood plain and land use data. In almost every case in the Army, these CADD systems were from Intergraph, and the installations were directly aided in their implementation of the technology by a local Corps of Engineers District. A significant cadre of CADD expertise was developing at the Corps districts, and the technology was clearly paying back the high cost of hardware, software, and training in the engineering drafting and design arena. Success related to the use of CADD for automated mapping and facility management (AM/FM) was mixed, however, for many of the same reasons that users struggled with GIS implementation.

Installations were spending considerable sums in developing their AM/FM data themes, usually starting with new photography supplemented with extensive ground data collection/verification. Maps were generated for the built-up (urbanized or cantonment) areas at 1:1000 or 1:2000 scale and for the training lands or noncantonment areas at 1:5000 or 1:10,000 scale. Corps of Engineers Districts provided the support to build these data sets, usually through a contractor such as Mid-States Engineering, Chicago Aerial Survey, or GRW Aerial Surveys. Sometimes contractors fielded support staff at the installation to help with the implementation of the technology. But many of the installations floundered with the new technology, unable to keep the data up to date (utility and building data were constantly needing modification) and unable to dedicate adequate staff to get over the hump in gaining competence with these complex systems. While installations were sometimes able to secure the extensive up-front funding for building databases, they seldom had success in obtaining approval to hire even low-graded staff as system operators. And those who were trained as CADD operators often left these positions for greener pastures once they realized the market value of their new skills.

Initially, several Corps of Engineers Districts (Savannah, Kansas City, St. Louis, and others) took the lead in exploring this technology and supporting the military installations. As momentum grew and more districts became active in CADD and AM/FM technologies, a critical mass developed for a standard contract through which all districts (and the installations they support) could acquire hardware, software, training, and other services. In 1987, Intergraph Corporation was awarded a five-year contract to provide the Corps of Engineers (through the Huntsville Division) with CADD and AM/FM hardware, software, training, and other services.

When this contract was awarded, the Corps of Engineers decided to establish a CADD Center within the Information Technology Laboratory of the Waterways Experiment Station, another Corps lab located in Vicksburg, MS. The purpose of the CADD Center was to (1) help districts and installations to implement CADD technology, (2) address issues relating to data, cartographic, and state-of-the-practice standards, (3) bring technology users together in forums to advance knowledge and understanding of this technology, and (4) serve as Contracting Officer Technical Representatives for modifications and interpretations of this five-year contract (Stephens 1988).

The CADD Center established a structure of field advisory groups, called single discipline task groups, that tackled application development and standards issues in specific disciplines (such as architecture and mechanical or electrical engineering) and served as a network for users to gather and exchange ideas and issues. One of these single discipline task groups was established in 1990 to address GIS technology (Costanzo 1990). Members attending the initial meeting of this group immediately protested to the CADD Center organizers that GIS was multidisciplinary in nature, and "single discipline" was a misnomer. Despite these protests, the Corps was making efforts to link the GIS and CADD communities and to discuss both technical and cultural differences between users of the two technologies.

While many of the engineering disciplines in the Corps of Engineers were exploring the potential of automated design and drafting technology, other laboratory researchers and district and division staff were tackling new digital data streams from *in-situ* sensors and remote data platforms. To explore the potential applications of data from these new sensors, the Corps of Engineers established a Civil Works Remote Sensing Research Program which provided an opportunity for the laboratories to evaluate application to the Corps' water resources and waterways management mission. Over the years, Corps researchers and cooperating district staff analyzed data streams from *in-situ* soil moisture probes to evaluate ground water saturation, explored the potential for digital radar data on storm location and intensity for use in hydrologic models, and used satellite and *in-situ* sensors to model snow temperature and depth in rugged terrain to predict spring runoff and potential downstream flooding.

Each of the Corps laboratories participated in this research program, which was managed by Harlan (Ike) McKim of the Cold Regions Research and Engineering Laboratory (CRREL) in Hanover, NH. As a natural outgrowth of this program, and the need to help district and division staff translate the results of this research into their daily operations, McKim and CRREL leaders proposed that the Corps of Engineers establish a remote sensing center to be located at CRREL. In this proposal, GIS was identified as a key technology to help integrate and manage these incoming data streams. Proposed in the late 1980s, the center idea was the subject of some debate prior to July 1990, when Corps headquarters granted approval and budgeted for future funding for the Remote Sensing/GIS Center. In 1992, McKim became the first center chief. Key functions of the center include: (1) a central point of contact for remote sensing and GIS, including management of the Corps' civil works remote sensing and development program; (2) a source of information, dissemination, and direct assistance for the development of applications of these technologies; (3) training and development of standardized approaches for specific applications; and (4) the coordination of and emphasis on sharing lessons learned with remote sensing and GIS technology (Bruzewicz 1992).

In addition to serving the Corps water resources programs, the center has also been active in the Corps emergency management activities. On several occasions, the center organized efforts to help analyze oil spills, and Andy Bruzewicz, one of the lead scientists at the center, assembled teams (with members from other labs and districts) that analyzed remote imagery and developed GIS databases after Hurricane Andrew and during the 1993 floods in the midwest and along the Gila River in the southwest. As a natural outgrowth of these efforts, Bruzewicz has initiated efforts, in coordination with the Federal Geographic Data Committee (FGDC), to better plan for multiagency coordination of GIS logistics and data access in emergency situations.

Corps of Engineers Headquarters Tackles GIS Technology

In the 1960s, the Chief of Engineers established an independent environmental advisory board to review Corps practices and policies in environmentally related issues. In spring 1987, this advisory board met in New Orleans with a topical focus of environmental data. As this meeting progressed, staff from Corps labs and districts made presentations highlighting the use of GIS for data management, analysis, and presentation. These presentations raised concerns from the Chief of Engineers (at that time, Lieutenant General E. R. Heiberg) and the advisory board that the Corps was charging rapidly down several different roads implementing GIS technology while no one in the agency was directing traffic. As a result, the advisory board advised the chief to tackle this issue (Chief of Engineers Environmental Advisory Board 1987). A working group, chaired by Dr. William Klesch, from the Civil Works Directorate of Corps Headquarters, Washington, D.C., was created.

Dr. Klesch sought membership from throughout the Corps of Engineers, but the assembled committee (more than 30 strong) consisted of staff from labs, districts, divisions, and headquarters, without any installation staff involvement. The group tackled many issues (user needs, hardware, software, data requirements and quality, costs, technology transfer, and intra- interagency coordination) and, after a year of study, reported back to the Chief of Engineers and the advisory board. The group's primary recommendations were to establish an agency GIS coordinator and a headquarters-level steering committee and to require all divisions and districts to develop GIS implementation plans (Klesch 1989).

The Chief of Engineers (by then, Lieutenant General Henry Hatch) accepted the recommendation and assigned the responsibility for Spatial Data Systems Management (including GIS) to the Engineering Division, Civil Works Directorate. However, Corps headquarters was down-sizing, and creating a new headquarters-level position proved difficult. To help fill the need, Colonel Sam Thompson, at the time a professor of geography at West Point, volunteered to devote a sabbatical year to this task and moved to Washington, D.C., in 1989. Colonel Thompson convened a group again, this time involving more military installation staff, to try and chart a course of GIS actions for both the Corps of Engineers Civil Works programs and the Army installations. The Army and the Corps were pursuing different GIS directions, however, and the only clear consensus was that a headquarters-level coordinator was needed for GIS. Before Colonel Thompson returned to West Point, he completed a job description and left the paperwork for a recruitment action with M. K. Miles, who held responsibilities for the surveying and mapping activities within the civil works program.

Miles vigorously pursued attempts to fill this position, but he also bumped against the brick wall of staff reductions at headquarters. Not to be discouraged, he recruited help from the labs. As time passed, hope for permission to advertise the coordinator position dwindled, so Miles accepted the responsibility himself and began charting a course that he hoped would serve all parties in the Corps of Engineers.

The Federal Geographic Data Committee (FGDC) was by 1990–91 beginning to emerge as an important voice for coordinating and framing GIS activity throughout federal agencies. The only Department of Defense (DOD) participation, however, was from the Defense Mapping Agency, which represented the military mission activities but not the total range of geodata usage and activities within DOD. Miles established ties between the Corps of Engineers and FGDC and gained another DOD seat on the committee (and various subcommittees) representing the interests of the Civil Works program and the DOD facility and environmental management communities. In the years since, Miles and other Corps and DOD facility managers have been active participants in many FGDC and National Spatial Data Infrastructure (NSDI) activities, including the Geospatial Data Clearinghouse, the Metadata Standard, and the Spatial Data Transfer Standard.

As Corps Spatial Data Systems coordinator, Miles also began participating in the GRASS Inter-Agency Steering Committee forums, for the first time bringing a consistent voice from Corps of Engineers headquarters to this loosely assembled gathering of agencies. The U.S. Department of Interior, National Park Service and Geological Survey, SCS, and the U.S. Army Corps of Engineers represented the key agencies participating in this forum (although several other groups usually attended), and Miles joined the committee just as these agencies were debating the value of formalizing their relationship with a Memorandum of Understanding (MOU). Miles accepted the task of guiding this MOU through the long path of acceptance and concurrence at Corps headquarters. By this time, GRASS was being used by several military installations and Corps districts, but GRASS was also the subject of considerable debate, as GIS software choice had become one of the issues of fragmentation within the emerging GIS user community. As the GIS software marketplace matured, voices of concern were being raised regarding the long-term commitment for a government developed and maintained software package that initially looked very different from any marketplace option but now appeared to overlap with the commercial marketplace at least in part (Goran 1992).

In his role as Corps of Engineers Spatial Data Systems coordinator, Miles pursued another major effort. The CADD Center established in Mississippi in 1987 was facing a major milestone, the expiration of the five-year CADD contract with Intergraph, and the Corps needed a new option for continuing to acquire CADD products and services. The Naval Facilities Engineering Command (NAVFAC) had already drafted a comprehensive contract for both CADD and GIS technology and approached the CADD Center staff with the idea of teaming on this contract and expanding the existing CADD center to be multiservice and multitechnology. The Air Force Civil Engineer (AFCE) expressed interest in participating in this proposed center but also insisted on including both CADD and GIS technology.

Figure 12.2 Sites visited for the Tri-Service GIS Implementation Study.

While this multiservice idea received general support from the Corps of Engineers CADD community, debate erupted in the Corps laboratories (Figure 12.2). The Remote Sensing/GIS Center had just been established at CRREL for Civil Works applications, and CERL was already providing GIS technology assistance to dozens of installations. Further, the Topographic Engineering Center (TEC), another Corps laboratory located at Fort Belvoir, VA, which primarily focuses on tactical military requirements, was concerned about a potential overlap with its mission area.

With support from both the Navy and Air Force, and despite voices of opposition for the GIS component of this center from the Corps laboratories, the plan proceeded and the Tri-Service CADD/GIS Technology Center was chartered in summer 1992 as a:

> "multiservice vehicle to set standards, coordinate facilities of Computer Aided Design and Drafting and Geographic Information Systems within the Department of Defense, promote systems integration, support centralized acquisition, and provide assistance for the installation, training, operation and maintenance of CADD/GIS systems" (Stephens 1992).

The focus of the center was facilities engineering, not tactical operations (thus addressing TEC's concerns), and the charter was signed by the highest ranking facilities engineer in each of the three services. Tri-Service Executive Steering and Working Groups were established to oversee the center, and the Corps of Engineers was assigned as executive agent, responsible for the planning, budgeting, staffing, and execution of the center's programs.

In 1994, due to the increased attention of the Clinton Administration to GIS and digital spatial data (evidenced by the chairmanship of the FGDC being elevated to the Secretary of Interior and Executive Order 12906 on the National Spatial Data Infrastructure), Miles finally obtained approval to hire a GIS specialist to help him coordinate Corps of Engineers geospatial activities. Penny Capps from the Topographic Engineering Center was selected for this position.

With this expanded staff, Miles and Capps focused on developing an Engineering Circular to implement the Executive Order on the National Spatial Data Infrastructure (NSDI) and on establishing a Corps of Engineers Geospatial Data Field Advisory Group. They also plan to establish an Army-wide geospatial data committee, which will potentially use the Corps Engineering Circular as a model for an Army-wide (and perhaps DOD-wide) guidance document. The Corps Geospatial Advisory Group, which includes members from each of the labs and a representative from each Corps division, has pioneered implementation of the NSDI metadata standard, with internet accessible servers that maintain information on the Corps' geospatial data holdings.

GIS within Corps of Engineers Civil Works Programs

The Corps of Engineers District offices, with support from the laboratories, have developed many applications for GIS technology in support of navigation, water control, wetlands permitting, coastal zone management, and related programs. In the 1970s, the Hydrologic Engineering Center (HEC) developed the Spatial Analysis Methodology (HEC-SAM) as a planning tool for flood forecasting and water management studies. With SAM, the HEC software tools included both hydrologic models and spatial expressions of these models, thus representing an early approach to interfacing modeling and GIS technologies.

As was common in the GISs of the era, the "spatial expression" from the HEC models was an integer matrix, which could easily be produced on conventional dot matrix printers. The HEC-SAM software was written to run on the standard Harris minicomputers used throughout the Corps of Engineers, but performing the analytical functions and generating these number matrices was computationally intensive and slow on this host hardware. To speed computation, low spatial resolution was accepted. But at either low or high resolution, dot matrix printed integer maps lack locational features and need to be registered against conventional paper maps for interpretation. Despite these limitations, the HEC-SAM software was a very advanced approach to spatial analysis, the capability was used in many Corps of Engineers studies, and the software was ported to additional hardware environments and adopted by others throughout the world.

As GIS technology evolved, Corps researchers and project managers decoupled the HEC analytical models from the primitive spatial tools and created data exchange links to other GISs. By the late 1980s, Fort Worth District was running damage models for the often flooding Upper Trinity River, using a HEC hydraulic model interfaced with GRASS to predict water boundary delineations and depths for specified flood events, plus a structure damage model to predict building damage potential from these specific flooding events. Output from these models (then translated to ARC) was used by the Corps and local governments to help plan flood control strategies and projects (Walker 1993).

Building on the Fort Worth District efforts (and using the same contractor), Omaha District linked a suite of HEC models to GIS databases for their Readiness Management System (RMS) which was designed to help the Corps of Engineers and local governments understand the impacts of significant water releases from (or failures of) reservoirs and dams. In 1992, Omaha District conducted its first exercise of RMS by simulating failures of the Oahe Dam in South Dakota, along the Missouri River. Both CRREL and CERL supported this exercise and have since combined talents to create an RMS decision support capability (Frederickson 1994).

In autumn 1986, the Rock Island District, in cooperation with CERL, installed the first district implementation of GRASS on a MASSCOMP computer. One of the earliest applications of this technology involved an evaluation of dredged material placement site selection at a frequently dredged area near Keithburg, IL, on the Mississippi River. A National High Altitude Photography (NHAP) color infrared image was used as a base for this site selection model, in combination with soils, aquatic and terrestrial habitat, and stream depth data (Abrahamson 1987). An effective model was developed which promised to reduce field requirements for the interagency teams that made final site evaluations, but the model never evolved from prototype to full application. However, Rock Island District found many other applications for the technology and continued to build upon this early GIS start to develop a robust set of GIS tools (later including ARC), data, hardware, and expertise. The district GIS expertise and capabilities proved invaluable during the 1993 Midwest floods, when Rock Island District, with support from the Remote Sensing/GIS Center, CERL, and WES, provided mapping services to assess flood risks and damages and to help focus both district and Federal Emergency Management Agency (FEMA) efforts.

Starting in the late 1980s and continuing through the 1990s, hydrologists at the Detroit District have gathered diverse types of lake shoreline data into a common database called the Great Lakes Shoreline GIS. Water depth, historical shoreline, topography, bathymetry, land use, and socioeconomic data, along with imagery at multiple scales, have been gathered through this multistate and multinational effort to help understand the effects of current and proposed human activities on lake coastal zones and the potential effects of lake level variations on coastal zone land uses. This Great Lakes effort involves some daunting GIS technology challenges. Gathering, integrating, storing, and efficiently retrieving and exchanging the massive data for this vast region pushes even the fast-moving boundaries of scanning, storage, software access, and networking technologies. Additionally, system users need to shift focus dynamically from local to regional to multiregional concerns, forcing interim programming approaches to still unsolved spatial data generalization problems. Other Great Lakes Districts (Chicago, St. Paul, Buffalo) participate in this study, and the Remote Sensing/GIS Center and CERL have teamed efforts to develop a decision support capability for district staff using this system.

The Corps of Engineers waterways and wetlands permitting program provides another important GIS application area. To process applications for activities requiring permits (such as building a pier or draining a wetland), several districts partnered during the 1980s in the development of the Regulatory Analysis Management System (RAMS). Running primarily in UNIX workstation environments, RAMS provides permit application tracking functions and links to relevant databases. Permit evaluations inherently involve an analysis of spatial issues, such as a delineation of wetland boundaries, and regulatory decision makers can benefit from GIS capabilities, provided the use of these capabilities doesn't slow the permit evaluation process. But building a GIS database is often time-consuming. To avoid delaying permit evaluations, some districts have prebuilt spatial databases in areas with frequent applications, and other districts have tested aircraft-mounted video collection procedures to obtain data quickly on current conditions which can be easily added to spatial databases. Researchers from CERL, TEC, and the wetlands research program at WES (Waterways Experimental Station) have worked with district regulators to link RAMS to GIS and to help develop spatial analysis models for permit application evaluation and performance monitoring (Peterson 1994).

LUMS for Camp Lejeune

Like the Army, the Marines Corps' training is land use intensive, and continuous military pressure on the land, coupled with increasing environmental constraints, gave rise to similar interest from the Marines in GIS for land use planning. Marlo Acock, the civilian natural resources management officer at Marine Corps Headquarters, became an early champion for the potential of a technology based system to help improve Marine Corps land use planning. Camp Lejeune in North Carolina was selected as the pilot site for this concept, which was titled the Land Use Management System (LUMS). Staff from the Naval Civil Engineering Lab (NCEL), which has a technology infusion role for the Navy and Marines similar to the role of CERL for the Army, were recruited to help define the technical requirements for LUMS (Foresman 1993).

In 1983, Tim Foresman of NCEL assembled a team to develop specifications for an eventual LUMS procurement. Over the next year, this team conducted comprehensive evaluations and bench tests of available GIS technology, leading to the eventual selection of ESRI's ARC/INFO software. Once ARC/INFO was selected, work was initiated on two major applications: a weapons firing and maneuver area scheduling system and a natural resources inventory and mapping database.

The U.S. Geological Survey's 7 1/2 minute quadrangle was used as the spatial base for LUMS. This same 1:24,000 scale was the standard widely used by CERL for Army installations, although there is currently a shift towards the 1:12,000 quarter quadrangles used by USGS for digital orthophotoquads, as these digital rectified photographic products, where available, provide an excellent base for natural resources management, digital elevation model production, and various interpreted themes.

LUMS was originally conceived and continues to be implemented as a multiuser, multioffice system that blends together many Camp Lejeune land users and land managers with a common spatial database and a common computing environment. While LUMS was originally fielded on Prime minicomputers, Camp Lejeune staff has made several hardware upgrades, and the system now supports more than a dozen nodes based upon workstations from Sun Microsystems.

In 1987, Camp Lejeune hired a GIS specialist, Rick Slader, to coordinate system usage and help build applications. Slader's diligent efforts made a success of the troubled range scheduling application, and then he tackled the complex communication and server requirements necessary for remote site usage. Other LUMS applications, supported by Slader and domain experts, include a land acquisition EIS, siting of a new fire station and PX (on-post store), and management of wildlife and forestry programs. Slader and associates have also implemented an aggressive training program to help overcome some technology resistance of Camp Lejeune staff.

Certainly, LUMS suffered from some GIS implementation problems: long learning curves for staff to acquire software competence, a shorter than anticipated lifespan for the original minicomputer host hardware, and problems with the custom range scheduling component of the software. But the installation staff has successfully sustained its GIS initiative, upgrading systems components and plans as needed, and now enjoys a very effective and productive system. The Marine Corps and Camp Lejeune's corporate approach to LUMS planning and implementation have helped build an enduring framework for success (Cullis 1994).

Technology champions at Camp Lejeune enjoyed the early support and hands-on involvement of Headquarters Marine Corps and NCEL staff, creating both a push and a pull for the success of the technology. Despite the success of Camp Lejeune, the Marine Corps has been conservative in expanding GIS beyond this pilot. However, by the mid '90s other Marine Corps installations were in the midst of implementing LUMS-like systems (with some ITAM components).

Decision Support Systems

There were both failures and successes in implementing GIS technology on military installations, and clearly many lessons had to be learned in "getting smart" about future implementations. One consistent concern, regardless of the type of system or tasks to be accomplished on the system, was the long learning curve to become proficient. A significant investment was required to develop GIS expertise, and many installations struggled for months and years without achieving a sufficient level of expertise to benefit from the system's potential.

CERL tried fielding "experts" with the system who were either responsible for passing the hat to an on-staff designee or who would themselves migrate to an on-staff position. In the passing-the-hat case, however, the clock would be ticking on the process, and the hat was often still ill fitting when the laboratory expert headed home. Likewise, there were many hiccups in the staff migration process, as federal slots were often subject to cuts, freezes, delays, or competitive constraints.

Another approach to this learning curve issue was to isolate users from the complexities of the GIS system but still open the door to their specific application requirements. This approach involved the development of decision support frameworks that embedded GIS and other functions, packaged in a customized graphical user interface, with extensive on-line help facilities. This essentially became the dominant strategy for CERL (and many others) in the late 1980s and early 1990s.

In pursuit of this strategy, CERL researchers conceived and wrote a fast track developer toolkit called "Xgen" that captured a limited but critical subset of capabilities from the X-Windows graphical user interface (GUI) environment. X-Windows had become the dominant GUI for UNIX workstations, so it was a logical choice for a GUI tool source to be interfaced to GRASS. But X-Windows itself was complex and presented a learning obstacle to many programmers and GUI developers. Xgen, on the other hand, was easy to learn and did not require formal programming skills to implement. With Xgen, GIS users became developers, creating custom environments for their own or colleagues' specific applications (Poulsen 1992).

Once completed, Xgen was tested at Fort Lewis. While many other installations were still struggling with GIS basics, Fort Lewis had a staff of three GIS expert users, a custom-built dedicated GIS laboratory, and a history of numerous successful GIS applications. Within a week, all three of the Fort Lewis GIS experts had mastered the use of Xgen and created their own customized applications, each with a different look and feel and sound (some examples utilized voice options). These three applications—a radar tower siting analysis (Goran 1991b), a spatial database of historic settlements in the Fort Lewis area (Riche 1991), and a GIS generated map used by Fort Lewis trainers to identify sensitive locations and seasons that constrain training activities (Hansen 1992)—were packaged into a traveling GIS road show that was demonstrated at numerous meetings and workshops.

The concept of GIS embedded inside a decision support environment offered many advantages from both the GIS end users' and the laboratory developers' points of view. Users only needed to learn the specific GIS tasks pertinent to their day-to-day requirements, so that the GIS input to and output from these tasks would fit into a work flow of information management, collaborative working relationships, and report generation. Besides these user advantages, the design and development of decision support systems offered a rich field for various "value added" firms. On-line tutorials, linked data sets and software libraries, multimedia links, network connections for collaborative interactions, and embedded analysis models could all be incorporated into decision support offerings.

Staff at the Corps of Engineers labs vigorously pursued the integration of GIS and decision support technologies. Working originally with GRASS GIS, Xgen GUI, and database management technologies as the essential building blocks, CERL began to delve into a host of application specific decision support systems. Results included the following:

- XCRIS—The X-Windows version of the Cultural Resources Information System (Cole 1994). This decision support system provides a comprehensive work flow environment to DOD cultural resource managers, linking GIS, multimedia, database management, and CADD with a GUI interface. It was funded through the Legacy Resource Management

Program, which was initiated by Congress in fiscal year 1991 to insure protection of valuable natural and cultural resources on DOD lands. XCRIS was tested at Fort Riley, KS, then fielded at other DOD sites. Although Fort Riley staff previously had most of the XCRIS data and system component elements, integrating the pieces made a significant difference for the users. They were finally able to investigate adequately and still respond quickly to requests for actions by military trainers or installation facility managers.

- LCTA System—Land Condition Trend Analysis is a program, developed by CERL as part of ITAM, for inventory and monitoring of military lands. The LCTA System is the suite of software capabilities developed for analysis and management of the field collected data (Anderson 1995). The system is primarily hosted on PCs using standard commercial DBMS software packages, but usage of the system is linked to GIS capabilities and several ancillary products (such as soil erosion status maps or land capability maps) are generated with a GIS. Field sampling sites are also selected using satellite imagery, soils data, and a specially developed GIS model (Warren 1990). CERL, Colorado State University, Oklahoma Biological Survey, and others have developed a variety of GIS applications models that link LCTA and GIS software into decision support environments.
- TRAINER—This system (also developed by CERL and the University of Illinois, as part of ITAM) was designed to provide a resource for military training planners (Johnston et al. 1991). It includes spatial data sets providing both land feature and land use data on installation training lands, plus a variety of models and options for selecting training routes and scenarios, which balance least-cost environmental damage and the most desirable training option. In 1994, TRAINER was tested at Fort Stewart, GA, as a primary link between the land uses (training planners) and land managers.
- PRISM—Also funded by the Legacy Resource Management Program, PRISM represents an attempt to provide a fully-integrated resource management planning decision support system to support natural and cultural resource management requirements such as forestry, endangered species habitat management, and archeological site management. The system was custom developed for Fort Polk, LA, with extensive staff input and considerable custom database development and integration (Majerus 1994).
- ASAN—The Assessment System for Aircraft Noise was originally developed by Armstrong Laboratories at Wright Patterson Air Force Base in Ohio, to provide an integrated decision support framework for evaluating the impact of aircraft flight routes on the environment (primarily humans, animals, and structures). ASAN version 1 was tested at several Air Combat Command installations, and Armstrong Labs and CERL are partnering to develop a next generation of this system (Smyth 1991).

When developers fully coordinate with users, these decision support systems (DSS) deliver a valuable set of capabilities that link dispersed data sources into useful decision environments with greatly reduced learning curves. However, DSS are development intensive in two respects: Developers must provide DSS functionality for each type of anticipated application, and the required data sets must be populated for these applications to be successful and meaningful. At some point, sophisticated users find the DSS environment unduly restrictive and need to learn to interact directly with the DSS building blocks—the GIS, the DBMS, the GUI, the multimedia system (if separate), and the analytical models. Users who stay within the DSS boundaries often find they want to perform some function which is easily supported by the underlying software tools but not supported by the DSS.

One of the concepts of CERL in delivering DSS technology was to provide tiers of capabilities, to support different levels of users at a site or within an organization. The novice users or those who used the system infrequently would primarily stay within the DSS boundaries. More expert users would stretch the DSS boundaries by learning the GUI toolbox and enough about the underlying tools to add on to the DSS, if needed, to support the first tier users. These expert users might also just bypass the DSS and go straight to the underlying software and data. System developers represent the third tier of users who might add to these building blocks or even swap component software or data sets. This concept was highlighted in a *GIS World* article by Jim Westervelt (1990b), who was one of the original developers of GRASS and principal architect of this CERL DSS effort.

While these CERL DSSs were originally developed with public domain GRASS, they are being adapted to run (unchanged from the user's perspective) with commercial GIS software from ESRI, Intergraph, LAS, and other vendors, and the GUI and DBMS are also being changed in some cases to accommodate user configurations or preferences.

Tri-Service Spatial Data Standards

As GIS technology became more widely implemented in DOD, many technology proponents and users raised concerns about data standards. In response to these concerns, the Tri-Service CADD/GIS Technology Center undertook the ambitious effort to develop the Tri-Service Spatial Data Standards (TSSDS), to include a data dictionary, standard schema, and common symbology for use in all three services. The TSSDS efforts are being coordinated with the Defense Mapping Agency and with other federal agencies through the FGDC. A new FGDC working group, focused on facilities, is enhancing coordination of the TSSDS throughout the federal government and with state and local governments. The Tri-Service Center is also working with the Defense Information Systems Agency (DISA) to incorporate some of the data elements into the Defense Data Repository System (DDRS).

In developing the TSSDS, the Tri-Service Center staff attempted to capture various data schemes already in use within the services and to extend these to meet requirements across the services. In the Army, a standards effort was already underway, attempting to link master planning and environmental data through an exchange of GIS themes. This effort, called Master Planning Environmental Overlays, had been initiated by Army headquarters planning proponents and was incorporated into the TSSDS. Schemes from Edwards Air Force Base, Patuxent River Naval Air Station, Camp Lejeune Marine Corps Base, Fort Bragg, Fort Lewis, and others were used to help formulate the draft TSSDS.

The TSSDS was designed to be implemented on the most commonly used systems in the three services: Intergraph CADD and Modular GIS Environment (MGE), ESRI's ARC products, GRASS, and Autodesk's AutoCAD, with the use of either Oracle or Informix DBMS on either DOS or UNIX platforms. Formal tests of TSSDS are underway or planned for each of these systems and environments. A draft version of the standard was distributed to more than 1,000 sites in January 1994 (Smith 1994), and revisions to the standard continue through a sophisticated CD-ROM distribution of the software and comments solicitation.

Assessing GIS Implementation Success

In the spring of 1994, Brian Cullis, an Air Force Major and then graduate student in the Department of Geography of the University of South Carolina, conducted a comprehensive assessment of field-level (installation) organizational GIS programs across DOD. The aim of his doctoral thesis work was to identify and analyze the factors encouraging (or inhibiting) successful individual adoption of organizational GIS resources. The Tri-Service CADD/GIS Technology Center provided sponsorship for this work.

With help from DOD labs, service headquarters, and Corps of Engineers Districts, Cullis first sought to define the population of known, field-level organizations where a GIS investment has matured to at least a self-reported degree of effective use. A total of 38 organizations across

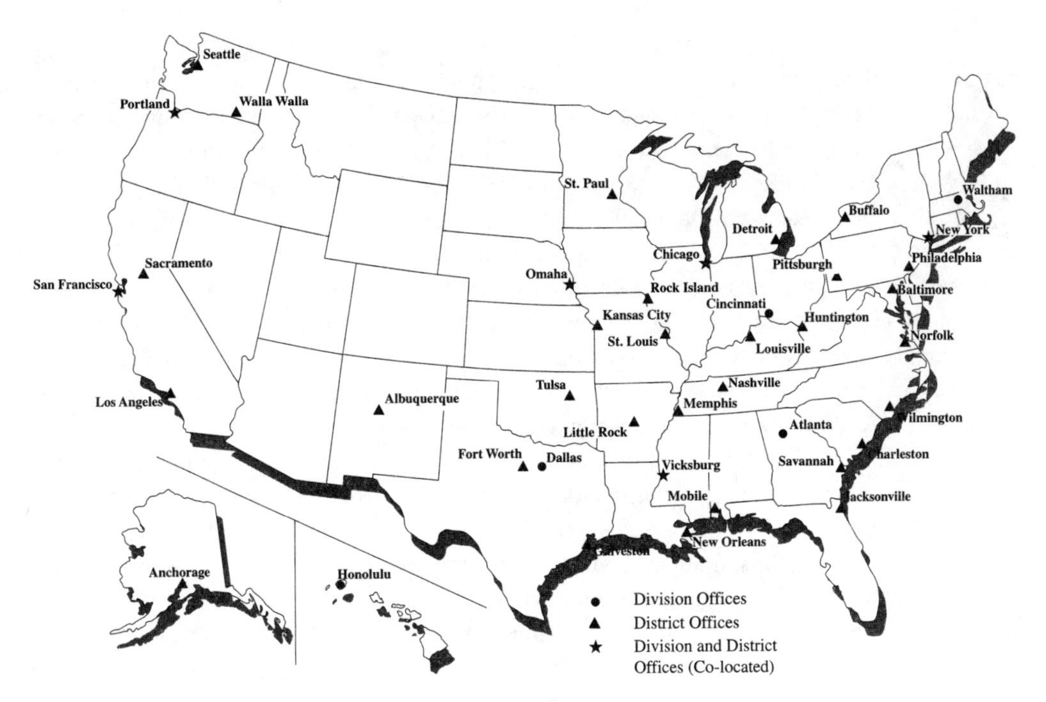

Figure 12.3 Corps of Engineers Districts and Divisions.

the services were identified (Figure 12.3). At each of these sites, Cullis administered personal interviews and quantitative surveys to organizational administrators, GIS program managers, and direct and indirect users of the GIS. While analysis of the data is still underway, several noteworthy trends and individual GIS model parameters have been defined.

In his analysis of factors affecting individual GIS adoption success, Cullis considered 12 key variables:

1. Organizational support for GIS
2. Ease of GIS mission integration
3. Benefits to GIS user
4. GIS linkages to other spatial data processing systems
5. Extent of GIS training
6. User confidence in GIS database
7. GIS use access
8. User knowledge of related spatial technologies
9. Facilitators to GIS adoption
10. GIS hardware procurement
11. Spatial data procurement
12. Relative advantage to GIS use

These influence factors were compared to three dependent variables: (1) Extent of GIS Use, (2) GIS User Satisfaction, and (3) Combined Extent of GIS Use and User Satisfaction. The major predictors of Extent of Use were found to be GIS Use Access, Extent of GIS Training, and Benefits to GIS User, while the major predictors of GIS User Satisfaction were facilitated by Benefits to GIS User, User Confidence in GIS Database, GIS Use Access, Extent of GIS Training, and Organizational GIS Support. Combined Extent of GIS Use and Satisfaction was best predicted by Benefits to GIS User, GIS Use Access, Extent of GIS Training and Organizational GIS Support.

GIS program managers and administrators were in remarkable agreement in their perceptions of the key obstacles hindering their success with GIS technology. In order, they listed (1) education, (2) manpower stability, (3) funding stability, (4) technical issues, (5) organizational politics, and (6) quality of the spatial database. All of the perceived obstacles were components of the factors determined to be significant in GIS use and satisfaction.

More than 60% of the sites surveyed implemented GIS technology with no increase in organizational manpower, and few installations developed structured implementation plans with specific objectives in advance of system acquisition. From this study, it is clear that GIS implementation on military installations could significantly benefit from better planning and a stronger organizational commitment from the implementing organizations. Of particular concern is adequate manpower allocation and training. At several installations contacted by Major Cullis, turnover or reassignment in the trained staff (or insufficient training in the first place) had resulted in abandonment or neglect of the GIS.

To improve success in individual adoption and organizational benefit, GIS technology cannot be viewed as another tool but rather as a process which involves critical social components including manpower, training, and a sustained organizational commitment to change (Cullis 1995).

GRASS Roots Grow, Blades Wither

After the initial release of GRASS software, the community using GRASS expanded rapidly, with the primary expansion being outside the federal agencies. On university campuses in North America and elsewhere, faculty and graduate students were discovering that GRASS was available at no cost, provided a wide range of GIS functionality, came with full access to source code (even a programmer's manual) and for public domain code, was relatively well documented and dependable. Segments of the architecture and engineering community also began experimenting with GRASS, often as a first low-step cost introduction to GIS technology. A 1986 award to the CERL GRASS developers from the Urban and Regional Information Systems Association (URISA) recognized the user friendliness and functionality of GRASS and helped draw attention to the software (Westervelt 1986).

Original distribution of GRASS outside of DOD was handled by letter request with nine-track tape as the distribution media. However, as letter traffic increased, CERL sought other options to answer the mail on public access to its software product. As early as 1986, a National Aeronautics and Space Administration (NASA) funded remote sensing commercialization program (the Space Remote Sensing Center) at Stennis Space Center in Mississippi offered to help with distribution. The Space Remote Sensing Center hosted the 1987 GRASS-GIS Users' Conference, began providing training courses and workshops for GRASS, created an interactive multimedia tutorial for GRASS, and also provided code distribution and support services. The Space Remote Sensing Center was operating as one of NASA's commercialization centers, and promoting a GIS that worked directly with imagery seemed an appropriate path to help broaden the user base for NASA's imagery products. In the next few years, several other private firms and academic institutions, hoping to find a niche in the expanding GIS marketplace, decided to offer GRASS related products, training, integration, database creation, and other services.

Starting in 1987, CERL also operated a GRASS Information Center, which set up shop to serve customers with information, publications, videos, and meeting coordination. Center tasks included the publication and distribution of the *GRASSClippings* newsletter (also started in 1987); coordination of the annual GRASS-GIS Users' Conferences; coordination and information distribution on training workshops; and management of GRASS related materials. These tasks were transferred to the Open GRASS Foundation (OGF), and the center at CERL closed, after the January 1993 signing of the Cooperative Research and Development Agreement (CRADA) between OGF and CERL. Later, OGF passed these responsibilities on to Rutgers University.

In 1990, CERL experimented with a file transfer protocol (FTP) hosting environment (Westervelt 1990), so that Internet literate explorers could obtain latest versions of software code contributions to GRASS. For academics and for the few Internet experienced government staff, this access provided an important resource. Not only was source code easily obtained (although unsupported), the Internet repository for GRASS also provided a place for software developers to present their new GRASS programs for testing and review. Because GRASS offered both source code and a growing user community, many students and professors elected to test their ideas and algorithms by writing extensions to GRASS. These contributions were encouraged and sometimes evaluated by Dr. Fred Limp of the University of Arkansas, who authored the column "Growing GRASS" featured in each issue of *GRASSClippings.*

This two-way exchange of code led to the next phase in exploiting the Internet, the on-line exchange of ideas (Knauerhase 1992). In late 1991, two Internet e-mail forums were established to serve the GRASS programmer (grassp) and GRASS user (grassu) communities. The forums were established on the mail exploding principle. A user sends mail to the forum address and that message is then "exploded" to the addresses of all forum subscribers. This principle results in a low maintenance communication mechanism but a high volume of traffic for all subscribers. The forums, like the FTP site, proved very popular, and soon hundreds of subscribers were exchanging messages and ideas about GRASS and GIS topics.

Despite a following of loyal users both inside and outside DOD, CERL struggled with the complexities of maintaining a major software system and the appropriateness of this effort in light of an increasingly robust commercial marketplace. While GRASS excelled in certain capabilities, those military installations with GRASS software generally requested better cartographic output, tighter database coupling, and more polished releases.

The Installation Spatial Technology Advisory Board (ISTAB), established in 1993 by CERL and the Army's major commands, was intended to provide these users a voice through which to communicate their GIS requirements. At the first meeting, in 1993, CERL staff asked the ISTAB members: Should we continue with public domain GRASS development? At that time, the ISTAB answer was an unequivocal yes. Two years later, CERL staff simply announced plans to discontinue GRASS software development, and the ISTAB members unequivocally concurred. In the intervening years, several of the installations with GRASS had switched to and/or added a commercial GIS.

To help installations transition to commercial GIS, CERL framed partnership agreements with ESRI, Intergraph, and a GRASS third party developer, Logiciels et Applications Scientifiques, Inc. (LAS) from Montreal, Canada. The ISTAB strongly endorsed these agreements and the shift in focus to commercial software, which allows CERL's technical staff to focus solely on military installation applications. Rich Manning of Dugway Proving Ground, UT, ISTAB chair during this 1995 meeting, wrote in summary: "Commercial-off-the-shelf software products are leading edge technology and applicable to many natural resources and environmental applications. . . . CERL can provide a support role to installations during conversion from (public domain) GRASS to COTS" (Manning 1995).

Although CERL is cutting loose from GRASS, the Internet forums continue to thrive, and many academics and federal researchers continue to rely on and expand the public domain software base. And LAS has done much to add value to the public domain base, with their commercial offering, GRASSLANDS, released in November 1995, providing an impressive World Wide Web smart Windows 95 and Windows NT implementation.

GIS and OGIS Open

The goals of OGF rapidly evolved in the early 1990s from a singular focus of serving and expanding the community of GRASS software users to a broader agenda of opening doors in the entire geodata technology industry. *GRASSClippings* matured from a newsletter to the *Journal of Open Geographic Information Systems,* and the organization changed its name from the Open GRASS Foundation to the Open GIS Foundation. In the autumn 1993 issue of *GRASSClippings,* David Schell, the executive director of OGF, explained that "In changing our name, we are emphasizing our commitment to identifying and promoting the basic platform technologies that define and drive the concept of Open GIS" (Schell 1993). In fact, OGF's expanded emphasis on Open GIS had evolved directly from members of the GRASS community, who had become very concerned about data access across different software environments and had begun debating and proposing approaches to interoperability.

In an opening plenary session address to the March 1993 GRASS GIS Users' Conference and Exhibition in Reston, VA, Kenn Gardels of the University of California, Berkeley, presented a tiered concept of interoperability, from simple data transfer to full access to data and tools from heterogeneous software environments (Gardels 1993). In this tiered model, Gardels was defining "levels of seamlessness" between various vendors products and GRASS, but the issue, as he presented it, was clearly not GRASS specific. In fact, dynamic sharing of data between different software environments is the core of Open GIS and is equally relevant to users of any GIS software environment (Buehler 1994).

CERL had established, in 1993, a new research effort focused on exploring emerging geodata models. One of the driving concerns in initiating this effort was bridging the data access and analysis gaps created by multiple GIS software and hardware environments within a single military installation, between multiple installations within a single major command, and between military installations and their neighboring communities. The Army was planning a new installation support module that was a key element of the Sustaining Base Information System procurement, and this module was intended to draw data from existing spatial data systems to support a suite of comprehensive master planning applications. At the core of this initiative was the need for dynamic interoperability, and CERL researchers were hoping to address this need with their new research effort.

The approach pursued by CERL in exploring this interoperability requirement was very different from the approach in designing and developing GRASS. Rather than immediately writing code and testing pilots, CERL tasked the University of Arkansas, through a contract, to initiate a consensus process, involving as many voices as feasible from the geodata processing industry, including current and emerging future GIS vendors. The University of Arkansas and OGF vigorously pursued this tasking, and a consensus process was soon defined in a series of meetings with an ever expanding list of persons attending. Emerging from this process was the definition for an interoperability specification, the Open Geodata Interoperability Specification, or OGIS.

In a *GIS World* article by CERL's Kurt Buehler, OGIS is defined as "whatever is necessary to realize interoperability among services that access geospatial databases and processes" (Buehler 1994). OGIS planners hope to provide end users and developers (working in an OGIS compliant environment) to draw upon any data sources or geodata process, either locally or across a network, without interruption of work flow, to complete their application tasks. This OGIS vision is closely aligned with the goals of the National Spatial Data Infrastructure (NSDI) to provide widespread access to spatial data and tools.

An OGIS design document has been completed and several organizations plan to participate in the testbed phase of this effort. NOAA, for example, is leading an OGIS tested effort related to an interagency study of wetlands and drainage in central Florida, and the University of California Berkeley will be leading another tested effort involving multiple agencies and databases relating to environmental and coastal dynamics in the San Francisco Bay Area. Intergraph and the University of Arkansas plan another test (Jim Farley of the university's Center for Advanced Spatial Analysis is the Tested Coordinator), as does the National Geographic Society and the Mitsubishi Corporation. The Army will be testing OGIS to draw from existing GIS, CADD, and other databases, both on post and in neighboring communities, that have data or capabilities to help evaluate stationing options, carrying capacity, and investment strategies.

Deepening GIS Roots within DOD

While many DOD installations have already implemented GIS, the Air Force Air Combat Command (ACC) is just getting underway with an extensive effort to field GIS at 18 air bases, aided by CERL staff, and armed with lessons learned from Major Cullis' study. One of the major applications of the technology will involve aircraft noise impact assessment, using the ASAN capability developed by Armstrong Labs, but these GIS will also be used for environmental and master planning, natural and cultural resource management, and general installation support. The ACC GIS implementation process is being strongly guided and funded from the major command level, and system proponents are casting their nets wide to insure proper coordination and full implementation planning. Costs for development of installation specific GIS databases at multiple sites run high, however, and will probably require staged implementation planning after the fielding of ASAN, which will be run by on-post environmental planners but will examine impacts along regional flight routes rather than on post data.

Many other organizations and installations within DOD are actively implementing GIS technology for facility and environmental management functions. The Major Range and Testing Facility Bases Environmental Coordinating Committee (MECC) has for several years been involved in a cross-service effort to implement GIS in support of testing range requirements. Each of the six installations that comprise the MECC (Edwards Air Force Base, CA; Eglin Air Force Base, FL; Patuxent Naval Air Test Center, MD; China Lake Naval Test Center, CA; White Sands Missile Range, NM; and Aberdeen Proving Ground, MD) has an active GIS capability supporting environmental requirements and base planning, and these bases actively collaborate on applications and approaches.

The Fort Lewis Northwest GIS Center's support for other west coast installations may prove a model for DOD. Hohenfels Training Center in Germany is providing similar GIS services to multiple installations in Bavaria. At smaller installations, the work force for environmental and base planning functions is insufficient for on-site GIS implementation, but a GIS database can be managed and maintained by a supporting center, preferably linked through electronic collaboration tools with the installation staff.

GIS Supports the Installation Enterprise

While GIS implementation throughout most of the DOD has been focused within specific stovepipes (e.g., within environmental, tactical military units, master planning, or safety organizations), there are several 1990s initiatives that involve a comprehensive, basewide approach to GIS as a component of installation management (Figure 12.4). Certainly, the Tri-Service Spatial Data Standard is an attempt to harmonize data fields across all DOD installation management stakeholders, and DOD laboratories are tackling the technical questions of data interoperability

Figure 12.4 Military installation GIS for facility and environmental applications.

and user collaboration necessary to support enterprise-wide use of spatial information across military installations, between installations and their neighbors, and between the various installations within a major command or service. However, expanding from stovepipe to enterprise, from project-based to databased GIS implementation requires installation and service-wide planning and effective and consistent management support.

Thus far, the track record for such planning and management support is mixed. In the Army, the Sustaining Base Information System (SBIS) was conceived as a comprehensive framework for the implementation of systems technology in direct support of base operations. Initially, spatial technology was omitted from the specifications, but the inclusion of target capabilities for base planning and training range scheduling resulted in a late addition of spatial capabilities into the scope of work. SBIS is designed to rely on data already contained in existing installation systems, but presumably, if SBIS is fielded service wide, these existing or legacy systems would eventually be replaced. The SBIS procurement action has suffered some of the usual woes of large government contracts, such as delays, reductions in funds, and decreasing scopes of effort, leaving planners at SBIS receiving installations uncertain about appropriate future directions.

But while service level efforts struggle, many installations are making great strides in broadening their spatial technology operations. Aberdeen Proving Ground, bordering the Chesapeake Bay in Maryland, has successfully expanded a GIS program, initially focused on installation restoration activities, to support a range of environmental and engineering functions. Fort Belvoir, VA, expanded its automated mapping/facility management system to environmental and other base support functions, replacing stovepiped environmental systems with a more comprehensive installation standard.

Edwards Air Force Base, a flight test center in the California desert north of Los Angeles, has evolved a very broad GIS implementation vision that emerged from efforts to develop a dynamic framework for its base comprehensive plan. Conceived as a living plan by engineering and planning staff, the Edwards GIS was also championed by Bob Wood and others within the Environmental Division. Applications such as the use of GIS to model subsurface features and flows from installation restoration drill logs, to reduce the total number of drill holes required, are now easily paying back Edward's original investment. The flight test center is setting the pace for installation-wide approaches for GIS implementation within DOD (Edwards Air Force Base 1995).

Conclusion

GIS has already become a normal and expected capability within DOD. While lessons are still being learned as to how to plan properly for implementation and maintenance of these systems, hardware advancements are continuously reducing costs and improving performance. Software advances are helping to shorten learning curves, extend functional capabilities, and provide more robust links to analysis models, report writers, and other tools required to fulfill specific tasks. Standards for technology acquisition (NAVFAC CADD II), database elements and definitions (TSSDS), data documentation and exchange (SDTS), and data and analysis interoperability (OGIS) all promise to facilitate expanded use and sharing of data and technology, and ambitious programs to install fiber optics cables (such as the Army's Sustaining Base Information System) are providing the network backbone to realize these promises.

The metadata standards element of SDTS has inspired the Corps of Engineers to undertake an ambitious program to fully document data holdings and provide network accessible metadata servers. Other elements within DOD will likely follow the Corps' lead in providing institutional guidance for spatial data holdings and technology implementation, now provided in Engineering Circular 1110-1-83. Major Cullis' implementation study promises to guide DOD managers and policy makers towards better planning and more consistent policies for technology implementation.

With new initiatives like the Tri-Service Spatial Data Standard and the next generation HEC models, DOD continues to maintain a strong leadership in GIS technology development. But technology developers are no longer lone voices in the laboratories, trying to market their vision and link with technology champions. Rather, DOD managers have embraced the technology; created centers and coordinators and boards and councils; instituted standards, protocols, and procedures; documented and digitized their facilities; and are aggressively charting an agency-wide course into the fast lanes on the spatial information highway.

Bibliography

Abrahamson, C. A., A. J. Bruzewicz, and M. O. Johnson. 1987. "Use of Geographic Information System (GIS) to Improve Planning for and Control of the Placement of Dredge Material." Unpublished paper presented at the GRASS-GIS User Conference, Stennis Space Center, MS.

Anderson, A., W. Sprouse, D. Kowalski, and R. Brozka. 1995. "Land Condition Trend Analysis (LCTA) Data Collection Software Users Manual," Version 1.0. CERL ADP Report 95/13. Champaign, IL: U.S. Army Construction Engineering Research Laboratories.

Bruzewicz, A. J. 1992. "U.S. Army Corps of Engineers Reaps Many GIS Rewards." *GIS World* (March), 60–63.

Buehler, K. 1994. "OGIS Augments Data Transfer." *GIS World* 7 (10): 60–63.

Chief of Engineers Environmental Advisory Board. 1987. *Report of the 41st Meeting of the Board, Environmental Data and Its Use in Decision Making.* New Orleans, LA: Unpublished.

Cole, J. M. S., and L. Corbe. 1994. "User's Manual for the Cultural Resources Information System (CRIS)," Version 2.0. CERL ADP Report EC-94/29. Champaign, IL: U.S. Army Construction Engineering Research Laboratories.

Costanzo, D. 1990. *Proceedings of the Geographic Information System (GIS) Single Discipline Task Group Meeting.* Vicksburg, MS: CADD Center, Information Technology Laboratory, Waterways Experiment Station.

Cullis, B. 1994. "GIS Implementation Success Study." Presentation to the Installation Spatial Technology Advisory Board. Urbana, IL.

Cullis, B. 1995. *An Exploratory Analysis of Responses to Geographic Information Systems Adoption on Tri-Services Military Installations.* Tri-Service CADD/GIS Technology Center, Contract Report CADD-95-1. Vicksburg, MS: Waterways Experiment Station.

Edwards Air Force Base, Environmental Management Division, 1995. "IRP Activities at Edwards Air Force Base, Kern County, CA." Champaign, IL: U.S. Army Construction Engineering Research Laboratories.

Foresman, T. W. 1993. "Tactical GIS Helps Marines Preserve Natural Resources at Camp Lejeune." *Geo Info Systems* (May): 44–47.

Frederickson, K. E., J. D. Westervelt, and D. M. Johnston. 1994. "A Geographic Information System/Hydrologic Modeling Graphical User Interface for Flood Prediction and Assessment." CERL Technical Report EC-95/03. Champaign, IL: U.S. Army Construction Engineering Research Laboratories.

Gardels, K. 1993. "What is Open GIS?: *GRASSClippings,* 7 (1): 40.

Goran, W. D., W. E. Dvorak, L. V. Warren, and R. D. Webster. 1983. "Fort Hood Information System: Pilot System Development and User Instructions." CERL Technical Report N-154. Champaign, IL: U.S. Army Construction Engineering Research Laboratories.

Goran, W. D. 1991. "Interagency Steering Committee Coordinates Development of Public Domain GIS Software." *Federal Geographic Data Committee Newsletter,* 1: 6–7.

Goran, W. D., and D. Finney. 1991. "GRASS GIS Critical to Army's Land Management Program." *GIS World,* 4 (9): 52.

Goran, W. D. 1992. "What's the Contribution and Appropriate Role of Public Domain GIS Software?" *GIS World,* 5.

Hansen, T. 1992. "GRASS Provides Historical Overview of Ft. Lewis." *GRASSClippings,* 6 (1): 7.

Johnston, D. M., R. B. Olshansky, I. Lee, and L. D. Hopkins. 1991. "Development of Algorithms for Classification, Spatial Allocation, and Tradeoff Analysis for Training Land Management." Urbana, IL: Department of Urban and Regional Planning, University of Illinois. Unpublished paper.

Klesch, W. L. 1989. *Final Report of the Ad Hoc Advisory Committee for Geographic Information Systems, U.S. Army Corps of Engineers Environmental Advisory Board Meeting.* Denver, CO: U.S. Army Corps of Engineers.

Knauerhase, R. 1992. "OGI Promotes On-Line Exchange of Ideas." *GRASSClippings,* 6 (1): 1.

Lozar, R. C., R. C. Belles, R. Gauthier, and M. S. Larson. 1986. "Environmental Early Warning System User Manual." U.S. Army CERL Technical Report N-86/06. Champaign, IL: U.S. Army Construction Engineering Research Laboratories.

Maizel, M. 1990. "National Center for Resource Innovation Farm Preservation Program." *GRASSCLippings,* 4 (2): 1, 7.

Majerus, K., and C. Rewerts. 1994. "PRISM's Kajan Kaleidoscope for Integrated Natural and Cultural Resource Management at Ft. Polk, LA." Paper presented at the 9th Annual GRASS GIS Conference and Exhibition. Reston, VA.

Manning, R. 1995. "Memorandum for the Director, CERL. Subject: Summary of 28 November 1995 Meeting, Installation Spatial Technology Advisory Board (ISTAB)."

Martin, M. V., W. D. Goran, R. C. Lozar, J. M. Messersmith, M. S. Ruiz, and W. D. Westervelt. 1989. GRASS GIS Implementation Guide, U.S. Army CERL Special Report N-90/02.

Peterson, T., and R. Gebhard. 1994. "Wetlands Stewardship Through Application of Geographic Information System Technology." *CADD/GIS Bulletin,* 94 (1). Vicksburg, MS: Waterways Experiment Station.

Poulsen, C. 1992. "Using Xgen in GRASS 4.0." *GRASSClippings,* 6 (1): 21.

Riche, B. 1991. "Consolidated Environmental Constraints." *GRASSClippings,* 5 (1): 1, 19, 23.

Ruiz, M. S., and J. M. Messersmith. 1990. "Cartographic Issues in the Development of a Digital GRASS Database." CERL Special Report N-90/02. Champaign, IL: U.S. Army Construction Engineering Research Laboratories.

Schell, D. 1993. "The Open GIS Foundation: A Forum for the Promotion of Open Geographic Information Systems." *GRASSClippings,* 7 (2): 4–5.

Severinghaus, W. D., and W. D. Goran. 1991. "Integrating Land Management and Training Through ITAM and GRASS." *The Military Engineer*, 83 (544): 42–43.

Sliwinski, B. 1989. "Methodology for Performing Return-on-Investment (ROI). Studies for Implementation of GRASS on Military Installations." CERL Technical Manuscript N-89/25. Champaign, IL: U.S. Army Construction Engineering Research Laboratories.

Smith, H. L. 1994. "Tri-Service Approach to CADD/GIS Data Standards." *The Military Engineer*, 86 (565): 27–29.

Smyth, J., and N. Reddingius. 1991. "Assessing Aircraft Noise with GRASS." *GRASSClippings,* 5 (1): 1, 11.

Stephens, C. 1988. "The Corps of Engineers CADD Center." Presentation given to a Corps of Engineers CADD User Forum. Seattle, WA.

Stephens, C. 1992. "Tri-Service CADD/GIS Technology Center Charter." Vicksburg, MS: U.S. Army Corps of Engineers, Waterways Experiment Station.

Walker, S., T. Nelson, and T. Betancourt. 1993. "Comprehensive Floodplain Management: A GIS Approach." Paper presented at the 9th Annual GRASS GIS Conference and Exhibition, Reston, VA.

Warren, S. D., M. O. Johnson, W. D. Goran, and V. E. Diersing. 1990. "An Automated Objective Procedure for Selecting Representative Field Sample Sites." *Photographammetric Engineering and Remote Sensing*, 56 (3): 333–335.

Webster, R. D., R. L. Welch, and R. K. Jain. 1975. "Development of the Environmental Technical Information System." CERL Interim Report E-52. Champaign, IL: U.S. Army Construction Engineering Research Laboratories.

Westervelt, J., W. D. Goran, and M. Shapiro. 1986. "Design and Development of GRASS: The Geographical Resources Analysis Support System, Exemplary Systems in Government." *Urban and Regional Information Systems Association*, 167–180.

Westervelt, J. 1990a. "Release News: GRASS 3.1 on Internet." *GRASSClippings,* 4 (2): 2.

Westervelt, J. 1990b. "The Two Classes of GIS Users." *GIS World*, (October/November): 111–112.

CHAPTER 13

A Quarter Century of GIS at Oak Ridge National Laboratory

Jerome E. Dobson and Richard C. Durfee

Introduction

When Tim Foresman asked us to write about the role of Oak Ridge National Laboratory (ORNL) in the history of GIS, we sent a memorandum to about 50 colleagues requesting reports and articles from the early years. Within days, we were deluged with more than 70 items published in the 1970s and another 50 from the 1980s. Neatly stacked by year, the reports tell a fascinating story. The early stacks are filled with concise reports documenting comprehensive, fundamental developments supported by the National Science Foundation (NSF). Modeling, statistics, analytical cartography, and data processing dominate the titles. The middle years, roughly the Carter and first Reagan administrations, are filled with ponderous documents assessing all sorts of policies, plans, technologies, and resources. The tallest stack by far is 1980, the final year of the Carter administration. Later stacks focus on GIS addressing major national themes, such as acid precipitation and lake acidification, hazardous waste management, environmental restoration, and climate change. Every stack contains a few items of fundamental development funded by the popular issues of the day.

Jerome E. Dobson has been a lead scientist at the Oak Ridge National Laboratory for approximately three decades. During his career he has led the development of many ecological projects which required innovation in the new technologies of remote sensing, GIS, and environmental modeling.

Richard C. Durfee works as a research scientist and colleague of Dobson at the Oak Ridge National Laboratory. During his career over the past three decades, he has been engaged in the use of advanced technology for environmental applications.

Authors' Address: Jerome E. Dobson and Richard C. Durfee, Oak Ridge National Laboratory, P.O. Box 2008, MS 6237, Oak Ridge, TN 37831.

What strikes us most in reading the earliest documents is how current they sound. Consider, for instance, that GIS modeling, a forefront activity of today, was the primary impetus behind the laboratory's initial NSF-sponsored GIS effort starting in 1969 (Baxter et al. 1972, Meyers et al. 1972, Olson et al. 1972, Schuller 1972, Voelker et al. 1972), and the integration of statistics and GIS was a major theme (Durfee 1972). Actually, the term, GIS, though first introduced in 1964, was not extensively used until the late 1970s. Hence, the ORNL group was called the Geographic Data Systems Section, and the first comprehensive geographic data management system developed at ORNL was called the Oak Ridge Regional Modeling Information System (ORRMIS) (Durfee 1974). Many early publications reporting technical developments that are commonplace today are remarkable only because of their dates. Examples include perspective and isometric drawing of cartographic surfaces (Tucker 1973), integration of remote sensing and statistics in GIS (Durfee et al. 1975), raster-vector transformation (Coleman et al. 1977, Edwards et al. 1977), viewshed calculation (Tucker 1976), polygon intersections (Edwards and Coleman 1976), transportation routing models (Joy 1982), and true 3-D imaging (Levy et al. 1983, Olins et al. 1983). Many reports belie the current notion that GIS has failed to deliver on its promise of supporting research and analysis beyond inventory and mapping. Prominent examples include decision support to various National Energy Plans (Berry et al. 1978, Dobson et al. 1977, Dobson and Shepherd 1979, Honea and Hillsman 1979, Honea et al. 1979a), wilderness area designation (Voelker et al. 1979, Oakes and Voelker 1983), aesthetic impact analysis (Petrich 1984), and military air traffic management (Harrison et al. 1992, Hilliard et al. 1992).

This is our heritage, and we take great pride in reporting it here. First, we briefly describe ORNL, its mission, and its mode of operation to help you understand why we respond as we do to national needs and trends. Second, we summarize our vision of GIS as an evolving societal capability for representing and analyzing real world phenomena and processes. Third, we summarize the development and use of GIS technologies at ORNL in three eras—regional modeling and fundamental development (1969–1976), integrated assessment (1977–1985), and issue-oriented research and analysis (1985–1994).

Oak Ridge National Laboratory (ORNL)

Harold Meyer, the eminent urban geographer, once told a story about Oak Ridge. He was at the University of Chicago during World War II when the Manhattan Project was underway. Week by week, faculty members were whisked away to Site X "somewhere down south." He and other faculty members suspected what was going on, but they kept the secret through tacit acceptance of national security needs. "When the nuclear physicists disappeared, we knew they must be experimenting with an atom bomb," Mayer recalled. "When the nuclear engineers left, we knew they were going to build it. And when the regional planners left, we knew it was big."

ORNL is still a big place, and most first-time visitors are surprised at just how big. The reservation itself covers about 140 square kilometers including three main complexes still known by their wartime codes: X-10, Y-12, and K-25. Except for its Biology Division and a few miscellaneous functions, the national laboratory itself is synonymous with the X-10 site. All of this is owned by the Department of Energy (DOE) and run by Lockheed Martin Energy Systems, Inc. (LMES). About 4,300 full time employees, including 1,800 professional staff members, work at the laboratory per se.

We know we aren't federal employees because we don't get Columbus Day off, and we know we aren't truly private because our benefits plan didn't change in 1984 when the contract transferred from Union Carbide to Martin Marietta Energy Systems or in 1995 when Martin Marietta merged with Lockheed. To be precise, we are a Federally Funded Research and Development Center (FRDC), defined as an institution "operated, managed, and/or administered by either a university or consortium of universities, other not-for-profit or nonprofit organization, or an industrial firm, as an autonomous organization or as an identifiable separate operation unit of a parent organization." Nationwide, there are about 100 FRDCs including, for example, NASA's Jet Propulsion Laboratory, the Environmental Protection Agency's (EPA) laboratories, and DOE's other national laboratories. There are ten DOE national laboratories designated as weapons laboratories (e.g., Los Alamos National Laboratory), special purpose laboratories (e.g., National Renewable Energy Laboratory), or multipurpose laboratories (e.g., ORNL). Our GIS relationships with the other laboratories have included standards development, information exchanges, and highly successful collaborative efforts, one instance of which is reported here in the "Integrated Assessments" section, but friendly competition would not be unprecedented.

LMES is actually a subsidiary of the Lockheed Martin Corporation, established expressly to maintain a corporate veil between public and private research interests. In the final analysis, our allegiance is to ORNL, much as academic faculties claim allegiance to their college or university rather than to the state, church, or private foundation that owns it. Many of us take a rather highminded view of our responsibility to conduct research in the public interest regardless of what may be trendy or well-funded in the public arena. There is, however, no line item for us in the federal budget. We depend on funded projects, but we're not allowed to bid on competitive requests for proposals (RFPs). In other words, we're not supposed to compete with universities and private companies for research funds. So how do we support those highminded goals? By keeping close working relationships with many federal agencies that need our advice and technical assistance. We have a broad research agenda, and we continuously optimize between what we want to do and what those agencies will fund. The system has worked well for many years.

As for our atomic roots, the next time you hear a television news anchor refer to "the government's Oak Ridge nuclear facility," please keep in mind that the label is about 20 years out of date. After World War II, ORNL was administered by the Atomic Energy Commission and its mission was primarily nuclear. In 1974, however, the parent organization changed to the Energy Research and Development Administration (ERDA), and ORNL's mission broadened to include all types of energy. Since 1969, that scope has included major work in environmental issues and energy conservation. In 1977, the parent changed again to the Department of Energy. Today, only 11% of ORNL's total effort is related to nuclear technologies. Most of the work is for DOE, but a substantial portion of our funding comes from a host of other federal agencies through interagency agreements between them and DOE.

Of the 16 research divisions, at least four have substantial GIS activities underway. The oldest continuous effort is that of the Geographic Information Systems & Computer Modeling (GCM) group, headed by Richard C. Durfee in the Computational Physics and Engineering Division. The Environmental Sciences Division (ESD) maintains a substantial GIS facility devoted to environmental research, including environmental modeling and landscape ecology. The Energy Division was a close partner in the early development of GIS modeling capabilities and in conducting national assessments. Its GIS focus today is primarily on transportation modeling. The Health and Safety Research Division (HSRD) employs GIS to study relationships among environment, health, and risk. In addition, GIS is employed heavily in facilities management and environmental restoration activities by other groups on the Oak Ridge reservation but outside ORNL.

Evolving Vision of GIS

Many people think of GIS as a computer tool for making maps, but this simplistic definition vastly underestimates the complexity of the technology, as well as its actual and potential contribution to science (Dobson 1993b, c). Geographic and other spatial analyses require special data structures and tools to represent and investigate complex geographic relationships and phenomena involving Earth's surface. Even the most basic definition in common usage by the GIS community today would include "capturing, storing, checking, manipulating, analyzing, and displaying data which are referenced to the Earth" (Department of Environment 1987). Most formal definitions still reflect this limited focus of GIS as a spatial database management system, but popular usage has undergone a dramatic transformation. Many people now routinely include spatial databases in referring to the GIS for a specific place, and GIS often is used as a covering term for the collective technologies of computer cartography, computer graphics, digital remote sensing, spatial statistics, and quantitative spatial modeling when used in combination to support analysis of a site, a region, or even the entire Earth. Dobson, for instance, has explored the role of spatial logic in GIS and the joint potential of spatial logic and GIS for proposing and testing hypotheses and formulating theories (Dobson 1993b, 1993c, 1992a; Dobson et al. 1990).

A discussion of GIS can focus on three major themes depending on the interests of the reader. One perspective concerns the types of real-world problems or applications that GIS can assist in solving. A second involves the technologies and methods required to develop and use GIS systems including hardware, software, algorithms, and models. A third focuses on the many different types and characteristics of data that are handled by GIS systems including their content and the digital data structures used to represent them in a computer. As GIS technologies and multitudes of geographic databases have spread to the desktop in the last decades, metadata—information documenting the description, specifications, and quality of geographic data elements—have become very important. Good metadata are essential in determining fitness for each intended use, i.e., determining which applications can be accomplished while ensuring the desired quality of results and decisions made from those results.

This paper focuses primarily on the first and second themes mentioned above: ORNL's role in the development of GIS technologies and their application to real-world problems over the past two and a half decades. Over this time, hundreds of projects and tasks involving GIS have been carried out by several organizations at ORNL. It would be impossible to mention them all. We have focused on several of the larger efforts that show the diversity of applications and techniques. We offer examples in which GIS has proved useful in research and decision support. A section is included describing early GIS systems development and summarizing current systems capabilities. We have included an extensive bibliography for those who desire more information on GIS-related technologies and databases developed or used by ORNL scientists. The bibliography will serve as the primary acknowledgement and credit for the large number of groups and individuals involved over two decades. In addition, we recognize and appreciate the contributions of many other scientists and managers who promoted and supported the vision behind these efforts.

Evolution of GIS Technology at ORNL

An overview of the development of GIS technologies and applications at ORNL during the last 25 years is presented in Figures 13.1–13.3. The first figure presents the evolution of hardware platforms and software systems used in GIS development. The top portion of the diagram shows the migration from mainframes to minicomputers to PCs to workstations. Work is now proceeding with parallel processors and supercomputers at ORNL. The bottom part of the diagram portrays the use of in-house and external GIS software tools. When ORNL became involved with spatial applications, commercial GIS software did not exist. GCM staff developed all the algorithms and software systems, initially using Fortran on mainframes. This has evolved over time so that in the 1990s more of the software development is done using C language tools and various libraries along with application programming languages (e.g., ARC/INFO AML, Avenue, MapBasic, ProC).

As shown on the lower part of Figure 13.1, PC software tools started being used in GIS systems developed at the laboratory in the early 1980s. A number of PC-based high-resolution systems were developed in the mid 1980s for Department of Defense sponsors. Included were a vector/raster based system for facility and resource planning and a 3-D hydrogeochemical system for managing, analyzing, and displaying hazardous waste data. In the mid-to-late 1980s, full-scale commercial GIS systems matured with capabilities to store and process efficiently large volumes of data, controlled with user-friendly interfaces. The use of ORNL mainframe resources dropped significantly. The GCM group and other scientists at ORNL began to use these commercial systems (e.g., ARC/INFO, Intergraph, and ERDAS) on minicomputers, PCs, and UNIX-based workstations. A few vendors began to provide robust programming languages and macro capabilities for their software. This allowed different packages to be combined and commercial toolsets to be integrated with in-house software. This has resulted in more generalized data management systems in which GIS is one of several capabilities provided.

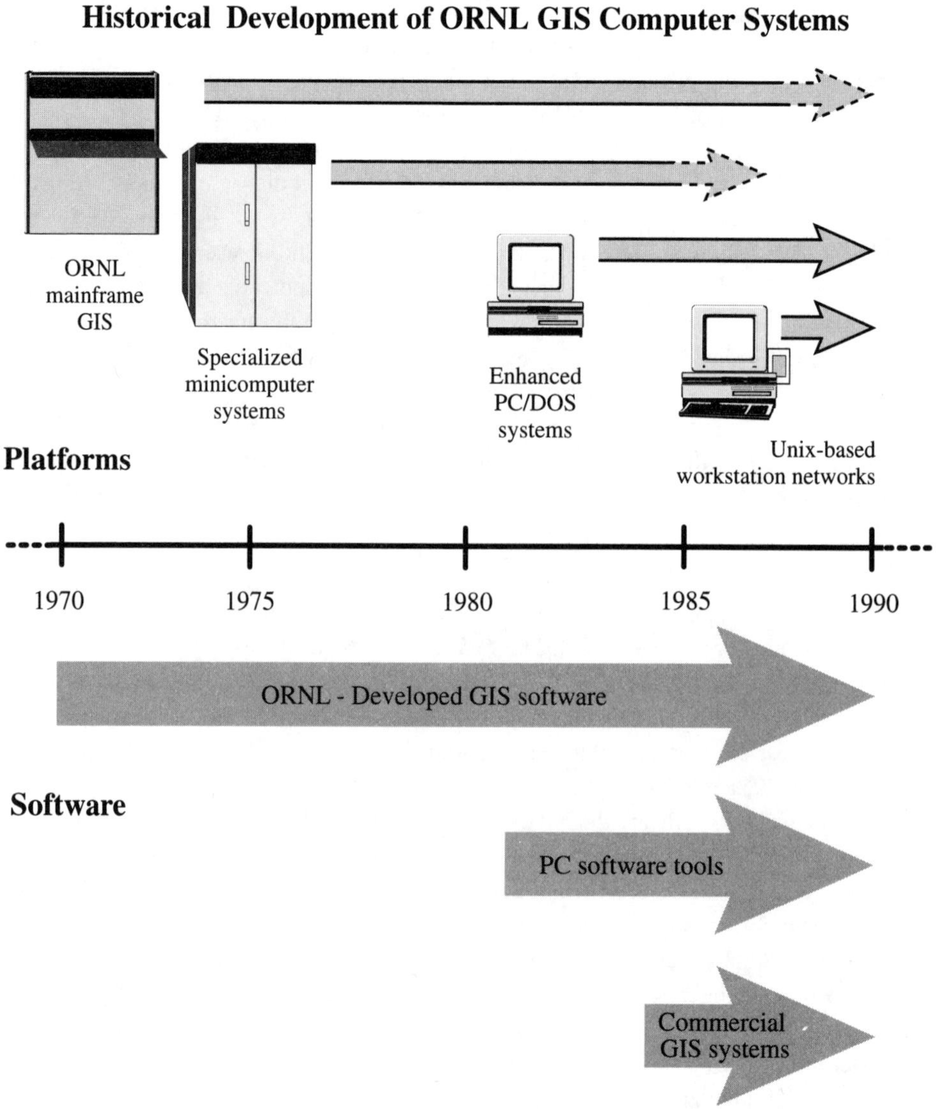

Historical Development of ORNL GIS Computer Systems

Platforms

ORNL mainframe GIS

Specialized minicomputer systems

Enhanced PC/DOS systems

Unix-based workstation networks

1970 1975 1980 1985 1990

Software

ORNL - Developed GIS software

PC software tools

Commercial GIS systems

Figure 13.1 ORNL developed most of its own GIS software and some specialized hardware until the mid 1980s.

At ORNL, GIS software and hardware have evolved together, each taking advantage of the other. Development at ORNL began on mainframes, including several generations of IBM 360 (e.g., models 75, 91) and IBM 3033 machines. These mainframes served as the workhorse for handling massive amounts of geographic data until the latter part of the 1980s. As interactive computing began at ORNL, early predecessors to the DEC PDP-10 were used in 1969–1970

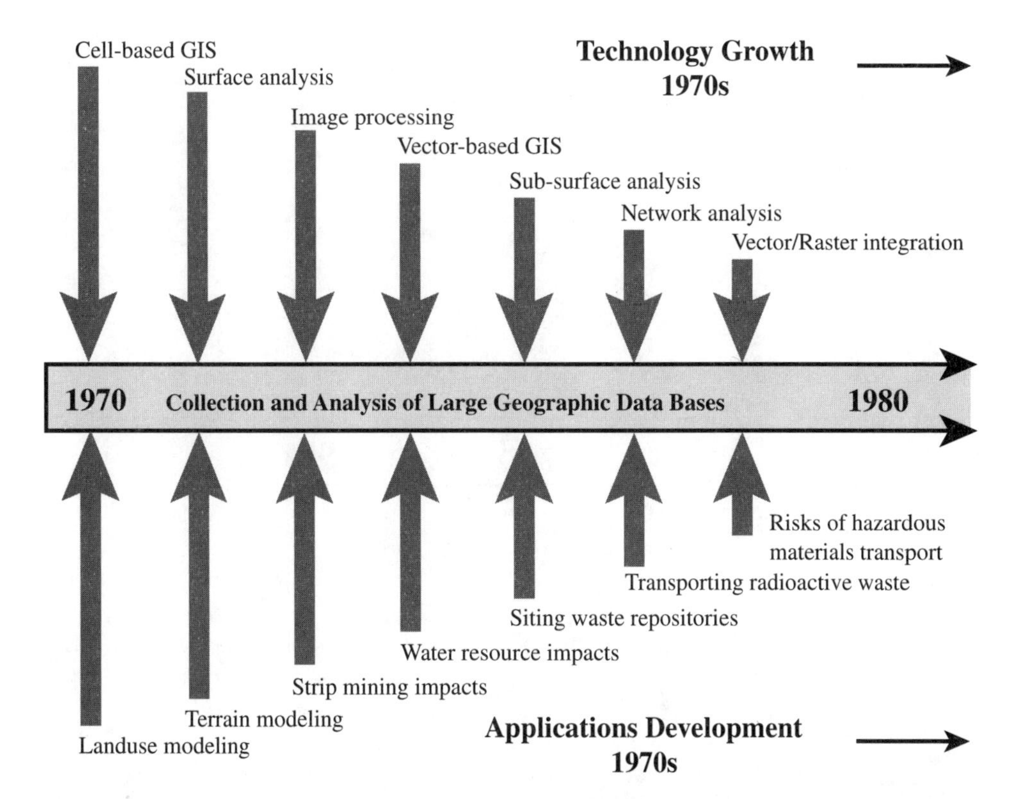

Figure 13.2 Methods and technologies evolved in response to changing applications in the 1970s.

to develop a county-based land use model. Teletype machines with paper tape were part of the initial development.

It quickly became apparent that special hardware was required to input and display map data. The first production use of specialized hardware was the adaptation of a Hough-Powell flying spot scanner used for bubble chamber photographs at the University of Pennsylvania to scan 35 mm film containing maps of land cover data for East Tennessee. A special camera assembly was built at ORNL to capture the maps on continuous rolls of 35 mm film. Software was written for the IBM mainframe to process the scanner data to reconstruct geo-registered gridded data for features such as rivers, lakes, highways, rail lines, soils, and many other types of geographic data (Meyers et al. 1976, Voelker 1976, Wilson 1976). These data were used as input to various models being developed under the National Science Foundation's (NSF) Research Applied to National Needs (RANN) program which will be described later.

A digital 35 mm film recorder was installed at ORNL in the early 1970s that could accept input from the mainframes. An early technological feat was the development of a computer movie film that simulated terrain and population changes over a 40-year period viewed as a

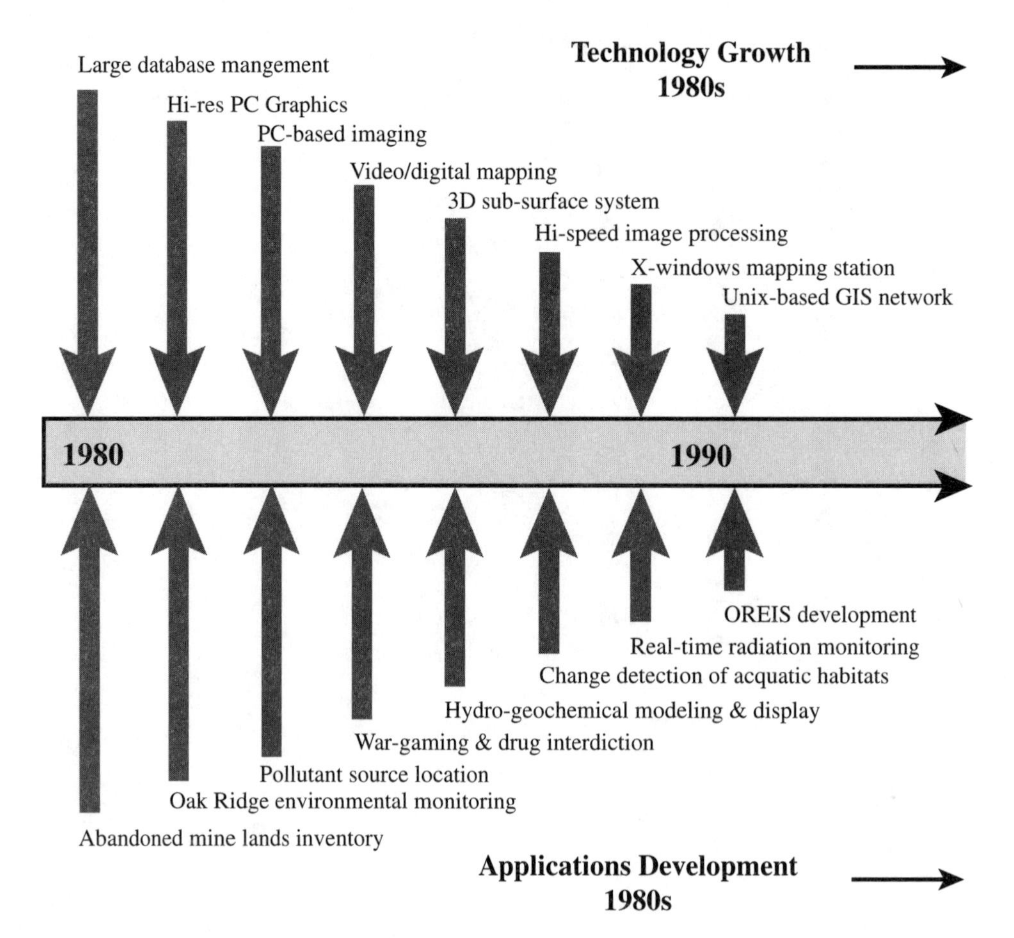

Figure 13.3 Methods and technologies continued to evolve along with new applications in the 1980s.

rotating 3-D perspective display for the Norris, TN, area as Norris Dam began its operation as a hydroelectric facility (Tucker 1973, 1976). The significance of this development was that all the algorithms, software, and data were developed at ORNL more than 20 years ago. It was 10 to 15 years later before this visualization/animation capability became available in a workstation environment for the environmental scientist.

From the mid 1970s to mid 1980s, minicomputers were linked with special graphics devices to handle the input, output, and processing of geographic data along with providing data linkages to the IBM mainframe computers (Durfee 1988). In the 1970s, the graphics stations evolved from Tektronix terminals with digitizing tablets to small stand-alone Terak desktop graphic stations. In a sense, the Terak systems were precursors to the PCs of the early 1980s. One of the most difficult problems in digitizing linear features from paper maps was the ability

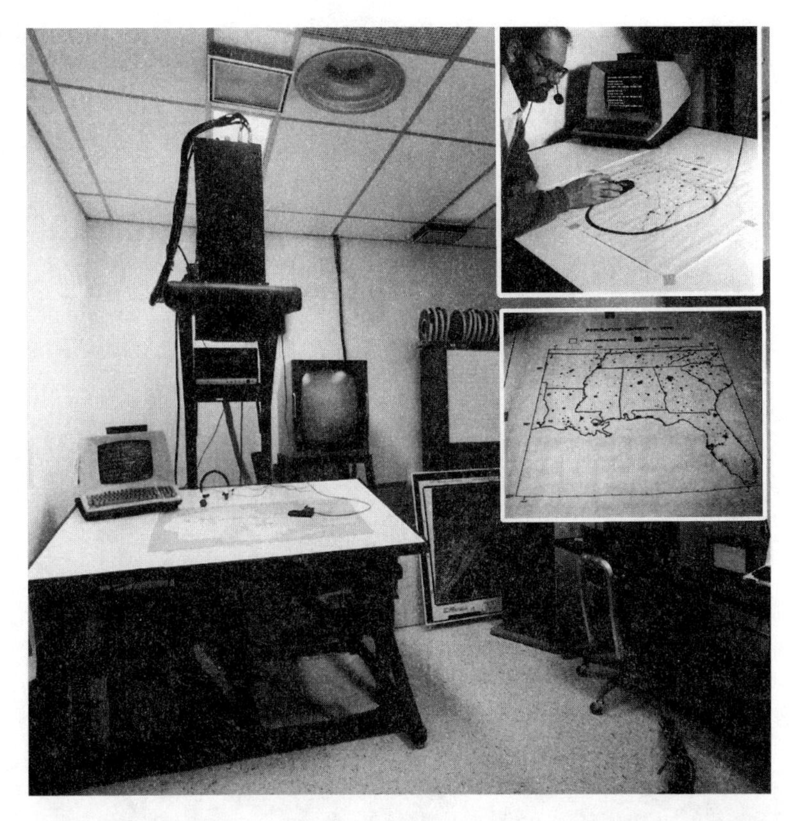

Figure 13.4 Richard Durfee demonstrates a video projection digitization system developed as one of ORNL's early (1970s) contributions to geographic information system (GIS) technologies. It projected a light beam that instantaneously followed the operator's mouse as linear features were traced on the map. This development addressed the problem of constantly having to compare digitized features from paper maps with their vector representation on a computer screen. The computer data directly overlaid the paper map in real time. Note the audio input headset.

to compare the paper map with digitized data on a computer screen. Thus, a video projection digitization system (hardware and software) was developed in the GCM Computer Laboratory to project a light beam that would instantaneously follow the operator's mouse as features were traced on the paper map (Figure 13.4). Audio input capabilities were tested with the system to determine if operators could "speak" geo-labels into the computer instead of typing them. Even though the audio input system was interactively trained for each operator, the results were not reliable enough due to variations in voice characteristics.

Two predominant data structures evolved to represent geographic data in the computer: vector structures and grid cells (raster-based systems). As more raster imagery became available and remote sensing began to expand (e.g., Landsat data, originally called ERTS), it became apparent that special image processing hardware and software were needed to allow image analysts to interactively control the computations with immediate display of image results. Thus in the mid-to-late 1970s, the I^2S imaging system with real-time video capabilities was acquired in the GCM Computer Laboratory. The software was adapted to allow interfacing with existing minicomputers. For the first time at ORNL, color satellite images could be manipulated and combined with geographic and other spatial databases interactively on the screen (Figure 13.5). In the 1980s, large-format hardcopy of these images and all types of GIS data were plotted on a 36-inch color Applicon ink-jet plotter that also served as the workhorse color plotter for all of ORNL for a number of years.

Over the years, the benefits of these spatial technologies have included their applicability to many different types of problems. For example, geographic techniques were applied to Electron Microscope Tomography (EMT) for 3-D reconstruction of DNA chromosomes (Olins et al. 1983). The I^2S imaging system played a major role in the interactive video input and registration of multiple micrographs, enhancement and transformation of the images, and stereo viewing of the results. When it was determined that more sophisticated true 3-D displays were

Figure 13.5 Richard Durfee working in the ORNL Geographics Laboratory in 1981.

needed, a special display unit, based on a prototype at the University of North Carolina, was built (Figure 13.6). Depth visualization was provided by a vibrating mylar mirror that was synchronized with a monitor mounted above the mirror. In the reflected image, data at greater depths were displayed when the mirror was at a greater deflection, thus varying the focal length to correspond to the appropriate depth. This occurred at a rate of 60 times per second, so the observer saw a continuous true 3-D image. A 3-D joystick controlled the rotation, zoom, and observer position for the data displayed. The display could be used for any x,y,z spatial data.

As shown on the upper part of Figure 13.1, enhanced PC hardware was used for small mapping applications with customized software developed in-house. By the mid 1980s, self-contained PC mapping packages were available, and by the late 1980s, UNIX-based workstations contained enough storage and memory capacity to be used for large GIS problems. The predominant UNIX-based hardware used for GIS at ORNL has been the Sun platform with a variety of other systems also used by the GCM group, the Environmental Sciences Division, and other organizations (e.g., Silicon Graphics, Intergraph, and DEC). The most widespread growth of GIS within ORNL is now occurring with desktop PCs linked to UNIX-based servers to support staff involved in all types of facility management and environmental work. The implementation of high-speed network communications is becoming a crucial resource both for the casual PC users and the specialists in the GIS Center. Groups of analysts can no longer afford to store and maintain duplicate volumes of data on each of their own workstations. Thus, we share common data and system resources across the network.

During the decade of the 1980s and continuing into the 1990s, a wide variety of graphic peripherals have been incorporated into the systems used in the GIS Computing and Technology Center of the GCM group. These have included various types of digitizing tablets; video and digital scanning cameras; color printers; film recorders; ink-jet, pen, sublimation, electrostatic and thermal plotters; projectors; stereo systems; video cassette and laser disc recorders/players; video digitizers; scan converters; and all types of storage media ranging from tape and disk systems to an optical jukebox. A schematic representing current facilities in the center is shown in Figure 13.7. All the PCs (486 and Pentium) and the UNIX workstations are linked by ethernet to Sun SPARCservers. Examples of some of the current spatial software systems include ARC/INFO and ARCView, Intergraph, MapInfo, AutoCAD and ARCCAD, ERDAS Imagine, PCI, EarthVision, and AVS. Database management and analysis is based primarily on Oracle and SAS, plus FoxPro on PCs. The Environmental Sciences Division operates GIS systems that include Sun, Hewlett-Packard, and Silicon Graphics workstations with appropriate input and output devices. The Airlift Deployment Analysis System (ADANS) maintains a network of more than 30 Sun workstations at ORNL linked across the U.S. with current installations at several U.S. Air Force bases.

Figures 13.2 and 13.3 present the evolution of GIS methodologies over two decades along with examples of applications that were developed requiring those technologies. The technologies and applications shown are examples from a much larger suite that evolved over the two decades. Before typical GIS functions could operate properly on geographic and other

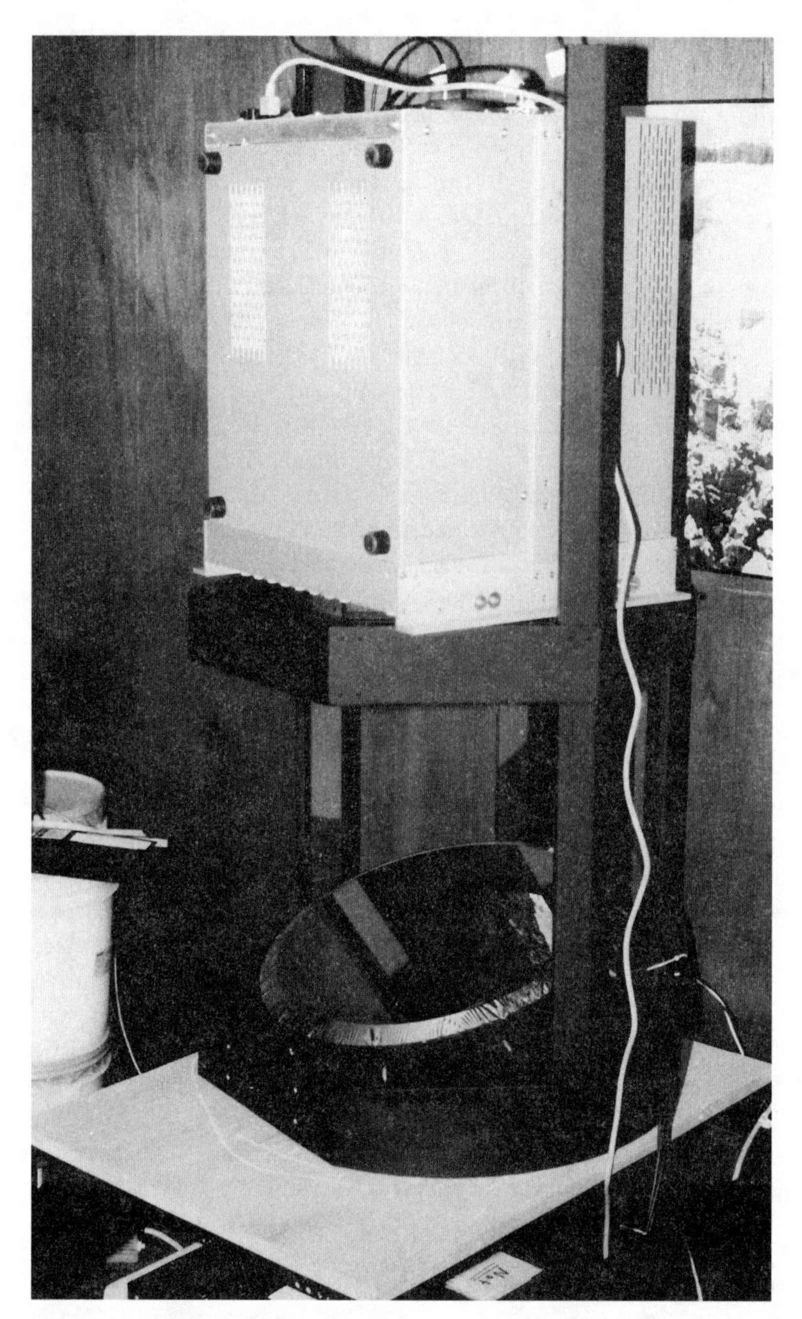

Figure 13.6 ORNL researchers, building on earlier work at the University of North Carolina, developed a special varifocal mirror system to provide viewers with a continuous three-dimensional image. The technology was used to display electron micrographs of chromosome structures and medical computer tomography scan images.

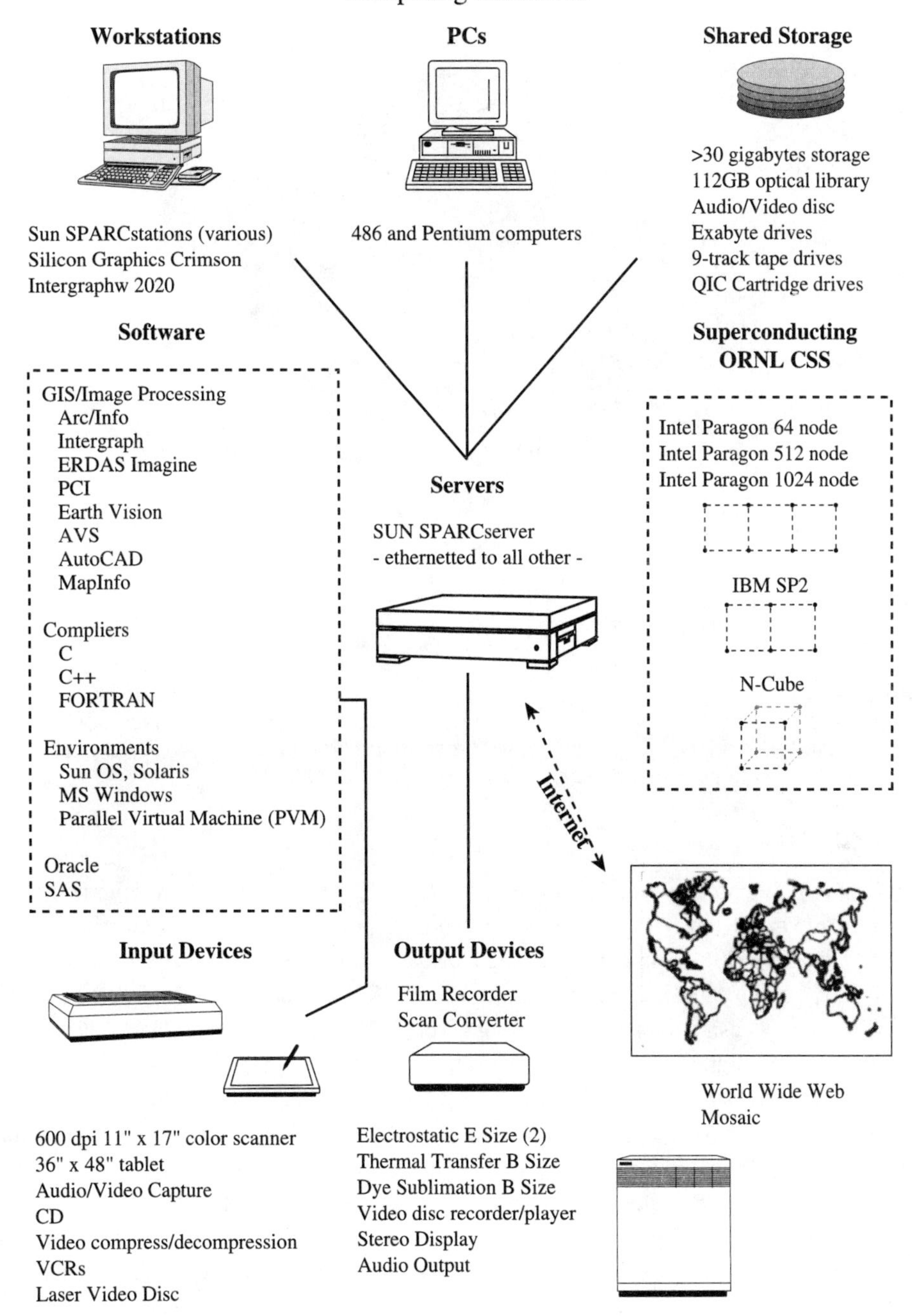

Figure 13.7 ORNL's GIS and Computer Modeling Group maintains a variety of computing resources to support efforts ranging from natural resource assessments to environmental restoration.

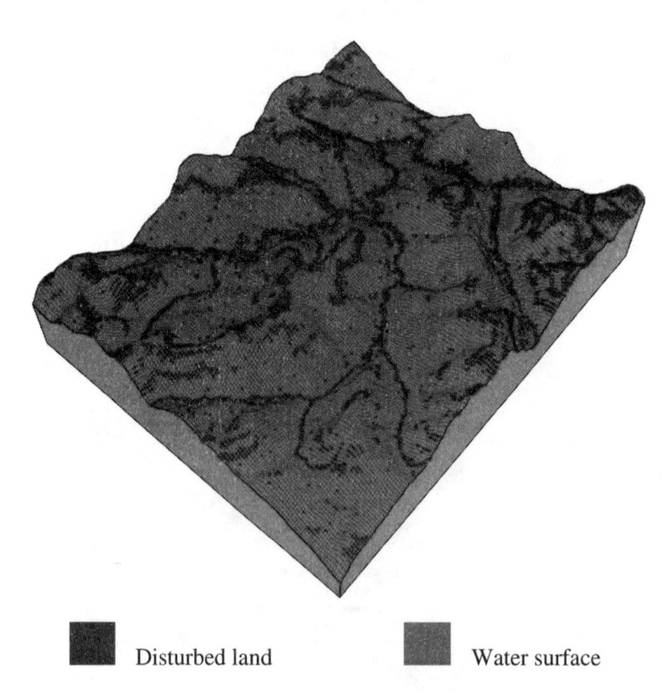

Figure 13.8 A three-dimensional perspective view of coal strip mines in the Cumberland Mountains north of ORNL. Streams are superimposed on the terrain. The spatial model computed the geographic relationship between disturbed land and nearby surface waters, allowing assessment of environmental and visual impacts of strip mines.

spatial data (e.g., input, analysis, output), it was necessary to implement cartographic and map projection software to represent the three-dimensional earth's surface as two-dimensional digital maps. The initial GIS software supported cell-based GIS with specialized surface analysis capabilities to support land use and terrain modeling. With the advent of the Landsat satellite, image analysis tools using statistical techniques were implemented. The tools were first used to assess strip-mining impacts north of Oak Ridge, TN (Durfee et al. 1975, Honea et al. 1979b). Figure 13.8 is a 3-D perspective display of strip-mined areas calculated from satellite imagery with nearby streams superimposed on the terrain. Spatial models were developed to determine the potentially most impacted streams and the visual impacts on Oak Ridge residents. Significantly, this work was conducted early in the history of GIS (i.e., mid 1970s). Another breakthrough occurred in the development of vector-based algorithms and their eventual integration with grid cell systems. These techniques were used for all types of water resource and energy-related studies with ORNL's Energy Division. Initial transportation databases and routing capabilities were developed in the late 1970s to aid in studying risks of hazardous waste transport (Figure 13.9).

Figure 13.9 This map illustrates the computation and display of possible truck routes for transporting radioactive waste material from nuclear reactors to a candidate repository site. These routes are superimposed on zones of computed population densities across the United States.

Another technical development at ORNL in the mid-to-late 1980s was the integration of video information with digital data in the computer. Raster digitization of video signals and the introduction of laser video discs opened up a whole new way of dealing with graphic and map data. In cooperation with the LMES Data Systems Research and Development organization (DSRD), one application was to demonstrate the feasibility of using video-scanned map images recorded on laser video discs to aid war-gaming exercises simulated on a high-resolution workstation. Our approach was to locate, computerize, and combine multiple video frames into large electronic maps that could be roamed and overlaid with lots of other geographic and military information in a real-time mode. As a natural extension to computerized video input, inexpensive video output techniques were soon implemented. By linking a scan converter with graphic output signals, screen images could be recorded as video frames on laser video discs under computer control. By computing and recording hundreds of successive frames, we were able to play back 2-D and 3-D terrain animations, even using PC systems.

Development has continued through the 1980s and into the 1990s with new hardware and software technologies continually offering opportunities to address new applications. Many of these efforts were sponsored by other federal agencies and carried out cooperatively with other divisions at ORNL or within LMES. As the applications evolved, the demand for more extensive databases increased along with the need for expanded computer resources. In some cases, more effort has been required to gather, process, and validate the data than to design and implement the models or systems. However, to solve real-world problems, both data and technology must be available.

The most advanced GIS research currently underway at ORNL involves algorithms that will take advantage of parallelism on large supercomputers. Recent work has shown around an 18,000-fold increase in speed for transformation and interpolation of large GIS data sets as compared to identical operations running on Sun SPARC systems (Hodgson et al. 1995). With the explosion in data collection from all types of earth sensor systems, integration of workstations and parallel processors will be a necessity to handle the massive data volumes. Recently, for example, sophisticated techniques have been employed to color balance and mosaic large numbers of orthographic aerial photographs to create very large composite images that are consistent in color and intensity across the entire array. Integration of high-precision, real-time global positioning systems (GPS) and video cameras with portable GIS capabilities is underway at ORNL for applications such as vehicle tracking. Airborne real-time GPS techniques, minimizing reliance on ground control points, have already been used in photogrammetric surveys on the Oak Ridge Reservation. As these technologies become commonplace, geographic data will be collected at an ever-increasing rate. To provide intelligent and efficient access to large amounts of geographic, multimedia, and other spatial data, work is also underway with integrating World Wide Web servers, data browsing and display tools (e.g., Mosaic and Netscape), and Internet through GIS systems and various programming techniques. These capabilities are important to the Oak Ridge user community as well as to the success of the National Spatial Data Infrastructure (NSDI) during the 1990s and beyond.

Regional Modeling and Fundamental Development (1969–1976)

In 1969, (a) the United States Congress passed the National Environmental Policy Act (NEPA), (b) NSF initiated the RANN program, (c) Ian McHarg (1969) published *Design with Nature*, and (d) ORNL delved headlong into regional modeling and GIS. All four events were prompted by growing concern over environmental impacts, and all four promoted multidisciplinary, integrated analysis. Prior to NEPA, research and development, infrastructural development, and resource management decisions had been based almost exclusively on engineering and cost/benefit considerations. Instantaneously, NEPA thrust all large enterprises, including the federal government itself, into a new legal and ethical milieu in which comprehensive, interdisciplinary analyses were absolutely essential. Alvin Weinberg, ORNL director from 1955 to 1973, immediately recognized the need and sought to diversify the laboratory's mission. Simultaneously, others in Washington, especially Senator Albert Gore, Sr., recognized the need for centers of excellence dedicated to the scientific advancements necessary to implement NEPA. RANN resulted from that movement; ORNL and 17 other institutions were designated as RANN research centers (Department of Civil Engineering, University of Washington, 1977).

ORNL's original RANN proposal did not turn up in our search, so there is uncertainty as to dollars and dates in the initial time frame. The 1972 proposal indicates an ongoing RANN project called Regional Environmental Systems Analysis (RESA) funded at $763,897 per year. That's about $2.5 million in 1994 dollars, clearly a substantial program to match ambitious goals. In the authors' words, its purpose was "to develop and communicate to the planning and management community a scientific basis for forecasting the environmental impacts of public and private decisions (such as land use) in order that we may more effectively manage our environmental resources." Its strategy was "to develop and validate a hierarchy of analysis models representing the relevant processes at each of several regional levels and to study and test their implementation processes."

The research was grouped into five major activities—socioeconomic, land use, ecological, sociopolitical, and implementation and communications—that "in combination, develop a capability to simulate the dynamic environment." From later documents, it is clear that the land use modeling effort became the principal impetus to GIS and remote sensing development. All topics focused on a 16-county study area in East Tennessee.

If NSF hoped to develop a cadre of expertise in the new sciences, it could hardly have done better. The vitae submitted with the 1972 proposal include, among others:

- John H. (Jack) Gibbons, later Director of the Congressional Office of Technology Assessment and currently President Clinton's Science Advisor (Director of the Office of Science and Technology Policy).
- James L. Liverman, later Assistant Secretary for Environment in the Department of Energy.
- Stanley I. Auerbach, who led ORNL's Environmental Sciences Division to become a premier research institution in ecology and related sciences.

The results of that initial effort range from thoughtful, concise discussions of conceptual issues, such as "the reciprocal impacts. . .between human activities and economics," to thoughtful, concise reports on fundamental algorithms, such as "a general subroutine for drawing graticules" (Stephenson 1972). It was a heady time when one concerted effort could tackle a complex problem in all its many aspects. By 1974, team members were able to report substantial progress in each activity. With regard to GIS, the most important development was ORRMIS developed by Durfee and the Geographic Data Systems Group (later renamed GCM). ORRMIS was described as "a geographically oriented information system . . . designed to operate either as a stand-alone inquiry-retrieval system or as a communications supervisor for use with simulation models." Its primary purpose was "to provide the data management capability for analysis models which forecast the spatial distribution and ecological effects of activities within a geographical region." One final note added that "the geographic orientation could be replaced with a more general adjacency relationship thereby extending its use to non-geographical models" (Durfee 1974). During the early part of this period, initial GIS software techniques were based on hierarchical grid cell systems. It became apparent that additional capabilities were needed for accurate cartographic representation and analysis of vector-based map data. By the mid 1970s, development of sophisticated polygonal-based GIS systems was well underway. Efficient storage and computational techniques for integrating grid cell and vector-based systems opened the door to larger and more complex national scale problems.

What influences did these developments have on the GIS community at large in software, standards, and institutions?

We know our contributions supported and accelerated the growth of a commercial GIS industry, but the extent of that influence is difficult to document. Many of the algorithms were published in open literature, primarily in government documents available through the National Technical Information Service (NTIS), and others were communicated through direct correspondence with a growing cadre of developers in government, academia, and business. In those days it was not possible for a national laboratory to obtain a patent or copyright on software, and we were not philosophically inclined to limit access anyway. We know, however, from personal recollections and from a few gracious letters of acknowledgment that leading commercial vendors relied heavily on our algorithms and concepts.

Our influence on geographic data standards began with an ad hoc effort by several national laboratories to facilitate data exchange among themselves. In 1978, Durfee, Bob Edwards, and Al Brooks, then head of Computing Applications at ORNL, were members of the Interlaboratory Working Group for Data Exchange (IWGDE) that produced a prototype which later influenced the International Standards Organization's (ISO) data exchange standard (ISO #8211) and the more complex Spatial Data Transfer Standard (SDTS) (National Institute of Standards and Technology 1992). From 1985 through 1993, Dobson represented ORNL on the Steering Committee of the National Committee for Digital Cartographic Data Standards (NCDCDS), one of the two committees that jointly composed SDTS. In all of the federal standards effort, Brooks holds the longest tenure, having served as a key member of the original IWGDE, the ISO committee, and NCDCDS.

Institutionally, we promoted the concept of a national geographic information and analysis center to serve as a resource throughout the scientific and user community and to provide a focal point for advancing research and education (Dobson 1984, 1992b, 1993a; Dobson and Durfee 1987). Dobson's original concept led directly to the establishment of the National Center for Geographic Information and Analysis (Abler 1987) and continues to influence formation of the new University Consortium for Geographic Information Science (UCGIS) (Mark and Bossler 1995) of which ORNL is a founding member.

Perhaps, our most important legacy is that we served as a testbed and proof of concept for integrated geographic technologies and systems, informing and educating many government agencies of the potential held by GIS, at a time when GIS was poorly understood and little appreciated. We privately cherish a few cases in which agencies specifically heeded our message and became enthusiastic supporters of GIS development and application. Ultimately, however, we were but one of many research and development organizations nudging society toward greater acceptance of GIS and none of us will ever know who played the greater part.

Integrated Assessments (1977–1985)

In the mid 1970s, a shift in federal policy greatly reduced NSF funding of the DOE national laboratories. From then on, hardly another penny was received to support pure research, development, or operation of the ORNL GIS systems, per se. GCM and Energy Division shifted to applications-driven research and continued to cover fundamental development and operations out of the funded projects. Like the original modeling goal, the applications tended to be comprehensive in scope. Hence, we never had the luxury of focusing on a particular technology (remote sensing or computer cartography, for instance) to the exclusion of other technologies. We were then and are still comprehensive integrators with analytical purposes paramount in everything we do. In many respects, this approach has been advantageous since (1) the integrated GIS technologies were then applicable to a wide range of spatial problems, and (2) the applications-driven development minimized "ivory-tower" research looking for a problem to solve.

The first seeds of the new order were sown when Durfee and Honea used ORRMIS for predictive modeling of coal strip mining and associated environmental problems (Durfee et al. 1975, Honea et al. 1979). Results of this work were presented to Robert Seamans, head of the Energy Research and Development Administration, ERDA, predecessor to the DOE. Soon afterward, we became heavily involved in siting analysis. In 1975 and 1976, ORNL systems were used, along with data from the Maryland Automated Geographic Information System, to support conflict resolution in power plant siting (Jalbert and Dobson 1977, Jalbert and Shepherd 1977, Dobson 1979). By the late 1970s, ORRMIS and related systems were heavily involved in decision support for federal energy policy and resource management. ORNL employed GIS extensively to evaluate the environmental impacts of various National Energy Plans (Berry et al. 1978, Dobson et al. 1977, Dobson and Shepherd 1979, Honea and Hillsman 1979, Honea et al. 1979a), to project coal production goals for federal leasing programs (Dobson 1979), and to evaluate the impacts of wilderness area designation on energy supply (Oakes and Voelker 1983).

Another national effort in which the laboratory played a major role was assisting the National Uranium Resource Evaluation (NURE) Program during the mid-to-late 1970s. The Computational Physics and Engineering Division (at that time called the Computer Sciences Division), in cooperation with the DOE Grand Junction Office, was the national repository for all data collected and was responsible for overall data management, GIS processing, spatial analysis, and mapping. Al Brooks was director of the Oak Ridge effort to support DOE in surveying the country for potential uranium resources and estimating possible reserves. Through a multitude of subcontractors, DOE conducted both aerial radiometric surveys and hydrogeologic ground sampling on a quadrangle by quadrangle basis across the whole U.S. Aircraft with special sensors were flown to detect radioactive elements such as bismuth, thallium, and potassium, as well as measuring magnetic information. One example of very specialized GIS work was the development, in cooperation with Bill Hinze from Purdue University, of spatial filtering, interpretation, and contouring techniques to convert one-dimensional flight line data into meaningful maps of regional magnetic data that would help identify magnetic anomalies (Tinnel and Hinze 1981). These maps were published and distributed by the U.S. Geological Survey. This was one of the earliest projects that required the handling of massive amounts of spatial, tabular, and textual information of many different types. During this time, specialized hardware GIS systems were implemented to provide new ways of digitizing and displaying large amounts of geographic data.

The energy assessments are early examples of policy analysis using GIS. A flurry of activity began each time President Jimmy Carter proposed a new National Energy Plan. Econometric models were run by the Energy Information Administration to project, as far as the year 2000, energy demand and fuel use by type in each major region of the country. These regional projections were passed to ORNL where energy demand was disaggregated to Bureau of Economic Analysis Regions and supply was allocated to counties. ORNL projected electrical generation from each existing plant and simulated plant construction or retirement to meet the President's goals by fuel type (Church and Bell 1981). Water quantity requirements were projected and compared to measures of water availability for energy in each stream (Dobson et al. 1977, Dobson and Shepherd 1979, Shepherd 1979) (Figure 13.10). Projections of electrical generation were passed to other national laboratories (Argonne, Brookhaven, Los Alamos, and the Solar Energy Research Institute) for evaluation of air quality, water quality, labor supply, and other impacts. Ultimately, the results were reported back to the Department of Energy where they contributed to policy analysis of the feasibility of each plan. The results of one GIS assessment for the Ohio River Basin were shown to President Carter in a live presentation using a graphics station when he visited ORNL in 1978 (Association of American Geographers 1978).

In short, as early as the 1970s the nation's energy system and many pertinent physical and cultural features were simulated through GIS linked with econometric models, location-allocation models, environmental assessment models, and spatial databases with national coverage. The principal output was by county, but many of the databases and computations were much finer than county level. The databases included, for example, population interpolated from Enu-

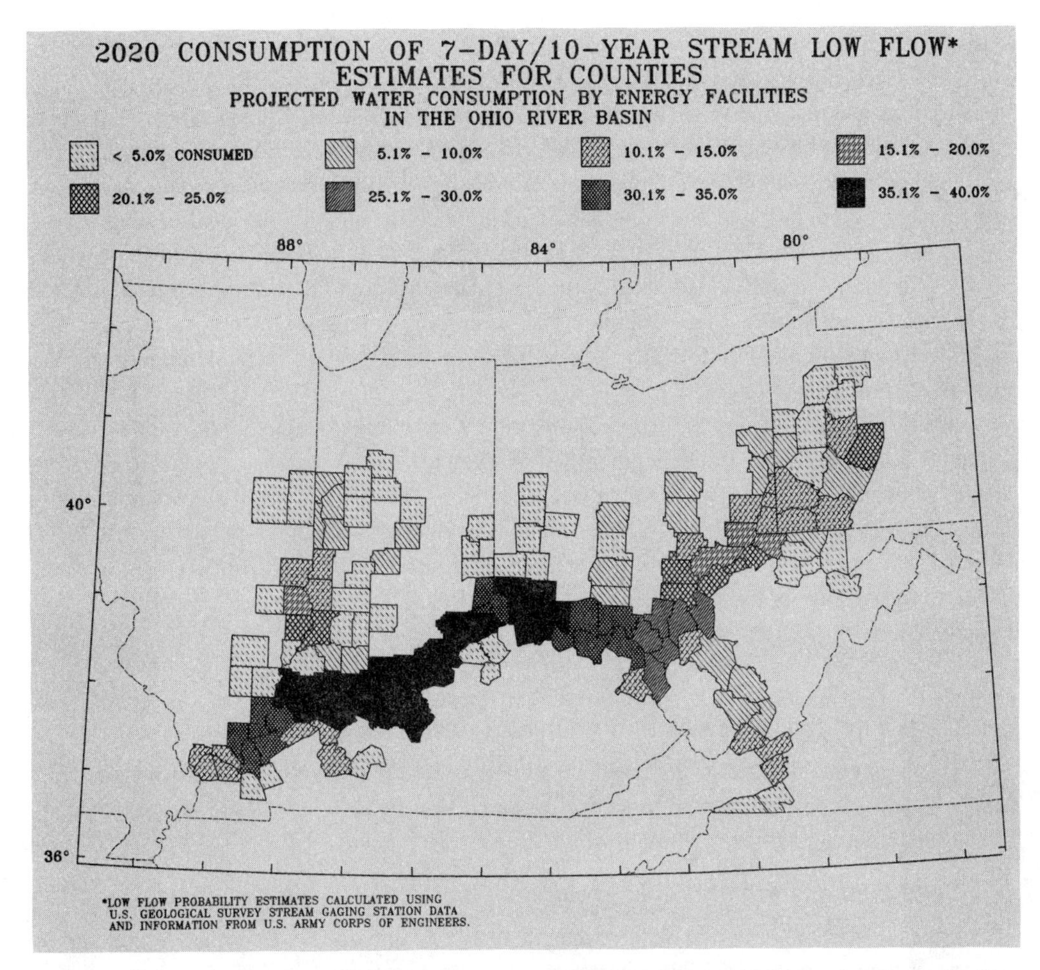

Figure 13.10 The results of this assessment of projected water consumption by energy facilities in the Ohio River Basin were shown to President Carter in a live presentation using a graphics station when he visited ORNL in 1978.

meration District counts (Durfee and Coleman 1983), all power plants over 10 megawatts in generating capacity, and all U.S. Geological Survey stream gauging station records. The models were as sophisticated as any in use at that time with or without GIS.

A substantial ORNL effort in the early 1980s involved the development of a national abandoned mine lands inventory (Honea et al. 1984) for the Office of Surface Mining (OSM). Federal legislation mandated reclamation of abandoned mine lands to protect the environment and human health, safety, and welfare to be funded by taxes on mining operations. A national inventory was necessary to assess the areas with the most significant problems and to establish priorities for reclamation. The effort was initially viewed as technology-based involving GIS, remote sensing, record-based information systems, and statistical tools. Major environmental impacts were addressed using Landsat satellite analyses. The legislation, however, focused on health and human safety, and many pertinent factors (e.g., open mine shafts, acid drainage, impacted water supplies) could not be determined from the available satellite data. Such impacts required more traditional field research methods. Thus, a major field survey effort, including on-site interviews with impacted populations, was conducted in conjunction with all the coal-mining states. Unique information-handling procedures were devised to standardize and computerize textual, tabular, temporal, and spatial data from forms and maps that could then be linked with GIS for spatial aggregation, statistics, and mapping. The results supported assessments at state, regional, and national levels to aid OSM in allocating reclamation funds and overseeing mitigation of the most severe problems.

Methodologies developed for one application were readily adapted and applied to other problems. For example, the initial demographic work of the late 1970s was extended to compute detailed population distributions for any place in the U.S. The technique which involved sophisticated spatial transformations (involving points, triangulations, Thiessen polygons, grid cells, and contours) was used to compute population distributions and exclusion areas around all nuclear power plants in the U.S. to aid the Nuclear Regulatory Commission in planning and licensing decisions (Durfee et al. 1983).

Issue-oriented Research and Analysis (1986–1995)

Starting in the mid 1980s, the emphasis shifted again, this time in a very positive direction as GIS became an important tool in topical research on scientific issues of national interest. Three illustrative examples are summarized in the following pages.

Lake Acidification and Acid Precipitation

Lake acidification and other environmental issues potentially related to acid precipitation were major themes in the late 1980s (Durfee et al. 1993b, Landers et al. 1987, Olson et al. 1980, Olson et al. 1983, Rosen et al. 1988). The Environmental Sciences Division (ESD) was involved prominently in the National Acid Precipitation Assessment Program (NAPAP), especially the National Surface Water Survey. Through extensive collaboration with EPA laboratories and numerous universities and private firms, ESD personnel collected, managed, and analyzed massive geographic databases for lakes and watersheds throughout the United States. The goal was to characterize contemporary chemistry, temporal variability, and key biological resources of lakes and streams in potentially sensitive regions.

Simultaneously, the Energy Division approached the same problem from a different perspective (Dobson 1990; Dobson et al. 1987a, b; Rush et al. 1985). While NAPAP focused on acid precipitation and investigated its impacts, the other project focused on lake acidification and investigated for causes. The approach combined GIS and digital remote sensing with the traditional field methods of geography. The methods of analysis consisted of direct observation, interpretation of satellite imagery and aerial photographs, and statistical comparison of two geographical distributions, one representing forest blowdown and another representing lake chemistry. Spatial and temporal associations between surface water pH and landscape disturbance were found to be strong and consistent in the Adirondack Mountains of New York. Evidence of a temporal association was found at Big Moose Lake and Jerseyfield Lake in New York and at the Lygners Vider Plateau of Sweden. The authors concluded that forest blowdown facilitated the acidification of some lakes by altering hydrologic pathways so that waters (previously acidified by acid deposition and/or other sources) did not experience the neutralization normally available through contact with subsurface soils and bedrock. Increased pipeflow was suggested as a mechanism which may link the biogeochemical impacts of forest blowdown to lake chemistry. Haring et al. (1992) cite this study as a prime example of GIS and remote sensing effectively employed in a traditional geographic research mode.

Both efforts illustrate an ORNL strength: the ability to assemble multidisciplinary teams and multiple organizations to attack complex problems. GIS, in itself, is an integrating technology that draws together different sciences needing common spatial data, visualization, and analysis capabilities. Such was the case in the acidification studies just described. While primary responsibility for these two efforts rested separately in the Energy and Environmental Sciences Divisions, GCM was heavily involved in both efforts, and there was considerable interaction between the two projects. As a follow-on to this effort, ESD, in cooperation with GCM and other groups, has continued to expand its GIS capabilities and resources. This allows ESD scientists to have hands-on access to GIS systems and databases to support their research efforts.

Coastal Change Analysis

For decades, the National Oceanic and Atmospheric Administration's (NOAA) National Marine Fisheries Service (NMFS) has been concerned with the problem of declining fish populations in coastal waters of the United States (Dobson et al. 1995). Suspecting that declines might be caused by losses of habitat, such as saltmarshes and seagrasses, and increases in pollution resulting from expanding urban development, rural development, and agriculture, NMFS initiated a research effort to solve the technical, institutional, and methodological problems of large-area change analysis. ORNL has led the technical effort to improve methods for satellite change analysis of uplands and wetlands and the prototype analysis of the Chesapeake Bay (Dobson and Bright 1991). The integration of these remote sensing and GIS methodologies in a laboratory environment, in field investigations, in workshop settings, and for external presentations shows how much this evolving technology is becoming ingrained in all phases of earth-sciences work.

The Coastal Change Analysis Program (C-CAP) is developing a nationally standardized database of land cover and land cover change in the coastal regions of the United States. As part of the Coastal Ocean Program (COP), C-CAP inventories coastal and submerged wetland habitats and adjacent uplands and monitors changes in these habitats on a one- to five-year cycle. This type of information and frequency of detection are required to improve scientific understanding of the linkages of coastal and submerged wetland habitats with adjacent uplands and with the distribution, abundance, and health of living marine resources. Satellite imagery (primarily Landsat Thematic Mapper), aerial photographs, and field data are interpreted, classified, analyzed, and integrated with other digital data in a GIS. The resulting land cover change databases are disseminated in digital form for use by anyone wishing to conduct geographic analysis in the completed regions.

Land cover change analysis has been completed for the Chesapeake Bay based on Landsat Thematic Mapper (TM) data. The resulting database consists of land cover by class for 1984, land cover by class for 1988/1989, and a matrix of changes by class from 1984 to 1988/1989. Although the Chesapeake Bay prototype focused on a single region, its purpose was to provide a technical and methodological foundation for change analysis throughout the entire U.S. coast. Four regional workshops (Southeast, Northeast, Great Lakes, and Pacific) addressed a full range of generic issues and identified the issues of special interest in each major coastal division of the United States. Ultimately, the protocol development effort (Dobson et al. 1995, Ferguson et al. 1993, Klemas et al. 1993) involved more than 250 technical specialists, regional experts, and agency representatives.

During the summer 1994 field season, fieldwork was conducted in the Gulf of Maine, the Oregon/Washington Coast, and Alaska. The Alaska effort included field verification of a 1986 land cover classification in the Yakutat Foreland and Russell Fiord, AK. Current fieldwork employs a hand-held Global Positioning System (GPS) linked directly to a color laptop computer. Commercial software integrates the live GPS location coordinates with raster images representing land cover and with vector images representing other features such as roads. The device has more than doubled productivity in the field. We are currently designing a modeling approach that will link GIS, transport models, and process models to address the linkage between land cover change and fisheries.

Environmental Restoration

Another major effort in the 1980s was the development of geographic workstations, algorithms, 3-D subsurface modeling techniques, and database systems for handling hazardous waste problems for Air Force installations (Durfee et al. 1989, Coleman et al. 1992). This work provided a foundation for supporting environmental restoration activities at DOE facilities. Since the late 1980s, environmental restoration has become a major theme of GIS development at ORNL. GIS has been recognized as an important technology supporting policy formulation, hazardous waste assessment and management, and remediation operations for environmental cleanup at Department of Energy facilities (Durfee et al. 1992). To conduct successful cleanup efforts and meet regulatory requirements at these facilities, many types of questions must be investigated, including:

- the types and characteristics of contaminants,
- the location of possible pollutant sources,

- previous waste disposal techniques,
- the spatial extent of contamination,
- relationships among nearby waste sites,
- current and past environmental conditions including surface, subsurface, and groundwater characteristics,
- possible pollutant transport mechanisms,
- efficient methods for analyzing and managing the information,
- effective cleanup strategies, and
- mechanisms for long-term monitoring to verify compliance.

Three programs involving significant GIS activities for Environmental Restoration (ER) in Oak Ridge include the Oak Ridge Environmental Information System (OREIS), the Remote Sensing and Special Surveys Program (RSSS), and the GIS and Spatial Technologies Program (GISST). The OREIS effort is designed to meet environmental data management, analysis, storage, and dissemination needs for all DOE facilities (McCord et al. 1992). The focus of this effort has been fourfold: (1) development of a consolidated database, (2) development of an environmental information system, and (3) development of data management procedures that will assure the integrity of environmental and geographic data throughout the facilities, and (4) 3-D subsurface modeling and visualization for hazardous waste studies.

The RSSS Program supports ER site characterization, problem identification, and remediation efforts through the collection and analysis of remotely sensed data from airborne sensors (Durfee et al. 1993a, c; King et al. 1993). One example has been helicopter radiometric surveys to determine gamma radiation for DOE facilities. GIS also aids in the interpretation and visualization of airborne electromagnetic and magnetic survey analyses to assist in the location of underground structures pertinent to hazardous waste burial and migration. Another example has been the delineation of unknown waste trenches in burial ground areas to aid in effective remediation of problems from hydrologic transport of contaminants.

The GISST Program is designed to promote the development, maintenance, and application of GIS technology, databases, and standards throughout the ER Program (Durfee et al. 1993a, c). The largest activity currently underway is the development of base map data, digital orthophotos, and elevation models for all the facilities using advanced photogrammetric techniques. This project is being carried out in cooperation with the Tennessee Valley Authority. Desktop mapping systems are being integrated into the daily operations of many Oak Ridge personnel dealing with the Oak Ridge Reservation. To support these activities, the resulting data from this project are being made available to users networked into a local file server.

Through these and other ER programs, environmental and facility databases have been developed and modeled extensively to improve understanding of relationships among sources, pathways (surface and subsurface), and receptors of environmental contaminants (Voorhees et al. 1988). Three-dimensional modeling, data management, and contaminant analysis have been enhanced through integration of image processing systems, relational database management systems, statistical analysis systems, and GIS. Policy formulation and operational management have been supported through telecommunications networks linking decision makers with analytical software and databases.

Transportation Modeling and Analysis

Transportation systems and networks are crucial to the U.S. economy and way of life. GIS is used increasingly to plan, develop, and manage transportation infrastructures with the goal of improving efficiency in construction and operation. Two main centers heavily involved in transportation modeling and geographic networks are the Energy Division and the Chemical Technology Division (ChemTech), both with support and collaboration by GCM. ChemTech has been primarily supporting DOE transportation needs, while Energy Division has been supporting the Departments of Defense and Transportation. Collectively the three divisions have developed detailed representations of highway, railway, and waterway networks for the U.S. and military air transport networks for the entire world. Energy Division, for example, is the developer and proprietor of the National Highway Planning System and the Interline railway routing model (Peterson 1984), while ChemTech has a major responsibility for routing hazardous materials on the nation's highway and rail systems (Joy 1982).

Operations Desert Storm and Desert Shield involved the largest airlift of personnel and equipment from region to region ever accomplished. The U.S. Air Force's Military Airlift Command (MAC) was responsible for this movement from the United States and Europe to the Persian Gulf Region. Prior to that event, ORNL had worked with MAC to develop the Airlift Deployment Analysis System (ADANS)—a series of scheduling algorithms and tools that enabled MAC to schedule missions to and from the Persian Gulf more rapidly and efficiently. ADANS is currently being used by MAC to schedule peacetime, exercise, and contingency missions (Harrison et al. 1992, Hilliard et al. 1992).

The ADANS architecture is based on a relational database management system, which operates on a network of powerful, UNIX-based workstations stretching across the U.S. with current installations at Scott Air Force Base (AFB), IL; McGuire AFB, NJ; and Travis AFB, CA. The configuration includes a database management system, a form generation tool, graphical display tools, a report generation system, communication software, a windowing system, and more than 500,000 lines of ADANS-unique code. All modules exchange data and run asynchronously. Thus, schedule planners can use the windowing system to keep track of and to modify multiple pieces of information. The three main components of the user interface are movement requirement and airlift resources data management, schedule analysis, and algorithm interaction (Harrison et al. 1992).

All data and algorithms are geographically explicit. The user inputs data on a station-by-station basis with the textual network editor, while the graphical network editor allows the user to establish a network and to enter or to edit information directly on a world map. Analysis functions facilitate determination of such information as the number of cargo and passenger requirements moved, how they are moved, to what aircraft they were assigned, and how station resources were employed (Harrison et al. 1992).

Summary and Future Vision

It is fair to ask what all this effort has done to advance science and to serve the national interest. The primary focus of the laboratory's efforts have been: (1) developing GIS technologies, (2) defining and implementing geographic and other spatial databases, (3) solving complex real-world problems through application of GIS, and (4) advancing the use of GIS within our national infrastructure involving government, industry, and academia. Major contributions include:

- Fundamental development of early geographic computational techniques which supported and accelerated the growth of a commercial industry.
- A testbed and proof of concept for geographic technologies and integrated systems, many of which now enjoy widespread use.
- Decision support to numerous federal agencies and assistance in solving their GIS-related problems.
- Development and integration of key methodologies and databases.
- Improved understanding of various scientific issues based on geographic and other spatial analysis.
- Improved standards for exchange and use of geographic and other spatial information and metadata.
- Promoting the concept of a national geographic information and analysis center to serve as a resource throughout the scientific and user community and to provide a focal point for advancing research and education.

In summary, we at ORNL believe we have played an instrumental role in the GIS revolution by establishing and implementing a coherent vision that has been welcomed by scientific, policy, and management communities. Today, commercial GIS products address many of the technical needs that required so much of our effort in the past, and the research frontiers have moved on to more complex methodological issues. However, no single commercial product today will handle all the current needs for GIS. One of our on-going roles will continue to be the integration of multiple products with in-house technologies to best meet real-world needs that arise.

Our vision of future GIS encompasses the design and implementation of (a) methods and structures to link GIS closely with environmental transport models and process models traditionally used by scientists in other topical fields such as biology, ecology, and economics; (b) migration of GIS and digital remote sensing techniques to supercomputers; (c) three dimensional visualization and analysis; (d) temporal analysis in a spatial context; and (e) improved statistical analysis capabilities for geographic and other spatial data. The use of supercomputers will become even more important as new data collections (e.g., the next generation of high-resolution satellite imagery) inundate the scientific community with terra-bytes of information. Justification for collecting and using these data will depend on the ability to extract meaningful information for which supercomputer technology is critical. We are currently addressing these and other technological issues, such as GIS animation, telecommunications, and real-time GPS and video linkage with GIS. We hope to maintain a leadership position through continued advancement of hardware and graphics systems, GIS software, and databases that will solve complex spatial problems more effectively. Fast server/client technologies with unified relational database management systems that store and integrate both tabular and spatial data will provide a major impetus to large enterprise GIS operations. We believe knowledge-based expert systems will play a role in advancing future development and use of GIS technologies. We hope to assist the GIS community in improving standards and quality assurance procedures, and we look forward to assisting in enhancement of the National Spatial Data Infrastructure.

Ultimately, we view GIS as an integrating technology with the potential to improve all branches of science that involve location, place, or movement. Consider, for example, that most of the advances in medical imaging have been based on visual analysis. Imagine how much greater the potential would be if the images were enhanced by data structures and analytical tools similar to those employed in analysis of the three-dimensional Earth. We envision that certain technological thresholds will open the door to entire fields. For example, true three-dimensional analysis (beyond visualization) and temporal GIS will open the door to geophysicists and paleogeographers studying plate tectonics and the dynamic forces operating beneath Earth's surface. The single advancement of linking GIS with transport models and process models will suddenly enable biologists, ecologists, sociologists, economists, and numerous other disciplines to incorporate spatial logic and geographical analysis alongside their traditional approaches. We look forward to a truly revolutionary new form of science as these and other developments take place.

Bibliography

Abler, R. F. 1987. "The National Science Foundation National Center for Geographic Information and Analysis." *International Journal of Geographical Information Systems,* 1 (4): 303–326.

Association of American Geographers. 1978. *AAG Newsletter.* Washington, D.C.: Association of American Geographers.

Baxter, F. P., W. C. Barr, T. R. Crow, F. G. Goff, and W. C. Johnson. 1972. "Ecological Analysis." Memo Report #72-4. Work supported by the National Science Foundation RANN Program under National Science Foundation Interagency Agreement No. AASA-R-4-79.

Berry, L. G., D. J. Bjornstad, F. D. Boercker, R. M. Davis, J. E. Dobson, W. R. Emanuel, R. B. Honea, D. D. Huff, J. R. Hyndman, C. R. Kerley, J. T. Kitchings, J. M. Klopatek, B. D. Murphy, D. C. Parzyck, R. J. Raridon, and H. H. Shugart. 1978. "National Coal Utilization Assessment: A Preliminary Assessment of Coal Utilization in the South." Report to Assistant Secretary for Environment Department of Energy, ORNL/TM-6122.

Church, R. L., and T. L. Bell. 1981. "A Comparison of Two Baseline Approaches to Regional Energy Facility Siting." *GeoJournal,* 3: 17–36.

Coleman, P. R., and R. C. Durfee. 1992. "An Overview and Tutorial for the Oak Ridge Hydro-Geochemical System (HGCS)," Version 2.2. Project report for Kelly Air Force Base. Unpublished.

Coleman, P. R., R. C. Durfee, R. G. Edwards. 1977. "Application of a Hierarchical Polygonal Structure in Spatial Analysis and Cartographic Display." *Proceedings of Advanced Study Symposium on Topological Data Structures for Geographic Information Systems.* Boston, MA.

Department of Environment. 1987. *Handling Geographic Information.* London: Her Majesty's Stationery Office.

Dobson, J. E. 1979. "Applied Geography: A Regional Screening Procedure for Land Use Suitability, Analysis." *The Geographical Review,* 69 (2): 224– 234.

Dobson, J. E. 1984. "Statement of Dr. Jerome E. Dobson." *The Role of Information Technology in Emergency Management.* Official Record of Hearings Before the Subcommittee on Investigations and Oversight of the Committee on Science and Technology, U.S. House of Representatives. Washington, D.C.: U.S. Government Printing Office.

Dobson, J. E. 1992a. "Spatial Logic in Paleogeography and the Explanation of Continental Drift." *Annals of the Association of American Geographers,* 82 (2): 187–206.

Dobson, J. E. 1992b. "The Long Hard Battle for NCGIA." *GIS World,* 5 (5): 22–25.

Dobson, J. E. 1993a. "A Rationale for the National Center for Geographic Information and Analysis." *The Professional Geographer*, 45 (2): 207–215.

Dobson, J. E. 1993b. "The Geographic Revolution: A Retrospective on the Age of Automated Geography." *The Professional Geographer,* 45 (4): 431–439.

Dobson, J. E. 1993c. "A Conceptual Framework for Integrating Remote Sensing, GIS, and Geography." *Photogrammetric Engineering & Remote Sensing,* 59 (10): 1491–1496.

Dobson, J. E., and A. D. Shepherd. 1979. "Water Availability for Energy in 1985 and 1990." ORNL/TM-6777. Oak Ridge, TN: Oak Ridge National Laboratory.

Dobson, J. E., and E. A. Bright. 1991. "CoastWatch—Detecting Change in Coastal Wetlands." *Geo Info Systems*, 1: 36–40.

Dobson, J. E., and R. C. Durfee. 1987. "Automated Geography: Status and Prospects." In H. W. Windhorst, ed., *The Role of Geography in a Post-Industrial Society, Vechtaer Arbeiten zur Geographie und Regionalwissenshaft,* 5. Vechta, F.R.G.: Department of Geography, Universitat Osnabruck, abt. Vechta.

Dobson, J. E., R. M. Rush, and R. W. Peplies. 1990. "Forest Blowdown and Lake Acidification." *Annals of the Association of American Geographers*, 80 (3): 343–361.

Dobson, J. E., R. W. Peplies, and R. M. Rush. 1987b. "Lake Acidification." *Environment,* 29 (5): 2–3.

Dobson, J. E., A. D. Shepherd, R. G. Palmer, and S. Y. Chiu. 1977. "A Nationwide Assessment of Water Quantity Impacts of the National Energy Plan," vol. I. ORNL/TM-6098, ORNL/OEPA-3. Oak Ridge, TN: Oak Ridge National Laboratory.

Dobson, J. E., R. W. Peplies, E. C. Krug, and R. M. Rush. 1987a. "Versauerung von seen in Neu-England." *Geographische Rundschau,* 39: 492–497.

Dobson, J. E., E. A. Bright, R. L. Ferguson, D. W. Field, L. L. Wood, K. D. Haddad, H. Iredale III, V. V. Klemas, R. J. Orth, and J. P. Thomas. 1995. "NOAA Coastal Change Analysis Program, Guidance for Regional Implementation," Version 1.0. NOAA Technical Report NMFS 123. Washington, D.C.: National Oceanographic and Atmospheric Administration.

Durfee, R. C. 1972. "The Use of Factor and Cluster Analysis in Regional Modeling." ORNL/TM-3720. Work supported by the National Science Foundation RANN Program under NSF Interagency Agreement No. AASA-R-4-79.

Durfee, R. C. 1974. "ORRMIS Oak Ridge Regional Modeling Information System," Part I. ORNL-NSF-EP-73. Work supported by the National Science Foundation RANN Program under NSF Interagency Agreement No. AASA-R-4-79.

Durfee, R. C., and P. R. Coleman. 1983. "Population Distribution Analysis for Nuclear Power Plant Siting." NUREG/CR-3056, ORNL/CSD/TM-197. Oak Ridge, TN: Oak Ridge National Laboratory.

Durfee, R. C., R. A. McCord, and J. E. Dobson. 1993. "The Application of GIS and Remote Sensing Technologies for Site Characterization and Environmental Assessment." *Proceedings of the 1993 Federal Environmental Restoration Conference and Exhibition*, Washington, D.C., Hazardous Materials Control Resources Institute.

Durfee, R. C., R. A. McCord, and J. K. Thomas. 1992. "GIS Applications in Environmental Restoration at Federal Facilities." *1993 International GIS Sourcebook.* Fort Collins, CO: GIS World, Inc.

Durfee, R. C., R. G. Edwards, M. J. Ketelle, and R. B. Honea. 1975. "Assignment of ERTS and Topographical Data to Geodetic Grids for Environmental Analysis of Contour Strip Mining." Unpublished.

Durfee, R. C., et al. 1988. "Review of Geographic Processing Techniques Applicable to Regional Analysis." Report to the Environmental Protection Agency. ORNL/CSD/TM-226. Oak Ridge, TN: Oak Ridge National Laboratory.

Durfee, R. C., S. M. Margle, L. E. Till, E. P. Tinnel, P. R. Coleman, and B. C. Zygmunt. 1989. "Development of an Integrated Graphics System (IGS) for Spatial Data Base Management and Analysis." Project report for the Air Force IGS Program and Energy Systems Technology Transfer.

Durfee, R. C., P. R. Coleman, F. E. Latham, S. M. Margle, R. B. Honea, D. W. Jones, D. S. Shriner, R. J. Olson. 1993a. "A Geographic Data Processing Demonstration: A Planning Aid for the EPA National Surface Water Survey Project." Project report for DOE/Office of Environmental Assessment. Washington, D.C.: U.S. Department of Energy.

Durfee, R. C., et al. 1993b. "Guidelines for the Creation and Management of Geographic Data Bases within a GIS Environment," 2 volumes and appendices, Version 1.0. ES/ER/TM-56.

Edwards, R. G., and P. R. Coleman. 1976. "IUCALC—A Fortran Subroutine for Calculating Polygon-Line Intersections, and Polygon-Polygon Intersections, Unions, and Relative Differences." ORNL/CSD/TM-12. Oak Ridge, TN: Oak Ridge National Laboratory.

Edwards, R. G., R. C. Durfee, and P. R. Coleman. 1977. "Definition of a Hierarchical Polygonal Data Structure and the Associated Conversion of a Geographic Base File From Boundary Segment Format." *Proceedings of Advanced Study Symposium on Topological Data Structures for Geographic Information Systems,* Boston, MA.

Ferguson, R. L., L. L. Wood, and D. B. Graham. 1993. "Photogrammetric Detection of Spatial Change in Seagrass Habitat." *Photogrammetric Engineering & Remote Sensing,* 59 (6): 1033–1038.

Haring, L. L, J. F. Lounsbury, and J. W. Frazier. 1992. *Introduction to Scientific Geographic Research.* Dubuque, IA: William C. Brown Publishers.

Harrison, I. G., M. R. Hilliard, I. K. Busch, C. Liu, R. S. Solanki, I. R. Moisson, and R. D. Kraemer. 1992. "Automated Airlift Scheduling, a Geographical Perspective." Paper presented at the Annual Meeting of the Association of American Geographers, San Diego, CA.

Hilliard, M. R., R. S. Solanki, C. Liu, I. K. Busch, I. G. Harrison, and R. D. Kraemer. 1992. "Scheduling the Operation Desert Storm Airlift: An Advance Automated Scheduling Support System." *Interfaces,* 22 (1): 131–146.

Hodgson, M. E., Y. Cheng, P. R. Coleman, and R. C. Durfee. 1995. "Computational GIS Burdens; Solutions with Heuristic Algorithms and Parallel Processing." *Geo Info Systems,* 5 (4): 28–37.

Honea, R. B., and E. L. Hillsman. 1979. "Regional Issue Identification and Assessment (RIIA): An Analysis of the Mid-Range Projection Series C Texas, Oklahoma, and New Mexico." ORNL/TM-6941. Oak Ridge, TN: Oak Ridge National Laboratory.

Honea, R. B., E. L. Hillsman, and R. F. Mader. 1979a. "Oak Ridge Siting Analysis: A Baseline Assessment Focusing on the National Energy Plan." ORNL/TM-6816. Oak Ridge, TN: Oak Ridge National Laboratory.

Honea, R. B., D. L. Wilson, C. A. Dillard, R. C. Durfee, C. H. Petrich, and J. A. Faber. 1979b. "Computer Software to Calculate and Map Geologic Parameters Required in Estimating Coal Production Costs." ORNL/TM-6548. Oak Ridge, TN: Oak Ridge National Laboratory.

Honea, R. B., F. P. Baxter, R. C. Durfee, R. G. Edwards, N. S. Fischman, R. W. Peplies, C. H. Petrich, R. M. Rush, K. A. Taft, C. E. Tanner, and D. L. Wilson. 1984. "A National Inventory of Abandoned

Mine Land Problems: An Emphasis on Health, Safety, and General Welfare Impacts." ORNL-6070. Oak Ridge, TN: Oak Ridge National Laboratory.

Jalbert, J. S., and A. D. Shepherd. 1977. "A System for Regional Analysis of Water Availability." ORNL/NUREG/TM-82. Oak Ridge, TN: Oak Ridge National Laboratory.

Jalbert, J. S., and J. E. Dobson. 1977. "A Cell-Based Land Use Screening Procedure for Regional Siting Analysis." ORNL/NUREG/TM-80. Oak Ridge, TN: Oak Ridge National Laboratory.

Joy, D. S., P. E. Johnson, and S. M. Gibson. 1982. "Highway, a Transportation Routing Model: Program Description and User's Manual." ORNL/TM-8419. Oak Ridge, TN: Oak Ridge National Laboratory.

King, A. D., R. C. Durfee, D. T. Bell, S. R. Conder, and B. W. Moll. 1993. "A Proposal to Vice President Gore's Environment Task Force—Use of Department of Energy Waste Sites from the Oak Ridge Reservation to Assess the Application of National Technical Means for Environmental Restoration." Item No. 66569, ORNL/TM-2972. Oak Ridge, TN: Oak Ridge National Laboratory.

Klemas, V. V., J. E. Dobson, R. L. Ferguson, and K. D. Haddad. 1993. "A Coastal Land Cover Classification System for the NOAA CoastWatch Change Analysis Project." *Journal of Coastal Research,* 9 (3): 862–872.

Landers, D. H., J. M. Eilers, D. F. Brakke, W. S. Overton, P. E. Kellar, M. E. Silverstein, R. D. Schonbrod, R. E. Crowe, R. A. Linthurst, J. M. Omernik, S. A. Teague, and E. P. Meier. 1987. "Western Lake Survey Phase I Characteristics of Lakes in the Western United States: Volume 1 Population Descriptions and Physio-Chemical Relationships." Contribution to the National Acid Precipitation Assessment Program, EPA/600/3-86/054a. Washington, D.C.: Environmental Protection Agency.

Levy, H. S., S. M. Margle, E. P. Tinnel, R. C. Durfee, D. E. Olins, and A. L. Olins. 1986. "The Varifocal Mirror for 3-D Display of Electron Microscope Tomography." *Journal of Microscopy,* 145 (2): 179–190.

Mark, D. M., and J. Bossler. 1995. "The University Consortium for Geographic Information Science." *Geo Info Systems,* 5 (4): 38–39.

McCord, R. A., R. C. Durfee, L. D. Voorhees, L. J. Allison, R. J. Olsen, M. L. Land, and J. K. Thomas. 1992. "OREIS Technical Brief: Oak Ridge Environmental Information System." *Environmental Restoration,* 1 (1).

McHarg, I. L. 1969. *Design with Nature*. New York: Doubleday.

Meyers, C. R., Jr., A. H. Voelker, and R. C. Durfee. 1972. "Land-Use Spatial Activities Simulation Model." Memo Report #72-2. Work supported by the National Science Foundation RANN Program under NSF Interagency Agreement No. AAA-R-4-79.

Meyers, C. R., Jr., D. L. Wilson, and R. C. Durfee. 1976. "An Application of the ORRMIS Geographical Digitizing and Information System Using Data from the CARETS Project." ORNL/RUS-12. Oak Ridge, TN: Oak Ridge National Laboratory.

National Institute of Standards and Technology (NIST). 1992. *Spatial Data Transfer Standard.* FIPS Publication 173. Springfield, VA: National Technical Information Service.

Oakes, E. H., and A. H. Voelker. 1983. "Wilderness Designation of Bureau of Land Management Lands and Impacts on the Availability of Energy Resources." ORNL/TM-8310. Oak Ridge, TN: Oak Ridge National Laboratory.

Olins, D. E., A. L. Olins, H. A. Levy, R. C. Durfee, S. M., Margle, E. P. Tinnel, and S. D. Dover. 1983. "Electron Microscope Tomography: Transcription in Three Dimensions." *Science,* 220: 498–500.

Olson, R. J., C. J. Emerson, and M. K. Nungesser. 1980. "Geoecology: A County-Level Environmental Database for the Conterminous United States." Prepared for Regional Impacts Division Office of Environmental Assessments, Assistant Secretary for Environment, United States Department of Energy, Washington, D.C. ORNL/TM-7351. Oak Ridge, TN: Oak Ridge National Laboratory.

Olson, R. J., C. R. Kerley, and G. W. Westley. 1972. "Socioeconomic Analysis." Memo Report # 72-5. Work supported by the National Science Foundation RANN Program under NSF Interagency Agreement No. AAA-R-4-79.

Olson, R. J., J. R. Krummel, and R. C. Durfee. 1983. "Review of Data and Analysis Techniques for Addressing Regional Acid Rain Issues." Draft report prepared for Corvallis Environmental Research Laboratory, Corvallis, OR, sponsored by the U.S. Environmental Protection Agency. Unpublished.

Peterson, B. E. 1984. "Interline: A Railroad Routing Model." ORNL-TM 8944. Oak Ridge, TN: Oak Ridge National Laboratory.

Petrich, C. H. 1984. "EIA Scoping for Aesthetics: Hindsight from the Greene County Nuclear Power Plant, Increasing the Relevance and Utilization of Scientific and Technical Information." In S. L. Hart, G. A. Enk, and W. F. Hornick, eds., *Improving Impact Assessment.* Boulder, CO: Westview Press, Inc.

Rosen, A. E., R. J. Olsen, G. K. Gruendling, D. J. Bogucki, J. L. Malanchuk, R. C. Durfee, R. S. Turner, K. B. Adams, D. L. Wilson, and P. R. Coleman. 1988. "An Adirondack Watershed Database: Attribute and Mapping Information for Regional Acidic Deposition Studies." Environmental Sciences Division Publication No. 2820, ORNL/TM-10144. Oak Ridge, TN: Oak Ridge National Laboratory.

Rush, R. M., R. B. Honea, E. C. Krug, R. W. Peplies, J. E. Dobson, and F. P. Baxter. 1985. "An Investigation of Landscape and Lake Acidification Relationships." ORNL/TM-9754. Oak Ridge, TN: Oak Ridge National Laboratory.

Schuller, C. R. 1972. "Sociopolitical Analysis." Memo Report # 72-6. Work supported by the National Science Foundation RANN Program under NSF Interagency Agreement No. AAA-R-4-79.

Shepherd, A. D. 1979. "A Spatial Analysis Method of Assessing Water Supply and Demand Applied to Energy Development in the Ohio River Basin." ORNL/TM-6375. Oak Ridge, TN: Oak Ridge National Laboratory.

Stephenson, R. L. 1972. "CATCH: Computer Assisted Topography, Cartography, and Hypsography, Part II ORGRAT: A General Subroutine for Drawing Graticules." EDFB-IBP 72-7 International Biological Program UC-48 Biology and Medicine, also issued as ORNL-TM-3790. Oak Ridge, TN: Oak Ridge National Laboratory.

Tinnel, E. P., and W. J. Hinze. 1981. "Preparation of Magnetic Anomaly Profile and Contour Maps from DOE-NURE Aerial Survey Data." ORNL/CSD/TM-155. Oak Ridge, TN: Oak Ridge National Laboratory.

Tucker, T. C. 1973. "CATCH: Computer Assisted Topography, Cartography, and Hypsography Part III, PERSPX: A Subroutine Package for Perspective and Isometric Drawings." ORNL-TM-3790. Oak Ridge, TN: Oak Ridge National Laboratory.

Tucker, T. C. 1976. "CATCH: Computer Assisted Topography, Cartography, and Hypsography Part IV: SEESIJ-A Subroutine Package for Determining the Visibility of Objects Throughout a Region." ORNL/TM-3790. Oak Ridge, TN: Oak Ridge National Laboratory.

University of Washington, Department of Civil Engineering. 1977. "Regional Environmental Systems; Assessment of RANN Projects." NSF/ENV 76-04273.

Voelker, A. H. 1976. "INDICES, a Technique for Using Large Spatial Data Bases." ORNL/RUS-15. Oak Ridge, TN: Oak Ridge National Laboratory.

Voelker, A. H., C. R. Meyers, Jr., R. C. Durfee D. L. Wilson, and T. C. Tucker. 1972. "Model Data Acquisition and Display." Memo Report # 72-3. Work supported by the National Science Foundation RANN Program under NSF Interagency Agreement No. AAA-R-4-79.

Voelker, A. H., H. Wedow, E. Oakes, and P. K. Scheffler. 1979. "Data Report: Research Ratings of the Rare II Tracts in the Idaho-Wyoming-Utah and Central Appalachian Thrust Belts." ORNL/TM-6885. Oak Ridge, TN: Oak Ridge National Laboratory.

Voorhees, L. D., L. A. Hook, M. J. Gentry, R. A. McCord, M. A. Fulkner, K. A. Newman, and P. T. Owen. 1988. "Database Management Activities for the Remedial Action Program at ORNL: Calendar Year 1987." ORNL/TM-10694. Oak Ridge, TN: Oak Ridge National Laboratory.

Voorhees, L. D., R. A. McCord, R. C. Durfee, M. L. Land, R. J. Olson, J. K. Thomas, and E. P. Tinnel. 1991. "Oak Ridge Environmental Information System (OREIS) Phase I—System Definition Document." Prepared for the Environmental Restoration Division. Unpublished.

Wilson, D. L. 1976. "CELNDX; A Computer Program to Compute Cell Indices." ORNL/RUS-14. Oak Ridge, TN: Oak Ridge National Laboratory.

CHAPTER 14

State and Local Government GIS Initiatives

Lisa Warnecke

Introduction

The United States is governed by a combination of local, state, and federal governments. Over time, states have had an increasingly important role in American government in general and regarding geographic information systems (GIS). Preparing this history of GIS in state governments is a formidable task because it requires analysis of my own extensive research on GIS in states over the last ten years with another approach, as well as analysis of others' work before and during this time.

This chapter is perhaps more challenging than some others because institutional memory is often lost in state and local governments since employees are usually not supported to maintain or publicize information about historical conditions. Accordingly, much information about GIS during the last 30 years has not been recorded or is not available. However, numerous individuals have been involved in state GIS for over a decade and provide insight in this regard. For example, Paul Davis and Karen Siderelis are close behind Al Robinette with almost 15 continuous years as the GIS manager of Mississippi and North Carolina, respectively. Other GI/GIS

Lisa Warnecke began her career in municipal government where she introduced information systems into the decision making process. She has become a leading authority on GIS evolution and implementation for the 50 United States. Her consultation services focus on policy implications of GIS technology.
Author's Address: Lisa Warnecke, Principal, GeoManagement Associates, 256 Greenwood Place, Syracuse, NY 13210

Author's Note: This chapter is dedicated to Alan Robinette, the first director of the Minnesota Land Management Information Center. Beginning in 1974, he served as a state GIS manager for a longer period than any other individual before he passed away in 1992. Al inspired many state and other GIS officials throughout the United States and abroad.

entrepreneurs have worked for states for a similar period, such as Hal Anderson in Idaho, Glen Daigre in Louisiana, Cliff Whitehead in Tennessee, Charles Palmer in Texas, Riki Darling in Utah, Larry Sugarbaker in Washington, and others. Several professors have influenced GIS and educated GIS staff in their states for similar or longer periods, including recently deceased Richard Dahlberg, formerly of Northern Illinois University; David Cowan of the University of South Carolina; Ken Dueker of Portland State University; and James Clapp, D. David Moyer, and Ben Niemann of the University of Wisconsin.

As our form of government in the United States, federalism encourages focus on the states because they are the key middle level of government. In addition, the states have increasingly crucial roles in implementing public policy and in developing, using, and coordinating GIS and geographic information among the 80,000 or more separate units of government in our country. Specifically, states (1) manage many of their own programs which use GIS; (2) direct, support, and request data from various regional and local activities using GIS; and (3) respond to several federal directives, incentives, and data requests that make use of GIS. Over time, each of these roles has served to accelerate state development, use, coordination, and institutionalization of GIS.

This chapter provides a review of state GIS developments and trends over time. The following section reviews state GIS developments chronologically, from the 1960s to the 1990s, and highlights societal, governmental, and technological trends impacted by GIS development in these decades. The third section describes the impetus for GIS development according to application, with GIS now applied to virtually every function of state government. This and the following sections review the earliest known and changing state GIS activities over time in each application area. The fourth section includes a detailed discussion of the history of GIS use in states for planning, growth management, and economic development. GIS use for environmental and natural resources applications is reviewed in the fifth section, followed by an overview of state GIS usage for transportation and utilities regulation in the sixth section. This is followed with discussion about several GIS application areas that have emerged in states in recent years. The final section includes a brief review of GIS coordination and institutionalization in states in the context of trends discussed throughout the chapter and concludes with some vision of GIS in states in the future.

GIS Development Timeline

As revealed in many chapters of this book, GIS is a relatively new technology which essentially began in the 1960s. However, many dedicated public servants in state and local agencies laid the groundwork for GIS by manually making maps and collecting data in previous decades. In many cases, their efforts and their vision laid the foundations for GIS and gave software inventors and experimenters something to test in the "real world." Increasing use of GIS may prove that data custodianship, though often unacknowledged, is a key responsibility of public employees. Without it, the axiom "garbage in, garbage out" holds all too true.

For example, an early effort that subsequently used GIS can be traced back to 1955. At that time, the state of Washington, Department of Natural Resources (DNR) staff developed a series of maps and overlays which showed land ownership, rivers and streams, roads, forestry, and other data. This resource was used by managers and field staff for more than 15 years and was painstakingly automated in the 1970s. Presently, Washington's DNR has one of the largest GIS installations any state agency in the country. It is likely that other state GIS initiatives had a similar origin, but these are not well documented.

State and local governments began to use the tools we now call geoprocessing or GIS in the 1960s (Cornwell 1982). During this decade, numerous states increased their manual mapping efforts level similar to Washington. Many states also responded to the growing public awareness of environmental problems and population growth by planning for use of individual natural resources or other lands. Planning was an early function for GIS during this decade, and it required information. Some early planning and mapping activities were complemented with investigation or experimentation in remote sensing and geoprocessing, two then-emerging technologies. For example, Connecticut, New York, and other states began to interpret and digitize aerial photography into land use categories in the late 1960s. While many states began planning efforts in the 1960s, and some developed automated information systems using emerging geoprocessing tools, only a few of these efforts have continued to the present time, such as in Minnesota and Texas.

The 1970s are well known as a decade of innovation, risk, and rapid technological change and experimentation with GIS. As in other sectors, entrepreneurial employees in various state agencies investigated GIS tools as a method to modernize and improve their operations. However, growth in use of GIS by state governments is also attributable to an increase in activism at all levels of government. The 1970s were a period of growth, investigation, and experimentation, both in terms of technology and government practices. Increased government activity characterized the decade, and this expansion resulted in more funding for technology development and transfer. In addition, the increasing recognition of environmental problems and population growth led to authorization and funding of state land use and natural resources planning. GIS was viewed as a new and improved way to understand conditions, mitigate problems, and plan for the future.

In the 1970s, planning was a more significant, integrating purpose for state GIS than for federal and local governments, as well as in states themselves in subsequent decades. Many states initiated efforts to coordinate and use multiple sets of data to meet planning and other needs, which became known as natural resources information systems (NRISs). About a third of the states had initiated a NRIS by the mid 1970s, and another eight states did so before the end of the decade (Cornwell 1982). Many of the NRISs developed some GIS capabilities. In addition, though not documented in NRIS inventories, GIS initiatives in at least ten states in the 1970s continue today, including in California, Colorado, Florida, Idaho, Maine, Michigan, New Hampshire, Oregon, Tennessee, and Washington (Warnecke et al. 1992). Combined, approximately two-thirds of the states had used some GIS capabilities for one or more applications by the end of the 1970s. Virtually all state applications up to the 1970s were for the support of one or more natural resources or land use planning functions.

State activism was complemented by federal direction and efforts that influenced state GIS activities during the 1970s. Several pieces of federal legislation facilitated GIS development in individual state agencies. The federal government also encouraged and provided funding for state planning, data integration, and the use of technology such as satellite imagery. Federal agencies promoted GIS development by providing grants, data, technical assistance, facilities, and expertise.

For several reasons, government expansion and promotion of technology waned toward the end of the 1970s. Following the recession of the early 1970s, the state land use planning movement lay dormant until the mid 1980s. The benefits of planning and new technologies were also tempered by increasing recognition of inadequate data and technical problems. In some respects, GIS benefits had been oversold, and many practitioners and policy makers realized that GIS tools had significant limitations. This, and some related financial and political constraints, led to the demise of some state NRISs and abandonment of GIS initiatives in some states and other sectors. Almost half of the states that had NRIS or GIS initiatives in the 1970s did not have them in the 1980s.

Significant governing changes in the late 1970s and early 1980s also influenced GIS development. While the 1970s were known for federal activism, states and other governmental entities experienced reduced federal initiatives and funding in the following decade. The 1980s became known for having "fend for yourself federalism." Little federal legislation that encouraged GIS use was adopted in the 1980s. Instead, a "states renaissance" emerged, and states defined and implemented much of the domestic agenda and assumed more prominent roles in several areas traditionally belonging to the federal government. State innovation since then has been manifested in several functions of government, including economic development, social welfare, natural resources management, and environmental protection (Ringquist 1993). State planning directives and programs also resurfaced in the mid to late 1980s after lying dormant for almost ten years.

Many of these state initiatives specifically encouraged and funded new and expanding GIS applications. Inventories conducted in the early 1980s document 16 to 19 functioning state NRISs in the early 1980s, most of which had some GIS capabilities (Caron and Stewart 1984; Cornwell 1982). Eleven of the NRISs with GIS in the early 1980s evolved as state or department-wide GIS today, including those in Alaska, Arizona, Illinois, Kentucky, Maryland, Minnesota, Mississippi, North Carolina, Ohio, Texas, and Utah (Warnecke 1995). Another ten states had GIS initiatives in the 1970s that grew through the 1980s and exist today.

The 1980s is frequently characterized as having significant growth in GIS technical capabilities, commercialization, and applications. In the states, applications growth accelerated dramatically during the mid to late 1980s, and 33 states had GIS activities in one or more agencies (Warnecke 1987). By the end of the decade, each of the 50 states had some GIS activity ranging from one to six different agencies, which included not only natural resources and planning but also transportation, public utility regulation, environmental protection, emergency management, economic development, and social services functions (Warnecke et al. 1992). Virtually all state GIS initiatives in the 1980s continued, either within existing program missions or through significant direction and support authorized by state legislation.

Another important aspect of the "states renaissance" that emerged in the 1980s is the improvement of the states in their fiscal and institutional capacity and capabilities. States established, strengthened, and professionalized their institutions with roles, responsibilities, and programs to accomplish state and federal agendas. This trend was also reflected in several state GIS activities, with increasing policy level interest, support, and coordination through directives, plans, policies, inventories, and standards for data, GIS, and related technologies.

From a historical perspective, coordination activities in geographic information began in the early 1900s, when states established Geographic Names Boards to establish the names of physical features. This was followed in the 1950s by the formation of state mapping advisory committees (SMACs) in about half of the 50 states to identify shared mapping priorities. Many state GIS users groups established in the 1980s were expanded, supplemented, and in many cases subsumed or overshadowed by broader groups, commonly known as GI/GIS councils by the late 1980s (Warnecke 1987).

While a few states previously had state cartographers and surveyors, the greatest growth in the incidence of coordinators, clearinghouses, and service centers to facilitate GI/GIS development and coordination occurred in this decade. For example, from 1985 to 1991 the number of statewide GI/GIS coordinators grew from 17 to 40 and those that were officially authorized tripled from ten to 30 (Warnecke 1995). Some states' GI/GIS coordination efforts were established specifically for GI/GIS, while other state efforts were created to help accomplish other government missions. However, by the end of the decade, all states had some form of GI/GIS coordination group, and 40 had some type of GI/GIS coordination organization within a state agency (Warnecke et al. 1992).

Many of the governing and GIS trends established in the 1980s continued and expanded in the 1990s. For example, worsening fiscal conditions since the 1980s, including a recession in the early 1990s, further constrained government activities, particularly at the federal level. The 1994 elections changed the political leadership of both houses of Congress for the first time in more than 40 years and indicated more support for limiting government. In particular, efforts increased to privatize numerous federal functions or devolve them to states and localities. Federal mapping and related GIS efforts have not been privatized or devolved, although some proposals have been advanced. Partially in response to these federal trends and also to respond to growing needs, the "states renaissance" has continued in the 1990s.

Federal encouragement of state GIS increased during the 1990s for several reasons. Legislation such as the Clean Air Act of 1990 and the Intermodal Surface Transportation Efficiency Act of 1991 (ISTEA) direct and fund state air quality and transportation planning. GIS capabilities represent a significant component of most of these efforts. The first federal executive order concerning GIS was signed in 1994. It directed federal agencies to coordinate their GIS activities with each other, as well as to work more closely with states, localities, and others (Clinton 1994). Individual federal agencies increasingly use GIS for internal purposes because it has become easier and more cost effective to use and because agencies have been modernizing their operations. Many federal agencies support GIS activities in their counterpart state agencies to help meet federal program missions.

By the 1990s, GIS were used to some extent by all states. In addition, it was the first decade with GIS applied in virtually all functions of government (Warnecke 1995). As discussed in the following section, GIS was originally initiated for environment and natural resources (ENR) or planning missions. However, its use expanded to several other application areas and agencies during this decade. While GIS was initially used on minicomputers, work stations and personal computers became the preferred platforms in the 1990s. Another important trend in this decade was that governments focused on providing better services to citizens. As technologies improved, attention grew regarding data and dissemination. An unprecedented level and formalization of state GIS coordination and institutionalization had occurred by the 1990s. An increasing number of informal efforts become official during this decade, reflected in more GIS authorizations, direction, funding, and activities. Moreover, many GIS users in and outside state government increased their involvement in state efforts because states are emerging as focal points for GI/GIS coordination among all levels of government and sectors in addition to state government. For example, state GIS coordination groups established in the 1980s expanded and strengthened efforts in the 1990s, indicated by broader participation, more attention to policy matters, and a wider range of data and technologies. In addition, personnel and financial resource commitments for statewide GI/GIS coordination also increased in the 1990s, along with the formalization and strengthening of the roles of GI/GIS coordination entities. This trend toward greater formalization and institutionalization of state GIS use and coordination is expected to continue.

Impetus for Development

GIS has now been applied to virtually every function of governance, as revealed in a 1994 inventory of GIS use within the 50 states (Warnecke 1995). Table 14.1 categorizes these findings according to a broad list of all government functions. For example, the table reveals GIS use in numerous states for essentially all types of environmental and natural resources (ENR) functions, including water resources (49), wildlife (42), geology (39), waste management (30), air quality (29), and forestry and agriculture (27 each), with additional uses including energy, public lands, and parks management.

As in other organizations, states initially used GIS independently for a variety of purposes, including automation of manual efforts such as mapping, analysis of conditions, or improvement of planning and management processes. More recent GIS efforts strived to accomplish a combination of these objectives and in many cases to integrate GIS as a part of managing and operating government. Early GIS development for several government functions is given in Table 14.1.

Table 14.1 GIS use in state government, classified by government function.

Number of GIS in Use	*Government Function*
General State Government: Planning, Administration, Finance, Revenue, Asset Management	
13	Revenue, including Property Taxation
13	Census Data Center
12	State Planning
9	Budget, Finance, Comptroller, State Property Management
4	State Surveyor, Cartographer, Geographer
3	Library
1	Banking Regulation
Environmental/Natural Resources	
49	Water-Quantity, Quality, Rights, or Drinking
42	Wildlife, Game, Fish or Biological Resources
39	Geological Survey
30	Waste Management, including Solid, Low Level
29	Air Quality
27	State Forestry Organization
27	Agriculture
24	Oil/Gas/Mining Regulation and Reclamation
22	Public Lands Management
22	Parks Management
20	Natural Heritage Program
18	Coastal Resources
12	Energy
Cultural Resources	
19	Historic Preservation
14	Archaeology
1	Other—Museum
Infrastructure	
50	Transportation
9	Utility Regulatory Commissions
Human Services	
25	Health (primarily epidemiology)
6	Social Services
5	Employment Security and Labor
3	Education
Other	
24	Public Safety, Emergency Management and Military
20	Economic Development
20	Community and Local Affairs

As described by Moyer and Neimann (this volume), land recordation in the United States is antiquated compared to other countries, and modernized land information systems (LIS) have been recommended for decades. LIS, sometimes referred to as a multipurpose cadastre, are large-scale information resources referenced by individual parcels of land. Improvements in information technologies and the demand for better information about land conditions and uses have encouraged some localities to develop automated LIS. Such systems were advocated for determination of equitable and efficient property assessment and taxation. At present, they serve as the basic conceptual model for multipurpose GIS efforts in local governments (Somers 1987). When comprehensively implemented, LIS can support multiple local services, such as utilities, public works, planning, building inspection, public safety, and others. Due in large part to the efforts of James Clapp, David Moyer, and Ben Neimann, Wisconsin continues to be the one state in the union with a land records modernization program.

Local GIS efforts began in various parts of the country, sometimes with federal or state support. The Census Bureau's Census Use Study evaluated computer mapping in New Haven, CT, in the mid 1960s and became a prototype for urban planning systems. Beginning in 1969, the Urban Information Systems Interagency Committee, chaired by the Department of Housing and Urban Development (HUD), provided funding to investigate the feasibility of and initiate automated information systems in six municipalities. The largest efforts were in Charlotte, NC, and Wichita Falls, TX (National Research Council 1980).

Early local GIS efforts sometimes focused on mapping automation, as in Houston, TX, and Milwaukee, WI, or planning, as in Clark County, NV, and San Bernardino, CA. Some early systems came to embrace additional objectives, thus becoming multipurpose LIS. Other systems suffered from lack of funding and support and were eventually abandoned. Some localities developed multipurpose LIS upon beginning to use GIS, such as Fairfax County, VA, and Pima County, AZ. A trend toward multi-participant systems emerged among some localities in the 1980s. In these cases, municipalities, counties and/or utilities shared GIS resources for the same or overlapping geographic areas, as in the metropolitan areas near Cincinnati, OH, Indianapolis, IN, Knoxville, TN, and San Diego, CA.

Over the years, some states assumed an active role in facilitating local LIS. In 1977, North Carolina established a program which provided funding and assistance for counties to modernize their land records systems and prepare new base and property maps. In 1990, the Wisconsin Land Information Program began to provide counties with funding and direction for LIS development in coordination with other local jurisdictions and state government. These are the only two states with programs that provided funding specifically for local LIS. However, other states have encouraged local GIS development to support other missions.

Planning, Growth Management, and Economic Development

Some of the earliest GIS applications were initiated to assist in land use planning. Since the formation of the United States, local governments have had the lead authority over land use, as institutionalized through zoning and subdivision laws and review. However, in the 1960s and 1970s, there was growing concern about environmental degradation and its impacts on humanity following publication of Rachel Carson's *Silent Spring* (1962) and similar work. Environmental problems, a growing population, and a seemingly imminent energy crisis stimulated action by state and federal governments. They responded by increasing their involvement in land use and planning. This period was also marked by a dramatic growth in federal funding for state and local activities. This governmental activism facilitated interest, funding, and incentives for GIS development that are still felt today.

Federal legislation not only provided strong incentives for planning efforts among federal agencies but also influenced state and local practices. The federal National Environmental Policy Act of 1969 (NEPA) encouraged the recognition of environmental impacts in land use planning practices and policies. Other federal legislation promoted and funded planning. For example, the Coastal Zone Management Act of 1972 helped establish coastal zone planning, and Section 208 of the Federal Water Pollution Control Act Amendments of 1972 provided direction and funding for areawide planning for waste treatment and water quality control. The adoption of a comprehensive federal land use policy was seriously debated and discussed in the late 1960s and early 1970s, and this helped lead to the land use and cover database and classification scheme developed by the U.S. Geological Survey.

The federal government provided significant funding, incentives and support for state and local land use planning activities. The Housing Act of 1954 and the Housing and Community Development Act of 1974 funded state and local planning agencies. The states used funds from the Comprehensive Planning Assistance Program (Section 701) to develop state land use plans and administered 701 grants to local and regional planning agencies. Grant recipients were directed to prepare a land use element as part of their comprehensive plan. This plan included an assessment of the condition, location, and future needs for facilities, land use, and open space, while considering their environmental impacts. Substantial amounts of data concerning natural resources, demographics, and economics were needed to meet these requirements. This facilitated experimental use of geoprocessing and related technologies.

Section 701 funding and incentives specifically stimulated development of GIS in some states. In 1974, the Maryland Automated Geographic Information (MAGI) system was initiated with these funds. It assisted in statewide land use planning and was used to prepare the state comprehensive land use and development plan. Maryland was one of the first states to use Harvard's GRID software. John Antenucci, president of PlanGraphics, Inc., and Jay Morgan, a professor at Towson State University, were among the early users of MAGI. Section 701 funds also supported development of the Mississippi State Data Base (MSDB) in the late 1970s. It used a more advanced GIS software from Harvard University, known as IMGRID, as well as imagery from the National Aeronautics and Space Administration (NASA) and soils data from the Soil Conservation Service. In the early 1980s, MSDB became known as the Mississippi Automated Resource Information System (MARIS) and was one of the first systems that was authorized by the governor and legislature as the statewide GIS. Paul Davis has served as MARIS director since then.

State governments also established planning direction during this period, beginning with the first statewide land use control law adopted in Hawaii in 1961. In addition, numerous states adopted environmental laws similar to NEPA and conducted related planning efforts. For example, Florida, Oregon, Vermont, and others imposed land use controls or required environmental permits during the early 1970s. These are often considered as major steps toward comprehensive land use planning. The U.S. Advisory Commission on Intergovernmental Relations, the American Bar Association, and the American Planning Association endorsed state-level land use planning and policy making. In 1976, the American Law Institute developed a Model Land Development Code which reflected this emerging interest in state planning of critical areas and state oversight of local planning.

Heightened planning activity and improving technologies since the late 1960s allowed states to develop land use data and technical capabilities to meet planning needs. The Minnesota Land Management Information System (MLMIS), one of the oldest continuing state GIS, was initiated in 1967 through state funding provided to the University of Minnesota. An early MLMIS product was a land use map which addressed the rapid development of lake shore areas. Subsequently, the state invested approximately $2 million in the development and analysis of a multipurpose statewide geographic database at the 1:250,000 scale. MLMIS was moved to state government and its name was changed to the Land Management Information Center (LMIC) in 1977. It is now one of the largest state GIS in the country. Its first director, Alan Robinette, served in this capacity from 1974 until his death in 1992. He was an inspiration to many GIS professionals within and outside Minnesota.

Another early system was New York's Land Use and Natural Resources (LUNR) inventory, funded by the Office of Planning Coordination (OPC) in the late 1960s. Don Belcher and colleagues at Cornell University, including Ernest Hardy and Ron Shelton, interpreted aerial photography and used early versions of Harvard University's GRID software to generate 1:24,000 scale acetate overlays for individual quadrangles in digital form. New York was the first state with a digital base which had one-kilometer grid cells. While the photos and overlays are still available, LUNR data was not maintained or upgraded beyond the early 1970s because OPC was eliminated, and the data had limited applicability.

Other states also developed or investigated potential for coordination, automation, and analysis of mapping and land use/cover data for planning during the 1960s and 1970s. For example, the Connecticut Interregional Planning Program, initiated in the late 1960s, facilitated development and interpretation of statewide aerial photography. These data, which were digitized into 55 land use categories, were subsequently used in Connecticut's statewide planning process in the mid 1970s. The Ohio Capability Analysis Program (OCAP) began in 1973 to assist in planning resource and land use. OCAP originally used grid cell-based software developed at Ohio State University, with data developed for individual counties at the 1:20,000 scale. Largely due to the long-term perseverance of Gary Schaal and others, OCAP is still in operation today.

Some state planning initiatives or investigations were authorized through state legislation. For example, the North Carolina Land Policy Act of 1974 authorized an information service which became known as the Land Resources Information Service (LRIS). While the legislation was eventually abolished, LRIS became one of the largest state GIS in the country. In many respects, this is due to the unique combination of political and technical skills of Karen Siderelis, who has served as its director since the early 1980s. In 1989, LRIS was renamed the Center for Geographic Information and Analysis (CGIA), and in 1991, it was elevated to one of the highest positions of any state GIS, in that it reports to the director of state planning. The Wyoming Conservation and Land Use Study Commission was created by legislation in 1973 to facilitate statewide land use planning and to investigate existing and determine the need for new data. Several other states conducted analyses and developed recommendations for integrated data and approaches for planning, but many of these recommendations were not acted upon.

Better remote sensing data and increased federal support in the 1970s furthered state planning sophistication. In particular, National Aeronautics and Space Administration (NASA) outreach programs for states, and data, studies, grants, and educational programs stimulated several state initiatives. Individuals including Alexander Toyahov, Joseph Vidale, Wayne Morayha, Phil Cress, and Dale Lumb conducted extensive outreach with state and local associations and individual states. For example, NASA funded a Satellite Remote Sensing Program at the National Conference of State Legislatures (NCSL) between 1976 and 1981. This program was managed during part of this time by Paul Tessar, who later served in GIS roles for the states of Arizona, Minnesota, and South Dakota, and now is GIS manager of the Wisconsin Department of Natural Resources. Paul Tessar, Loyola Caron, and others at NCSL conducted some of the earliest inventories of state activities related to GIS (Tessar and Caron 1980).

Individual states made use of emerging technology because of these NASA initiatives. For example, South Dakota's Planning Bureau decided to use satellite imagery since conventional data sources were inadequate to meet data and planning needs. Following a demonstration project with USGS's Earth Resources Observation System (EROS) office in Sioux Falls, Paul Tessar and others developed a Land Resource Information System (LRIS) using Landsat images for rural areas and aerial photography for urban and critical areas. Land cover maps and statistics for each county were generated at a detail of 1.1 acres, using Level II of the Anderson classification system. South Dakota was the first state to analyze Thematic Mapper imagery using state staff and funding resources. However, efforts ceased in the early 1980s when the Planning Bureau's Section 701 funding ended.

Other individual state and local projects were funded by NASA during the late 1970s and early 1980s. For example, NASA helped interpret Landsat imagery into a 17-category land use/cover classification for Pueblo County, CO. Results proved more accurate than those achieved by previous analysis of aerial photography. Georgia was another early user of satellite imagery and geoprocessing software. Similar to Tessar in South Dakota, Georgia's work was largely attributed to the efforts of Bruce Rado, then at the state Department of Natural Resources, and later a founder and now vice president of ERDAS Corporation in Atlanta. Efforts in these and some other states did not continue, often because federal and state support declined or GIS entrepreneurs departed for other positions.

An important thrust of several early state planning GIS efforts was analysis of land suitability and capability. Numerous state officials and others were motivated by Ian McHarg's *Design with Nature* (1969), which encouraged land use planning that is compatible with nature and existing resources. This book provided an important vision for application of GIS to planning. McHarg suggested two approaches. First, he suggested that the intrinsic suitabilities for land uses could be determined by overlaying maps and examining all the relevant factors and physical characteristics of a particular area, including natural, environmental, geological, and manmade characteristics. Second, he suggested that optimal locations can be determined for such a use by examining and weighing all the relevant factors for a particular land use. He envisioned that the optimal land use plan could be determined through synthesis of optimum suitability with optimum location. This methodology provided a new means for organizing, displaying, and synthesizing spatial data and was widely applied to regional and community development projects by the mid 1970s. It was facilitated by new technologies and remotely sensed data. McHarg's first approach was clearly reflected in some state planning offices, particularly where land suitability studies were prepared for specific regions or counties, as in Minnesota, Ohio, and South Dakota.

Additional applications of GIS to planning have evolved over time. The identification and protection of certain geographic areas of states began during the 1970s and advanced significantly in the 1980s and 1990s. Increasing state planning activism was reflected by the designation of certain areas or classes of land as "critical areas." This required some form of state intervention or oversight to ensure protection. New York's Adirondack Park Agency was created in 1971 to protect this mountainous region and became an early GIS user to help plan and regulate its growth. The California Coastal Commission was similarly established to protect coastal areas and also became a GIS user. States increasingly use GIS to identify and preserve such areas. Farmland preservation was an important use in the 1970s, with protection of other areas, such as wildlife habitat and drinking water supply, emerging afterwards.

State planning activism and federal technology transfer and funding opportunities facilitated much of the GIS experimentation in the 1960s and 1970s. However, many efforts struggled and suffered because of limited political and fiscal support, plus technical limitations. The 1973–1975 recession also limited growth, and the state land use planning movement lay dormant until the mid 1980s. As of the end of the 1970s, most states conducted land use planning on an incremental basis, with programs designed to protect particular resources, such as shore lands, rivers, or wetlands, or to regulate certain activities such as industrial facilities, surface mining, or other large developments. Virtually all states had some regulation of procedures for local planning, and sometimes of its content, but it was often limited to a one time only review (Healy and Rosenberg 1979). Localities in most states were required or encouraged to prepare and update community or land use plans. Accordingly, many local planning processes were institutionalized, and they evolved as GIS sites.

During the 1980s, planning attention grew from primarily addressing environmental concerns to also include economic interests, shifting in many respects to economic development or growth management. For example, planners becamc involved in developing office and industrial parks and ensuring that they had adequate infrastructure. Some planners initiated new GIS applications to determine optimal locations and wooed industries to locate in their areas. By addressing broader issues, such as housing and infrastructure finance, planning realized a broader constituency and strengthened support for the notion that local land use control is needed to maximize economic development and good government. The 1980s were also known as a time of growing state activism. This was reflected in a "second wave" of statewide planning and state regulation of urban growth in some states, as well as economic development efforts in virtually all states. Both thrusts provided fertile new ground for GIS applications.

While some early state experimentation with geoprocessing was for planning in the 1970s, use of GIS as a significant component of state planning was limited until the 1980s. One of the earliest and strongest examples of state legislation for statewide planning and growth management was in Florida. While statewide planning legislation had been adopted in 1972, Florida's 1985 act directed a strong top-down approach, requiring that development must be concurrent with available infrastructure. Florida was also a leading state regarding information policy at the time. The policy-level Florida Growth Management Data Network Coordinating Council was created to coordinate information to comply with the act. After initial focus on data processing and telecommunications issues and technology demonstrations, the council highlighted geographic information coordination as a primary objective in the late 1980s. This conceptual maturation at a policy level and related funding encouraged GIS and data development efforts already underway in numerous state, regional, and local agencies. It also provided an institutionalized framework for coordinated GIS development on a statewide basis.

Georgia, Maine, New Jersey, Vermont, Rhode Island, and Washington similarly adopted state planning legislation between 1986 and 1990, and all established GIS as a significant tool to meet state planning objectives. These acts also firmly established the legitimate role of states in local planning. State funding and assistance were provided to develop local plans in accordance with state goals, often including GIS and related data development. In some states, existing or new regional entities were empowered to review local plans or provide technical support. For example, Vermont's 12 Regional Planning Councils were equipped with state-funded GIS hardware and software and are responsible for technical assistance to localities and data applications development and management. Similarly, Georgia's Regional Development Centers were established to have significant data development roles for GIS.

The use of GIS for planning was established in most states with planning legislation; however, such direction was primarily limited to coastal states. A few other states have also facilitated local and regional use of GIS for planning. For example, Utah's Automated Geographic Reference Center has provided counties with a combination of state and federal GIS data for planning use. This is an important component of a state initiative to synchronize state, local, and federal planning activities. In 1994, 11 states were known as using GIS for planning (Warnecke 1995).

While essentially all states with planning initiatives use or encourage GIS, GIS has had far less penetration in state economic development. However, economic development was a leading example of state activism and leadership in the 1980s because virtually all states established some economic development activities during this time. Some of the early planning systems described above envisioned GIS as an aid to economic development. For example, GIS entrepreneurs in New York and South Dakota envisioned that GIS could be used to optimize siting decisions; however, these efforts did not come to fruition. More recently, Illinois, North Carolina, and others used their existing GIS to develop their proposals for the Superconducting Super Collider project in the late 1980s.

South Carolina has been the leading state to use GIS for economic development, due in large part to the vision of David Cowan at the University of South Carolina. State, HUD, and Appalachian Regional Commission funds were used in the late 1980s to develop statewide data at the 1:100,000 scale, and subsequent efforts facilitate GIS use in regional councils. Natural resources, transportation, utilities, and socioeconomic data available through GIS have helped attract businesses, such as Bavarian Motor Works' manufacturing plant now located in Spartanburg. As of 1994, 20 states had initiated similar GIS applications for economic development (Warnecke 1995).

The 1980s were also characterized by a significant reduction in federal support for planning. However, recent federal legislation, such as the Clean Air Act of 1990 and the Intermodal Surface Transportation Efficiency Act of 1991 (ISTEA), support state and local planning and encourage GIS use for air quality and transportation planning. During 1994, HUD began delivery of Map Info software to its state counterparts for use in their processing of community development grants. Grantees are also directed to use GIS as part of their planning process.

Political changes in the early 1990s, particularly the Congressional elections in 1994, minimized support for planning and growth management efforts. For example, efforts to limit development because of the location of endangered species on a site have been referred to as a "taking" of private property rights. It is claimed that in these cases government has deprived landowners of the use and full commercial potential of their land and that therefore, the landowner should be compensated for monetary loss. Some courts have supported these arguments, and various state legislatures and Congress are considering legislation to limit these takings. This trend may encourage GIS use, because it provides an incentive for localities to develop comprehensive, long-range land use plans and zoning laws to ensure the established right of localities to control land use.

Planning use of GIS encourages integration of data from multiple sources to analyze conditions and develop scenarios, while also facilitating commonality among multiple agencies in developing compatible data. However, history shows that government planning can be controversial and cyclical. Political trends in the 1990s served to minimize planning efforts, but many planning processes and GIS have become institutionalized and may not be reduced significantly or easily.

Environment and Natural Resources

Authority, roles, and responsibilities for environmental and natural resources (ENR) have traditionally been shared by federal, state, and local governments, with each level having a leading role for certain aspects of ENR. States have numerous ENR responsibilities, some of which were among the earliest, and are some of the most developed state GIS application areas today.

Early ENR applications were either initiated (1) to improve management practices for various individual natural resources, such as forestry, water, and wildlife, or (2) to automate, integrate, and provide common access to several sets of state natural resources data, usually known as natural resources information systems (NRISs). This section first includes information about systems initiated for individual resources. It follows with a discussion of more comprehensive ENR systems, including NRISs.

Forestry was one of the earliest ENR geoprocessing applications (Tomlinson 1987). State forestry organizations were early state GIS users, as they plan for and manage state-owned forests, conduct statewide forest inventories, mitigate and control fires, manage pests, and provide technical assistance for, and some oversight over, private forested lands. For example, the state of Washington's Department of Natural Resources (DNR) manages more than three million acres of land held in trust to support public institutions including schools. It began a manual mapping program in the mid 1950s for forest classifications to assist in this mission. Mylar overlays of several data sets were utilized until 1971, when DNR began to automate forest inventory and planning using GRID software. An orthophoto and soils mapping program for state forest lands was also initiated in the 1970s. The soils data were maintained on a Calmagraphics Mapping System, and other related information resided in different systems. In 1979, DNR was one of the first state agencies to begin a formal GIS needs assessment and requirements statement. Subsequent GIS installation provided that it was and continues to be an integral part of DNR's work. Their GIS has evolved to be one of the largest GIS installations of any state agency in the country.

Other agencies were also early users of GIS and remote sensing for forestry applications, often because staff learned of applications in other forestry agencies, and state forestry agencies sometimes have more funding available than other ENR agencies. Oregon's Department of Forestry tested a couple of mapping systems beginning in 1972. The Maine Forest Service was the first user of GIS in its state government, developing a system to assess landowners for their proportionate costs for pest management services in the late 1970s. California's Department of Forestry initiated a statewide vegetation classification program using some NASA funding and Landsat imagery in 1979. It also experimented with PIOS software for reforestation and fuel loading modeling. During the early 1980s, Arizona's Forestry Division and the U.S. Forest Service worked together to create a statewide vegetation database using Landsat imagery. Additional state forestry organizations, such as Alaska and Minnesota, initiated substantial GIS efforts by the 1980s, with more than half of these entities using GIS by 1993 (Warnecke and Herrington 1994).

One of the most common state forestry GIS applications was to manage forested public lands. However, some state agencies own lands that are managed for resource management and extraction other than forestry. Some early GIS efforts were initiated to support these needs. For example, Alaska's DNR began to use GIS in 1977 to track several activities on state lands, though efforts were abandoned because of the complexity and difficulty. However, by the early 1980s, multiple Alaska DNR divisions were using GIS, for such purposes as to evaluate oil and gas fields to determine leasing potential. Arizona is one of the most extensive users of GIS for public land management. Its state GIS center, the Arizona Land Resource Information System (ALRIS), has been located in the State Land Department since the late 1970s. ALRIS began use of PIOS, GRID, TOPO, and ELAS software in 1982.

Some states have recently expanded their use of GIS for land management to help determine and prioritize available and desired properties for acquisition. Public land acquisition has been one of the most significant state environmental policy actions at the state level in the 1990s. For example, voters in California, Florida, Maine, New Jersey, and Rhode Island approved land acquisitions to help protect wildlife habitat and other resources in addition to providing recreational opportunities. In Florida, "Preservation 2000" provided $3.2 billion for acquisition. GIS is being used there to inventory potential sites, set criteria for site selection, and monitor acquisition efforts.

Protection of resources was an important goal of some of the earliest GIS initiatives. For example, New York's LUNR, Ohio's OCAP, and South Dakota's LRIS were developed in part to identify valuable farmland and preserve it by providing this information to local planners. Many early land capability and suitability modeling efforts were developed to meet this objective. Additional agricultural GIS uses also evolved, including evaluation of various conditions, crop inventories and yield forecasting, and pest infestation and crop disease monitoring.

Many facets of water resources management were early drivers for GIS. One of the earliest state GIS users grew out of the Texas Water Oriented Data Bank. It was established in the 1960s to coordinate response to droughts that were followed by floods. The legislature authorized a centralized state data bank of all hydrologic data collected by several state agencies. It developed in-house mapping software and expanded to become the Texas Natural Resources Information System (TNRIS) in the mid 1970s. Utah's earliest GIS efforts were initiated to help develop the state's first water plan in the mid 1960s.

Other early GIS efforts were initiated to meet water needs. Authorization and funding in the Florida Water Resources Act of 1972 encouraged the South Florida Water Management District to became one of the earliest and most extensive users of GIS in Florida and the country. Idaho's Department of Water Resources (DWR) began to use satellite imagery and geoprocessing for analyzing and monitoring irrigated agriculture in the mid 1970s and extended to other water resources management applications. Aided by the efforts of Hal Anderson, Tony Morse, and others, Idaho's official state GIS center, now known as the Idaho Geographic Information Center (IGIC), began in 1978 and continues to be located in DWR. Arizona's similar DWR began geoprocessing and imagery applications to inventory irrigated agriculture and manage water rights in response to legislation in 1980. Wyoming's first GIS use was developed in 1985 by the State Engineer's Office to support the informational needs of a large water rights lawsuit.

Some states were also early innovators of wildlife GIS applications. GIS was used to determine, monitor, and analyze species, habitat, and land carrying, capacities. Colorado's first

GIS activities began in 1972 in the Division of Wildlife. It developed a wildlife inventory by digitizing species distributions to obtain computer-generated maps and tabular listings for each of 60 species and their habitats. Don Schrupp and others worked with the U.S. Fish and Wildlife Service's Western Energy Land Use Team (WELET) and the Cooperative Wildlife Research Unit at Colorado State University in the early development of the Map Overlay and Statistical System (MOSS) and the Systems Application Group Information System (SAGIS) software. While not used extensively in states, MOSS and SAGIS were employed by other federal agencies such as the Bureau of Land Management and the National Park Service.

Tennessee's Wildlife Resources Agency (TWRA) was the earliest user of GIS in its state government. However, it also became the largest agency user of GIS following a 1984 study which recommended that the statewide GIS should be located in TWRA. Through long-term efforts of Clifton Whitehead and others, TWRA was the first state wildlife agency in the country to use ARC/INFO and ERDAS software together. In addition, it has conducted several regional GIS projects for federal agencies and others. Some wildlife GIS applications have been stimulated by The Nature Conservancy's Natural Heritage Program. Initiated in 1974, it has provided a focal point in each state for the management of biological data inventories. The Gap Analysis Program, initiated in Idaho in 1988 to inventory, evaluate, and preserve biological diversity, has encouraged use of satellite imagery and GIS because it has been expanded with federal support to many states.

Another significant early GIS application was the determination of lands unsuitable for coal surface mining. The Surface Mining Control and Reclamation Act of 1977 encouraged states to conduct mining regulation. Accordingly, some states used extensive funding from the U.S. Office of Surface Mining to initiate and support their states' first GIS efforts. Funds also supported development of data resources that were used for additional purposes. For example, these early efforts in Illinois and Kentucky, beginning in 1982 and 1979 respectively, evolved to be the largest GIS in these states. PlanGraphics' vice president, Peter Croswell, headed up GIS efforts in Kentucky's Natural Resources Cabinet until the mid 1980s.

GIS applications have also increased for environmental protection, facilitated by federal actions and funding in many states. As described above, the National Environmental Policy Act of 1969 (NEPA) and similar state versions of NEPA encouraged land-use planning that was more sensitive to environmental impacts. This encouraged development and use of ENR data in planning. NEPA also required that all federal policies and actions be subject to an environmental impact assessment (EIS), which stimulated GIS development to conduct this work.

Subsequent federal legislation administered through the U.S. Environmental Protection Agency (EPA) encouraged and funded GIS efforts in numerous states and regions. Section 208 of the Federal Water Pollution Control Act Amendments of 1972 provided direction and funding for areawide planning for waste treatment and water quality control and encouraged early use of satellite imagery and GIS in Georgia, New Jersey, South Dakota, and Texas. In 1975, Ohio's Environmental Protection Agency was the first state environmental agency to use GIS, with funding provided by EPA and state government. EPA support for GIS grew in many states during the 1980s. GIS use was funded for individual media such as water and also to encourage integrated use of data in decision support and management through the State EPA Data Management (SEDM) and other programs. Several state GIS projects were funded under the Resource Conservation and Recovery Act in the late 1980s to coordinate hazardous waste permitting and compliance programs. EPA funded Montana's first use of GIS in 1987. It was initiated to help measure damage and manage four federally-designated Superfund sites on the Clark Fork River, and it later evolved to be the statewide GIS. Most recently, EPA funding has been used to support state GIS use for air pollution planning and control, in accordance with the Clean Air Act of 1990.

Early systems developed for individual ENR functions sometimes evolved to also serve as natural resources information systems (NRISs), or they were complemented by related NRIS efforts. A trend toward the development of state NRISs began in the 1960s and expanded in the next decade, with almost half of the states having NRIS initiatives by the late 1970s (Cornwell 1982). NRISs were usually initiated to provide integrated management, dissemination, and access to multiple sets of environmental and natural resources data to support planning and management of these resources. Several types of data were included in these systems, such as land use and cover, vegetation, topography, geology, hydrology, soil types, wildlife resources, climate, energy, environmental quality, and land ownership. Cell sizes ranged from 0.4 hectare (one acre) to one square kilometer (247 acres). NRISs had applicability for various projects, including inventory or assessment of forest cover, agricultural crops, and wildlife habitat, the monitoring of forest stand conversions and harvesting, water resources or suburban development, and the analysis of land use impacts and of mined land reclamation. NRISs were typically operated by one agency, sometimes on behalf of several agencies, and served as a clearinghouse or repository for automated, and sometimes non-automated, information.

NRISs were initiated as part of state government's response to growing public awareness and concern about environmental degradation, and to improve compliance with the array of new state and federal environmental legislation at the time. NRIS inventories in the 1980s included the early state systems developed for planning purposes that are described above. However, many NRISs had a narrower purpose, and they were housed in natural resource agencies which inherently have a narrower mission than planning agencies. Another important catalyst for the growth of NRISs was the rapidly improving data management, remote sensing, and GIS technologies that became increasingly apparent to states.

The federal government was an important catalyst for NRIS. It sponsored studies which identified state data management problems and needs, provided guidance and recommendations concerning these problems and new technological opportunities, and inventoried legislation and activities in all states. The Department of Interior sponsored a study of environmental data management which found uncoordinated data use and collection at state and federal levels, and it was concluded that most states are highly disorganized and decentralized in handling natural resource data (Council of State Governments 1978). It recommended that each state develop an autonomous data coordinating center with the sole objective of better data management, access, and analysis and that the center should be established by a mandate from the legislature or governor. In addition, NASA funded inventories and guides for state NRISs and satellite imagery use, as well as related projects in individual state governments (Tessar and Caron 1980; Seladones and Harwood 1981).

Most NRISs became early users of GIS, and often were the first GIS in their state. For example, of 16 functioning NRISs identified in the early 1980s, all but Hawaii and Nebraska had GIS capabilities (Caron and Stewart 1984). While many early systems no longer exist, some NRISs or other ENR systems, as in Arizona, Idaho, Montana, North Carolina, Oregon, and Utah, evolved to be official statewide GI/GIS centers. Other systems matured to service entire departments or to provide ENR data to others, as in Kentucky, Michigan, Ohio, and Texas.

These systems, and more recent efforts in other states, reveal a trend from independent use of GIS for individual natural resources to more comprehensive GIS use for ENR policy, planning, and management. A coordinated approach to GIS is being facilitated by a trend toward ENR coordination and organizational consolidation that is resulting from state and other government efforts to improve overall ecosystems planning and management. While use of the term NRIS faded, individual ENR agencies began to institutionalize internal GIS coordinators, clearinghouses, service centers, and groups in the late 1980s. Most of these GIS coordination efforts for ENR are coordinated with statewide GI/GIS initiatives. As discussed in the last section, some GIS coordination entities in ENR agencies serve as the formal or de facto statewide GI/GIS coordination entity (Warnecke 1995). Of all governing functions, ENR had some of the earliest applications and has had the most important role and influence on statewide GI/GIS development and institutionalization.

Infrastructure: Transportation and Utilities

An important distinction between GIS use for ENR from that for planning, growth management, or economic development is that through time the later applications increasingly required additional information. Planning missions that evolved in the 1980s include economic considerations and the physical as well as the natural environment, thus increasing the need for infrastructure data such as for highways and utilities. By this time, some transportation, highways, and utility departments had begun to investigate GIS and related technologies for their own needs, either (1) to automate various large-scale manual processes or (2) to analyze conditions and improve planning and management processes.

Most early transportation and utility users initiated use of computer aided drafting, design (CAD), or mapping (CAM) systems to automate preparation of facility or highway construction drawings and maps. As discussed in Chapter by McDaniels, Howard, and Emery, these efforts in utilities also became known as Automated Mapping/Facilities Management (AM/FM) systems. While some of the earliest users of these systems were large electric, gas, and telephone companies, water and wastewater utilities were among the earliest users of these technologies within municipal governments.

While states do not traditionally operate utilities, they have an important role in franchising and regulating utility activities within their borders. Several opportunities exist for state public utilities commissions (PUCs) to use GIS for internal and regulatory purposes. Eleven PUCs use GIS at this time and another six plan to use it in the future (Demers et al. 1995). Beginning in the late 1970s, North Dakota's PUC was probably the first one to use GIS. In 1985, the Ohio PUC started using GIS to help regulate utilities and has expanded operations for various purposes, including to monitor service areas and analyze demographics. Other states initiated similar efforts in the late 1980s, some in coordination with other agencies as in Kansas, New Hampshire, and Mississippi. In 1990, ten state PUCs were identified as GIS users; however, four of them ceased efforts while another five began using GIS in the 1990s (Warnecke 1990; Demers et al. 1995). Some PUCs and other state agencies began to use GIS to develop addressing data for statewide emergency 911 communications systems or to provide data or technical assistance for localities in this regard, such as Maine, Oregon, and Vermont.

Unlike PUCs, state departments of transportation have been extensive users of CAD for one or several purposes, including for highway mapping and the automation of preliminary and construction drawings. New York and Pennsylvania are acknowledged as the earliest DOTs to develop automated mapping systems. During the late 1960s, New York's DOT worked with LUNR, one of the first state GISs used for planning. Additional states began to develop automated mapping systems in the early 1980s, including Minnesota, New Mexico, Ohio, Texas, and Wyoming. A 1984 survey found that 19 state DOTs had Intergraph systems, four had other systems, 12 were planning to have a system installed by 1986, and 15 had no plans for a system at the time (Arizona Department of Transportation 1984). Some automated mapping installations in DOTs have been among the largest technology configurations of any GIS-related systems in state governments. For example, Wyoming's DOT had over 60 work stations in 1991.

Highway safety and pavement management were additional early applications of GIS in DOTs. Arizona's DOT began development of its Accident Location Identification and Surveillance System (ALISS) in 1970. It included a line segment database with attribute information concerning road center lines which were referenced by state plane coordinates. It constituted the first statewide digital spatial database in Arizona. Ohio's State Accident Identification and Recording System (STAIRS) was designed in 1980 to retrieve and display accident-related information for more than 112,000 miles of roads in the state. GIS capabilities enhanced the utility of STAIRS in road inventory, pavement and bridge management, skid resistance, and other applications. These and similar systems were expanded or separate GIS efforts were developed in the 1980s to utilize analytical GIS capabilities for transportation planning and management processes.

Beginning in 1983, Wisconsin's DOT was perhaps the first one to develop GIS as part of a comprehensive approach to fulfill operational, management, and planning needs. Largely through the efforts of Myron Bacon and David Fletcher and the leadership of DOT Secretary Lowell Jackson, the DOT was the first state agency to use GIS in-house. First concentrating on pavement management, a phased approach was conducted to institutionalize GIS and develop a foundation for multiple applications and technologies. For example, GIS was combined with optical-disc based photologging to view sections of roads that are selected via an automated map.

Colorado's DOT was another early user of GIS. Through the efforts of Greg Fulton and others, it purchased ARC/INFO software in 1986 for planning and pavement management, while a separate CAD effort existed in another part of the agency as in other DOTs. These agencies in California, Missouri, North Carolina, and Rhode Island developed GIS capabilities in the late 1980s. At this time, only a few state DOTs, such as Ohio and Oregon, which used CAD or automated mapping software expanded these efforts to incorporate analytical GIS capabilities (Petzold 1995).

While some state agencies learned about GIS from their sister agencies in the same state, many DOTs developed their CAD or GIS efforts in isolation from other agencies in their state. An exception is the North Carolina DOT which funded more than $1 million in ENR data development for highway planning that is integrated and used by other state agencies. However, DOTs often work with other DOTs to address common problems and needs. For example, they have had interstate committees addressing GIS, an annual GIS for transportation (GIS-T) meeting since 1987, and a set of studies sponsored by the national Transportation Research Board about GIS in DOTs beginning in 1990 (Vonderhohe, Travis, and Smith 1991). Communication and coordination between state DOTs helped accelerate GIS activities in DOTs to a greater extent than interstate relations for other functions of state government.

Since the 1980s, federal involvement, funding, and technical assistance from the U.S. Department of Transportation has also been an effective stimulus for GIS-T. In addition, the Intermodal Surface Transportation Efficiency Act (ISTEA) of 1991 established new transportation planning requirements and encouraged integration of systems that usually operated independently in DOTs. With additional funding and support, recognition grew that GIS could help DOTs accomplish these needs. This complemented multistate and other federal efforts. The use of personal computer-based GIS grew significantly in the 1990s, with a third of the DOTs using GIS on these platforms in 1995 (Moyer 1995). ISTEA also increased GIS efforts in the more than 400 metropolitan planning organizations (MPOs) across the country. While some MPOs and localities had used GIS for transportation in the past, virtually all MPOs are now using GIS or have plans to do so to meet ISTEA needs (Petzold 1995).

Emergence of Additional Applications

While mapping, planning, ENR, and infrastructure applications dominated most early development of GIS in states and substate governments, other governing functions have also benefited by GIS. In particular, applications emerged to improve delivery of several public services, including public safety, such as emergency management, law enforcement, and fire protection, plus social and human services.

An important component of these newer applications is that they often use demographic and socioeconomic data and analysis such as the Census Bureau's DIME file in 1980 and

TIGER database in 1990 (Chapter by Cooke). Perhaps the most significant GIS application impacting governance is the determination of district boundaries for officials elected to Congress, state legislatures and local councils. For example, Colorado's Joint Budget Committee funded the purchase of state government's first commercial GIS software in 1980 to meet reapportionment needs. More than a third of the state legislatures used GIS for this purpose after the 1990 Census, including Georgia, Mississippi, New York, Texas, and Wisconsin. Another use of GIS that promises to impact government is providing the public with access to data. Bob Peterski, formerly of the Colorado Department of Local Affairs, perhaps installed the first GIS technology in a library in the early 1980s. While later eliminated, this initiative proved that citizens could access useful data from GIS and paved the way for many related efforts in the 1990s.

GIS applications seem to accelerate and expand significantly as socioeconomic data is combined with other available data such as for ENR and infrastructure. New GIS uses are possible in planning, policy analysis, and management of several governing functions, as well as to improve operational efficiencies. For example, some agencies develop routing systems that combine traffic congestion and road repair activity to determine optimal routes for police, fire, or other service personnel.

Public safety applications make use of several sets of data and have been an important driver for GIS. For example, flooding after a severe drought in the 1960s caused Texas to develop what became one of the earliest state GIS in the country. In the early 1970s, NASA helped the U.S. Army Corps of Engineers by inventorying all dams in the country using satellite imagery. The National Flood Insurance Act of 1968 and the Flood Disaster Protection Act of 1973 established federal subsidization of insurance of property owners within designated floodway areas. Subsequently, the Federal Emergency Management Agency (FEMA) began a national $17 million program in 1980 to match addresses with flood zones and determine insurance rates for individual properties. While the system worked well and represented the first real time application of GIS, it was eliminated primarily because it threatened several mapping companies that had contracts to assess flood damage (Aangeenbrug 1992). However, the project help convince the federal government to pursue automated mapping, including getting the Census Bureau and USGS together (Aangeenbrug 1992).

Response to actual emergencies has been an important impetus for GIS activities and funding in some states. For example, while GIS efforts were previously underway in their states, agencies expanded their GIS efforts in response to major disasters including the Exxon Valdez oil spill, Hurricanes Andrew and Hugo, the midwest floods, and the west coast fires (Warnecke 1995). These disasters, particularly Alaska's oil spill, caused other states, such as Maine, New Jersey, and Rhode Island, to expand their GIS efforts to be better prepared for emergencies. GIS use has also expanded to address other types of emergencies. State GIS applications have been identified for certain types of emergencies, including flooding and floodplain management (30 states), hazardous materials/radiological incidents (27), earthquakes (13), fires (11), and hurricanes (10) (Warnecke 1995). FEMA is encouraging state GIS applications for emergency management by providing information about these activities to states and others.

Other public safety functions are also aided by GIS. Crime and fire analysis and prevention programs increasingly use GIS, as in Los Angeles, Phoenix, Seattle, and Tacoma. For example, crime statistics are analyzed geographically to help target optimal coverages and times. The California Department of Justice is among agencies also using GIS, including for narcotics tracking. Modernized communications systems, among them E-911, increasingly use GIS to optimize response. States such as Maine, Oregon, and Vermont are using GIS to develop statewide addressing data or provide assistance to localities in this regard.

While adopting GIS later than other parts of state governments, various state human and social services and labor and education agencies began to use GIS in the 1980s. For example, James Welsh, formerly at New York's Department of Social Services, pioneered social services GIS applications in the 1980s. It was used to support the state's neighborhood-based strategy for services delivered to children. South Carolina and New York City similarly used GIS to target services for children based on the locations of poverty, crime, teenage pregnancy, infant morbidity, and other conditions. Illinois and Washington use demographic and migration data with GIS to help locate health care and other facilities and services.

Some state labor developments have also been using GIS since the 1980s, among them Alaska, Arizona, and Washington. Various labor statistics and trends are analyzed geographically. These data are used for transportation and land use planning, including the targeting and locating of services, such as for job search and placement for welfare recipients. State education departments and schools have also utilized GIS. For example, state and local agencies in North Carolina and Washington use GIS to optimize bus routes and schedules and reduce fuel costs. A fast-growing GIS application in some state health departments is in epidemiology because ENR data can be analyzed with data about the occurrence of diseases to understand environmental implications on people. GIS applications in public safety and human and social services have been investigated and developed in an increasing number of state agencies during the 1990s. It is conceivable that these applications will experience the most dramatic growth in the industry in the future.

Concluding Remarks

Through time, as shown in Table 14.1, GIS has been utilized in virtually every function of government. As use matures, GIS also catalyzes planning direction, organizations, and coordinating relationships within and among states and others. In addition, many states and other organizations have evolved to adopt a comprehensive approach for all geographic information (GI) and related technologies. Institutionalized approaches have been evolving through the decades in various ways. As a result, the 1990s can be thought of as the decade of GIS institutionalization.

More analyses exist about state GI/GIS institutionalization than any other level of government. For example, an analysis of state directives funded by the Mapping Science Committee of the National Research Council identified 100 state GI/GIS directives among the 50 states (Warnecke 1993). State GI/GIS coordinating groups represent an important evolution in GI/GIS to facilitate the sharing data for use with GIS. A recent inventory identified 88 independent state GI/GIS groups among the 50 states and at least one group in each state (Warnecke 1996).

These groups have experienced increasing authorization, participation, strength, resources, and responsibility and influence over GI/GIS direction in states, though some are formal and others are unofficial. Localities, federal agencies, regional organizations, academic institutions, Indian tribal governments, utilities, nongovernmental organizations, the business community, and others increasingly participate in GI/GIS groups (Warnecke 1993). In addition, interstate relationships have emerged, such as through the National States Geographic Information Council that was organized by state GI/GIS coordinators and managers in 1991 to enable states to learn from each other and have a common voice to the federal government and others.

Maturing GI/GIS institutionalization is also evidenced by the increasing incidence of GI/GIS coordinators since the 1970s. As shown in Table 14.2, the incidence of these coordinators has increased from a third of the states in 1985 to more than 40 since 1995. The table also reveals stronger growth in the number of authorized coordinators than coordinators in general. Moreover, GI/GIS coordinators are increasingly located in central information technology agencies, rather than natural resources agencies where they were often located in the past (Warnecke 1995).

Many state GI/GIS groups and staffs have led and encouraged coordination through the development and adoption of plans, policies, procedures, guidelines, and standards. They have prioritized and implemented statewide data layers, promoted collaborative planning for future data development and other work, and served as clearinghouses of GI/GIS activities, data, and plans. Another example of GI/GIS institutionalization is the establishment of state programs providing assistance and resources to localities, while also providing incorporating local perspectives in state decision-making (Warnecke et al. 1992).

Figure 14.1 summarizes important trends in the evolution of state GIS use and institutionalization some 30 years. This short period in history has experienced a significant change in the use of GIS and related technologies and in the way state governments lead, manage, and coordinate these activities. Several trends in government and past experience with GIS that are discussed in this chapter suggest continuing state activism and influence upon federal, local, and

Table 14.2 Incidence and authorization of state GI/GIS coordinators.

	Authorized		*Unauthorized*		
Year	*Number*	*Percentage*	*Number*	*Percentage*	*Total*
1985	10	59%	7	41%	17
1988	15	52%	14	48%	29
1991	30	75%	10	25%	40
1994	31	77.5%	9	22.5%	40
1995	33	80.5%	8	19.5%	41

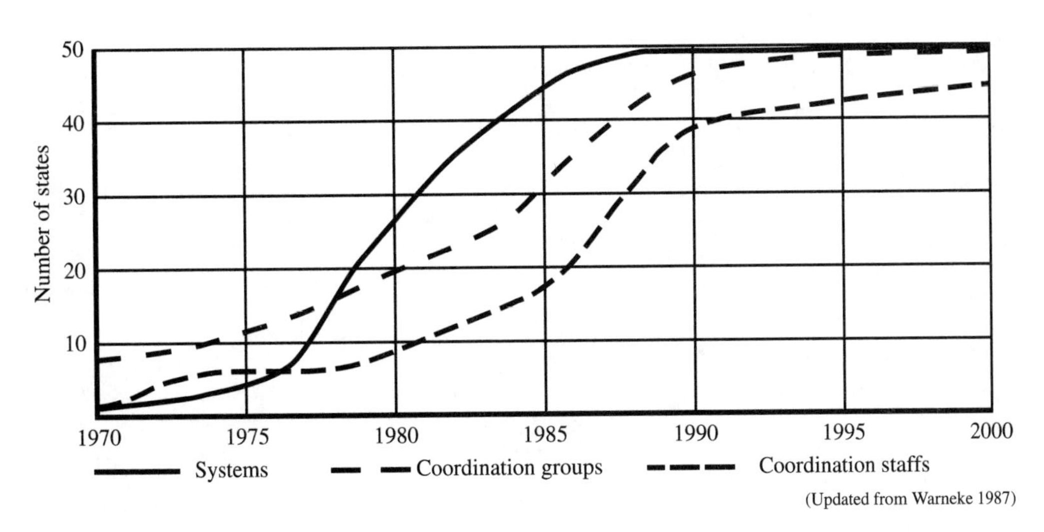

Figure 14.1 Trends in state GIS use and institutionalization.

other organizations. States are expected to have a stronger role in GIS, particularly because the definition of "state" is expanding beyond state government to encompass other organizations working within the geographic area defined by a state's borders. This broadening conceptualization has important consequences. As states expand their use and institutionalization of GI/GIS and provide external organizations with greater opportunities to shape direction and work together, benefits can be derived by all.

Bibliography

Aangeenbrug, Robert T. 1992. "Interview." *Geo Info Systems*, 2 (7): 22–30.

Arizona Department of Transportation. 1984. *Feasibility Report and Implementation Plan for Application of a Computer-Aided Design and Drafting (CADD) System.* Phoenix, AZ: Arizona Department of Transportation.

Caron, L. M., and D. S. Stewart. 1984. *An Inventory of State Natural Resources Information Systems.* Lawrence, KS: Kansas Applied Remote Sensing Program, University of Kansas.

Carson, R. 1962. *Silent Spring*. New York: Fawcett World Library.

Clinton, W. 1994. *Coordinating Geographic Data Acquisition and Access: The National Spatial Data Infrastructure.* Executive Order No. 12906. Washington, D.C.: U.S. Government Printing Office.

Cornwell, S. B. 1982. "History and Status of State Natural Resource Systems." *Computers, Environment and Urban Systems*, 7: 253–260.

Council of State Governments. 1978. *Environmental Resource Data: Intergovernmental Management Dimensions.* Lexington, KY: The Council of State Governments.

Demers, L., et al. 1995. *Survey of GIS Use by Public Utility Commissions.* Albany, NY: New York State Department of Public Service.

Healy, R. G., and J. S. Rosenberg. 1979. *Land Use and the States.* Baltimore, MD, and London: Johns Hopkins University Press.

Lester, J. P. 1994. "A New Federalism? Environmental Policy in the States, in Environmental Policy in the 1990s." In N. J. Vig and M. E. Kraft, eds., *Toward a New Agenda*. Washington, D.C.: Congressional Quarterly Press, 51–68.

McHarg, I .L. 1969. *Design with Nature*. Garden City, N.Y.: Natural History Press.

Moyer, D. D. 1995. Telephone interview.

Nathan, R. P. 1990. "Federalism—The Great 'Composition.'" In *The New American Political System*. Washington, D.C.: American Enterprise Institute for Public Policy Research, 231–261.

National Research Council. 1980. *Need for a Multipurpose Cadastre*. Washington, D.C.: National Academy Press.

Petzold, R. G. 1995. Telephone interview.

Ringquist, E. J. 1993. *Environmental Protection at the State Level: Politics and Progress in Controlling Pollution*. New York: M.E. Sharpe, Inc.

Seladones, S., and P. Harwood. 1981. *Earth Resources Data and the States: A Guide to Information Tools for Natural Resources Management*. Washington, D.C.: The Council of State Planning Agencies.

Somers, R. 1987. "Geographic Information Systems in Local Government: A Commentary." *Photogrammetric Engineering and Remote Sensing*, 53: 1379–1382.

Tessar, P., and L. M. Caron. 1980. *A Legislator's Guide to Natural Resource Information Systems*. Denver, CO: National Conference of State Legislatures.

Tomlinson, R. F. 1987. "Current and Potential Uses of Geographic Information Systems: The North American Experience." *International Journal of Geographical Information Systems*, 1: 203–218.

Vonderhohe, A., L. Travis, and R. Smith. 1991. "Implementation of Geographic Information Systems in State DOTs." National Cooperative Highway Research Program, Research Results Digest no. 180.

Warnecke, L. 1987. "Geographic Information Coordination in the States: Past Efforts, Lessons Learned and Future Opportunities." *Proceedings, Piecing the Puzzle Together: A Conference on Integrating Data for Decision-making*. Washington, D.C.: National Governors Association.

Warnecke, L. 1990. *Report on the Survey of State Regulatory Commissions' Interest and Experience in AM/FM/GIS*. Aurora, CO: AM/FM International.

Warnecke, L. 1992. "State Geographic Information Coordination Efforts and Groups." *Proceedings, Making Information Work*. Washington, D.C.: National Governors' Association.

Warnecke, L. 1993. "State of the States Regarding GI/GIS." Report prepared for the Mapping Science Committee of the National Research Council.

Warnecke, L. 1995. *Geographic Information/GIS Institutionalization in the 50 States: Users and Coordinators*. Santa Barbara, CA: National Center for Geographic Information and Analysis, University of California.

Warnecke, L. 1996. "Geographic Information/GIS Groups in the 50 States." Report prepared for the Federal Geographic Data Committee, Reston, VA.

Warnecke, L., et al. 1992. *State Geographic Information Activities Compendium*. Lexington, KY: Council of State Governments.

Warnecke, L., and L. Herrington. 1994. *Geographic Information/GIS in State Forestry Organizations: 1993 Survey Results*. Syracuse, NY: State University of New York College of Environmental Science and Forestry.

PART VI

International Experiences

Peter A. Burrough, Department of Physical Geography, University of Utrecht

The term Geographic Information System as such became known in the Netherlands in or around 1979–1980. Terms like Digital Cartography, automated soil mapping, and computer-assisted landscape information systems had already been in use for several years. The latter were actually the forerunners of current GIS, being created by adapting CAD-CAM systems such as Computervision. Considerable amounts of new software had to be written for data input, data coding, and retrieval, including Boolean operators that could be used for attributes of point, line, and polygon data. For me, these projects were a logical continuation of developments in computerized spatial analysis and mapping that started with the quantitative revolution in geology, soil science, and geography in the 1960s. My first isoline map was made by David Bickmore's Royal College of Art unit at Oxford in 1969, and, in the early 1970s, I reprogrammed John Davis's FORTRAN codes to create contour maps on a desktop computer.

Peter A. Burrough has been a leading figure in the European GIS scene. His early monograph, "Principles of Geographic Information Systems for Land Resource Assessment," by Oxford University Press became the first GIS textbook on both sides of the Atlantic.

Mark Monmonier, Professor of Geography, Syracuse University

The first time I used GIS was about a decade before GIS became GIS. In the late 1960s, I wrote a doctoral dissertation at Penn State on a method for relating digitized map overlays of crop production, soils, and topographic data. My project automated a manual overlay technique that George Deasy had used for years in his graduate seminar in crop ecology. I had learned about digital maps during a year (1964–65) at the University of Maryland's Computer Science Center, where I worked with image processing pioneer Azriel Rosenfeld and mathematician John Pfaltz on a technique for estimating surface area from a digitized contour map. Our system was called SAMP, for Surface Area Measurement Program. After I left, Rosenfeld and Pfaltz devised an overlay method based on a hexagonal grid! Data capture was messy and frustrating. At Maryland, we used image data, scanned at the U.S. Bureau of Standards. At Penn State, I tried two digitizers (one in Landscape Architecture and the other in Meteorology) before giving up, projecting my maps onto large sheets of fine-grained graph paper, and reading off coordinates to my wife Marge, who helped me convert data to punched cards. In the 1960s, GIS was part curiosity, part obsession.

Mark Monmonier has a distinguished career proselytizing the use of cartographic theory for GIS applications. His book, How to Lie With Maps, *has become a popular bestseller for technical and nontechnical alike.*

CHAPTER 15

The Incubation of GIS in Europe

David Rhind

Introduction

The story of how and when GIS developed depends crucially upon what definition is taken of Geographical Information Systems. Since the general understanding of what is a GIS has changed over the years and since the definition has also differed from country to country, it is vital to take a wide perspective. Digital mapping, computer cartography, AM/FM (Automated Mapping/Facilities Management), spatial information systems, geoscientific information systems, and much else are therefore subsumed in what follows. It does not, however, include processing of remote sensing imagery.

The evidence for the evolution of GIS in Europe is fragmentary. Most developments were only reported to a local audience and in "the grey literature." Where these remain in documentary form, it is chiefly in obscure and transitory technical reports usually written in the local language. By way of example, few of the 3,000 references collected on automated cartography alone by Karl-Heinz Meine in 1975 now seem to be generally accessible. But, on the available

---

David Rhind, director general and chief executive, Ordnance Survey, Britain's national mapping agency, directs a staff of 2,000. Prior to taking up this post in 1992, David Rhind was professor of geography in Birbeck College, University of London. He has been involved in GIS matters since 1968 and has worked extensively in Europe and in many GIS teams. *Author's Address:* David Rhind, Director General and Chief Executive, Ordnance Survey, Maybush, Southampton, U.K., SO16 4GU.

Acknowledgments: Thanks are due to many European friends and colleagues who, over the years (and in conjunction with others in North America and beyond), have provided the author with valuable material and numerous insights into GIS developments in their own countries. Despite the extent of these personal links, it is certain that this is only a very partial and incomplete view of how GIS evolved in Europe, inevitably biased towards developments by English-speaking individuals. To those who have been inadvertently left out, the author offers his apologies.

evidence and taking a wide definition of GIS, the development of GIS began in Europe as early as in North America and was for a period probably ahead of developments in that continent. GIS development has been characterised by great variation across Europe, initially with Swedish and British research groups leading the way and later significant contributions from the Netherlands and other countries. More recently, a substantial upsurge of interest and local developments in GIS have occurred in countries such as Italy and Portugal.

Most of the early developments were highly localised: there was no pan-European forum for GIS until the mid-to-late 1980s. Insofar as there was any exchange of information in the 1960s and early 1970s, the pioneers met with a few others from within their own disciplinary communities. However, a small number of individuals had close connections even at this early stage with counterparts in the U.S.A. One good example of this was the group of planners who saw the need to build information systems to underpin the statist planning concepts of the 1960s: their purview was not only Europewide but also embraced the U.S.A.

Europe was a pioneer in computing concepts and hardware. For instance, the first stored programme electronic computer, COLOSSUS, was built in Britain in the early 1940s and was used to decode German code messages. But Europe was not able to capitalise on its eminent early position in hardware, software, and (later) GIS. Commercial U.S. software, built on hardware from the same country, came to establish a hegemony over the global GIS market and has forced a harmonisation of sorts, with GIS constructs and even terminology being set by the leading U.S. vendors. This has been particularly marked in Europe. The U.S. invasion of the European GIS market has been fired by a realisation amongst U.S. vendors that the continent is relatively affluent, has a population more than 50% larger than the U.S.A. and contains many people who are both fluent in English and familiar with some aspects of U.S. culture. Centralised government structures ensured that a few key sales opened up the entire market. Set against these obvious attractions, however, were many hidden obstacles to the invasion. European culture differs from much of that in the U.S.A. The density of population (and hence of development) is typically much greater in Europe; many urban systems are intricate and bear the palimpsest of earlier structures. There are also few public lands and few national land agencies in Europe, at least in the same sense as in the U.S.A. Hence many of the commonplace applications in the U.S.A. are not significant in Europe. In these unpromising circumstances, it is surprising that the U.S.A. invasion has been as great a success as it has. Its success seems to be a tribute to the timeliness of North American GIS technology, the marketing skills of American evangelists, and shortcomings within the European industry and governments in the 1970s and 1980s.

This summarises the contents of the rest of this chapter. Case studies of GIS developments in three of the pioneering European countries provide greater detail. The analysis of GIS growth through a subdivision of the period from 1950 to the early 1990s into three sections provides a chronological perspective. A brief summary is given of the role of international linkages, and another outlines the importance of a few key personalities in the early days.

The Early Period: 1950 to Circa 1974

The earliest known processing of geographical data by computer is probably the production of meteorological charts from about 1950 onwards. Swedish work was reported by the Institute of Meteorology in Stockholm (1954) whilst Simpson (1954) reported some early British work, both using standard printing devices to construct embryonic raster maps. Within a few years, the first uses of cathode ray tubes for plotting maps were described in the literature (Döös and Eaton 1957; Sawyer 1960). One surprising characteristic of all this is that geographers and cartographers appear not to have played any role in the earliest developments.

The Swedish Contribution

This situation changed very soon thereafter in Sweden, notably within the Geography Department in Lund University and especially related to social science concepts and applications. The first published evidence of this seems to be Nordbeck's paper on geographical referencing in 1962. But Hagerstrand's famous paper of 1967 on the geographer and the computer—published by a man who was already a major international figure in geography after his work on the diffusion of innovation from 1950 onwards—had great influence worldwide. From 1967 onwards, a major programme of work on geographical data processing was funded by the National Swedish Council for Building Research. The range of work carried out can be judged from the book published by Nordbeck and Rystedt (1972) which described in 300 pages the conceptual and technical basis of the NORMAP programmes and their applications. The latter included calculating optimum locations for facilities on a hierarchical basis, examining the sensitivity of results to initial parameters, analysing flow information, coping with anisotropicity in space, and creating many kinds of maps. The connections between this Department of Geography and that at Northwestern University in the U.S.A. are noteworthy: some Lund publications like Bunge's (1962) *Theoretical Geography* were certainly strongly influenced by such linkages. Alongside these developments by social scientists, the electronic engineers in Lund developed what was probably the first large-format, mass-market raster plotter. Smeds (1973) described the Lund ink-jet plotter subsequently acquired by Applicon Inc. of the U.S.A.

Not all developments occurred in Lund, though its staff certainly fostered other developments elsewhere. Perhaps the most important practical development was the setting up of the Swedish Land Data Bank (Rystedt 1977; Andersson 1987). A Property Register Committee was set up by the Swedish Government in 1964 and, following its investigation, a Land Register Committee was established in 1966 to assess how "electronic data processing" could be used to assist in speeding the operation of a land register. Following various planning and experimental stages, the first phase of the on-line system was functioning for Uppsala by the end of 1975. The geographical referencing of individual properties consisted of centroids. Two years later, coordinate references for more than half of the properties in Sweden had been recorded.

A review published in 1977 of Swedish GIS described 12 systems which had evolved over the previous ten years (Wastesson et al. 1977). These were being run by the National Central Bureau of Statistics, the Central Board for Real Estate Data, the Ministry of Industry, the National Road Administration, the National Land Survey, the Nordic Institute for Studies in Urban and Regional Planning, and various local governments as well as university departments. It is important to recognise that all of this was facilitated by the long-standing Swedish commitment to open records and acceptance of a lack of privacy over many details which, in other countries, would be regarded as private information. For instance, a publicly accessible Swedish land record system has existed in map form since the early seventeenth century (Baigent 1990).

Work in Britain

In Britain, the earliest machine-readable geographical database was probably the mechanical one produced by Perring and Walters (1958). They compiled information on the distribution by National Grid square of 2,000 species throughout Britain and stored the results on 40-column punch cards. Initially, their plans were simply to use this for summarising and sorting the data, but they quickly realised that mapping of these data could also be produced on a mechanical tabulator. Another pioneer was Coppock who first came to appreciate the need for electronic aids to analysis after making attempts to analyse results in map form from his land use survey of the 1,600-square-mile area of the Chiltern Hills in 1954. Subsequently, he was involved in experiments with electronic area measuring devices. At the end of the 1950s, however, he analysed about half a million records from the Agricultural Census using an early computer in London University. The programmes summarised the data records and classified them ready for mapping by hand. Though the potential value of computer mapping was clearly appreciated at the time, the limitations of machine performance and output devices rendered such automation impossible (Coppock 1962). His work may be the earliest substantive GIS-based research.

Elsewhere, a number of other important developments had occurred. By 1960, for instance, routine use was made of computers in the Directorate of Overseas Surveys for projection change and other coordinate transformations, particularly for geodetic work in Africa (Windsor 1963). In addition, Bickmore (Rhind 1988) had suffered so many problems in manually producing the mammoth *Atlas of Britain* that by 1960 at the latest he had formulated plans for a sophisticated computer-based mapping system capable of matching the best quality that could be produced by hand. This system was subsequently built as funds were obtained and was described in Bickmore and Boyle (1964). Amongst the advances achieved by this duo was the world's first free cursor digitising table.

By far the best-funded and most comprehensive work of the period was that by the Experimental Cartographic Unit (ECU), founded by Bickmore in 1967 and directed by him for about a decade. Rhind (1988) has described the work of the ECU in some detail. Its roots lay in work in the Clarendon Press of the University of Oxford from 1960 to 1965 though the unit itself existed between 1967 and 1975. It was funded by the government's Natural Environmental Research Council, but its work spanned the full gamut of geographical activities. Thus, in 1969, for example, the unit's staff created computer programmes for converting digitised map coordinates to geographical ones, for projection change, for editing of cartographic features such as lines, for statistical summary of database characteristics, for the compression of data files, for generalisation of entities, for automated contouring, and for the production of 3D anaglyph maps. They defined and published standards for the exchange of geographical data in a standard format. Simultaneously, the ECU researchers were investigating digitising by automated line following and the accuracy of manual digitising of maps. They also created a 60,000 place gazetteer and produced various multicolour maps by computer. All this led in 1970 to the publication of a major work on the use of geographical data processing in planning (ECU 1971). The following year, ECU staff produced the first-ever multicolour map published as part of a standard map series—the geological map of Abingdon (Rhind 1971)—and took the first tentative steps in planning global databases. At the same time, they discussed with commercial organisations the packaging and marketing of its mapping software and planned an M.Sc. course in the handling of spatial data.

By no means did the ECU represent the entirety of the British efforts in GIS-related fields. A branch of central government produced the LINMAP system which provided statistical analysis and mapping capabilities within the Ministry of Local Housing and Government from about 1968 onwards (Gaits 1969). But a particularly interesting and important development was launched by one individual, T. C. Waugh, who was mainly based in the University of Edinburgh. The author of a line-printer mapping system (CMS) whilst an undergraduate student, Waugh set out to create in the Geographic Information Mapping and Manipulation System (GIMMS) a portable, high quality, vector mapping system with data manipulation and analysis capabilities (Waugh 1980). The first work was carried out in 1969–70 whilst he was a graduate student at the Harvard Computer Graphics Laboratory, and development has continued ever since.

In many respects, the high point of GIMMS use came in the late 1970s to the early 1980s when more than 100 sites worldwide ran it on mainframe and minicomputers. It was sold on a commercial basis from 1973 onwards (then at $250). Ultimately more than 300 sites in 23 countries ran GIMMS on a huge variety of computers ranging from PCs and Macintoshes to a Cray YMP. Other than the much simpler, batch-operated SYMAP line-printer package, GIMMS can be considered the first globally-used GIS. It pioneered the use of topological data structures, user command languages for interactive operations, macro languages, and user control of high quality graphics (including text placement and multiple fonts) in a widely-used and wholly-integrated system. In many respects, then, it is a prime antecedent of contemporary GIS and anticipated some key characteristics of the Harvard Odyssey system by nearly five years and ARC/INFO by a decade.

Though it was published just outside the period considered in this section, the book by Baxter (1976) has a good claim—in competition with that by Nordbeck and Rystedt (1972)—to be the first real text on GIS. Baxter's book is a comprehensive overview of the then hardware, software, data processing, and statistical aspects involved in what we would now regard as GIS. The section on data processing included detailed consideration of data coding and classification (including networks), spatial searches and transformations, data integration, and linkage. Numerous computer subprograms were included.

In no sense does this exhaust the work going on in Britain at this period. Following a major project carried out with the ECU between 1969 and 1971, the national mapping agency (Ordnance Survey) set up the world's first production line for converting its paper maps into computer form in 1973 and the following year carried out widely publicised experiments into how to restructure simple vector map data into several alternative forms suited to different applications (Thompson 1978); in effect, this was a prototype of later systems providing different views of a database. Until funding was withdrawn in 1974, a major central government initiative focused on the design and implementation of planning information systems to be used nationwide, fostered by the publication in 1972 of the report by local and central government on a General Information System for Planning (DoE 1972). Members of BURISA, a British variant of URISA (Urban Regional Information Systems Association), were catalysts in numerous attempts to build detailed descriptions of areas and their populations inside information systems in order to implement the 1960s and 1970s view of a planning-led society.

Though successful and long-lived systems such as that for Merseyside have resulted, most of this was thwarted by the poor technology of the time and by changing political ideologies. Yet, in the late 1960s and early 1970s, it can fairly be said that the U.K. was at the forefront of the use of GIS, even if the systems were not thus described for many years thereafter. Thereafter, the relative domination of GIS by Britain in Europe has decreased though the country remains a very strong player (see below).

Other Developments

An account such as that given above may mislead. Some GIS work was undoubtedly going on in many other locations across Europe at the time. For example, one of the first Ph.D.s in GIS was that of Brassel in 1973 in the University of Zurich. Many national mapping organisations such as IfAG in Germany had made their first experiments in computerised mapping by the mid 1970s. In general, however, such developments were highly individual and local and were constrained by limited access to expensive and unreliable technology.

The Middle Period: 1975 to 1985

The first use of the term "GIS" in a European publication is impossible to determine with certainty: Unwin (1991) has suggested that it may be as the title of a small conference run by this author in 1976! It is as well to realise that, with the exception of GIMMS (see above) and some expensive CAD/CAM systems (conveniently relabelled as being suitable for mapping), there was little or no commercial GIS software until near the end of the period. Minor successes were achieved by some GIS user organisations: Ordnance Survey, for example, sold its software to

Vienna and Cape Town. But such commercial products as existed—like Ferranti's mid 1970s CLUMIS—soon died through the dearth of any significant market.

The story of this period then is one of growing awareness of capabilities of GIS across the evolving continent of Europe. The changing European context certainly played some role. Though not initially a direct influence, the accession of Denmark, Ireland, and the U.K. to the European Community in 1973, followed in 1981 by Greece, by Portugal and Spain in 1986, and the accession of Austria, Finland, and Sweden to the European Union in 1995, has subsequently had quite profound structural, financial, and intellectual effects.

These effects have been most marked in recent years through EU support of pan-national research and encouragement of "mobility of human capital." The 1975–85 period, however, also saw a strong growth in the early links with North America and a rash of initial systems being set up by government bodies and individual projects within universities. One manifestation of the links with the U.S.A. was the appearance of "one-off" systems based on (and in some respects extending) the U.S. Bureau of Census DIME system from 1972 onwards. Examples of this included RGU in France, VEJKOS in Denmark, and TRAMS in U.K. This was fostered by SORSA—then an acronym for *Segment-Oriented* Referencing Systems Association rather than the current *Spatially-Oriented.* SORSA grew up as a collection of individuals from Europe (such as Owe Salomonsson from Sweden, Jean Salmona from France, and Jef Willis from U.K.) and North America who met in congenial places. Their distinguishing characteristic was an opposition to the consensus of the day amongst the planning fraternity that point-based geocoding systems formed the best practicable solution for describing space.

As it has transpired, the most significant national story of the period from 1975 to 1985 is the initial growth of GIS skills, entrepreneurial flair, and institutional influence in the Netherlands. At a time when the British central government decided to cease investment in what we now call GIS, other governments were becoming more receptive, and migration of key players was also to play an important role. Many individuals contributed to the "GIS business" in the Netherlands, and hence it is invidious to select only a few. But two examples will serve to illustrate the importance of international migration. Both Stein Bie (a Norwegian) and Peter Burrough (a British national) gained their doctorates in the University of Oxford working in Soil Science under Dr. Philip Beckett, and all three came into contact with Bickmore (see above) and his activities. The Oxford group had strong interests from the earliest period in data quality and the statistical inferences which could be derived from existing and new data. A 1972 paper by Beckett, Tomlinson and Bie, for instance, described a seminar on map quality.

Bie played a formative role in the development of a Landscape Information System at the Agricultural University of Wageningen from late 1976 onwards, drawing upon the experience of the U.S. Soil Conservation Service. When Bie returned to Norway to work on GIS in the National Computing Centre, Burrough—then returning for family reasons to Europe from Australia—took over his post and work (Burrough 1980) but subsequently moved to the Department of Geography in the University of Utrecht. In the latter location he published his influential 1986 book on GIS, originally as part of a soil monograph series edited by Beckett and published by the University Press at Oxford! The crucial need to sustain life in a land much of which is below sea level, the availability of government support at a formative period, and the Dutch ability and willingness to attract top quality international "stars" has ensured that the Netherlands became the preeminent centre of GIS research in earth sciences in the late 1980s and early 1990s (see below).

Developments elsewhere in Europe became more commonplace but were still typically based on home-grown software, often shared selflessly with friends and colleagues in the same discipline. Wolf Rase, working in a federal planning agency in Germany, made widely available a major library of FORTRAN subroutines, just as David Douglas did in Canada. At that time, the common perception of need was for such libraries of software modules, rather than complete and general-purpose programmes, which computer-literate experts could embed in whatever control programme suited their own purposes. Thus, the Geographical Algorithm Group's library in Britain was created in a project modelled upon the highly successful Numerical Algorithms Group approach. It may have been the first geographical software written to defined, professional standards of quality, consistency, and documentation (GAG 1977; Milne 1983). It is noteworthy that the great majority of those working in GIS at this period were geographers or cartographers plus researchers who required tools to carry out their work. An important exception was the French computer scientist Francois Bouille whose contributions included the hypergraph-based spatial data structure (Bouille 1977). The involvement of other computer scientists in this field was generally much later in Europe than in the U.S.A.

Sometimes developments which began for another reason became important for GIS. CERCO was set up as a sort of club of national mapping agencies in Europe in 1979/80 to facilitate discussion of common concerns, notably management issues. Over the years, however, this has grown into a group of 32 nations (including the Russian Republic) and has spawned a commercial group (MEGRIN) of 17 nations which is producing pan-European data.

The Recent Period: 1986 to the Early 1990s

This decade has seen the growth of use of GIS in most parts of Europe (Koshkariov et al. 1989). Most of the early national systems created by user organisations (e.g., FINGIS: Keisteri 1986) have been superseded in whole or part by the use of commercial GIS software. Clearly, a necessary condition for this to occur was the ready availability of commercial systems at acceptable cost. The first installation of a true commercial GIS in Europe was almost certainly ARC/INFO v2 in Birkbeck College, University of London, in 1983 (but see the GIMMS story above). This single copy of ARC/INFO was initially run on a DEC Vax 11/730 machine with less processing power than a 386 PC. Since then the situation has evolved almost out of recognition: growth rates in terms of installed systems have averaged about 20% per annum though the start point for growth has differed between countries, and the largest markets have varied considerably. In Germany, for instance, the utility market has played a key driver role. In general terms, the situation in Europe is now similar to that in the U.S.A. except where influenced by the presence or absence of local statutes or different government policies. One example of the latter is in regard to data pricing: a number of governments in Europe now require government organisations to price data above the costs of copying in order to offset ongoing costs of data maintenance. Currently, the extreme case is the U.K. where the national mapping organisation (Ordnance Survey) achieves nearly 80% recovery of all its costs when calculated on a full commercial basis, despite being tasked with various uncommercial activities.

In recent years, then, the range of GIS applications has widened and the number of systems has grown rapidly in Europe. This situation has been fostered by improvements in communications both within and between countries and across and between disciplines. Several factors caused this, including the creation of a global GIS research journal (the *International Journal of GIS* which grew from a meeting of British GIS workers but was consummated by marriage with colleagues in the U.S.A. and elsewhere), the popularity of trade magazines (such as *GIS Europe* and *Mapping Awareness,* with the British publishers buying up the American *GISWorld*) and the growth of a GIS book publishing industry comprising such major players as Longman/GeoInformation International, Taylor and Francis, and Oxford University Press. Also highly influential were the foundation of national GIS organisations (such as the Association for Geographic Information in the U.K. and CNIG and AFiGeo in France), national research initiatives in GIS in the Netherlands, U.K., and Portugal, pan-European initiatives such as the GISDATA project funded by the European Science Foundation plus funding of multilateral GIS product development by the European Commission under its DRIVE and IMPACT initiatives. The creation and location within the Institut Géographique National in Paris of a multinational team of employees drawn from national mapping agencies has materially aided Europeanwide thinking and progress in the geographical information field.

One other very significant factor has been the growth of education in GIS. It is possible that the first GIS course in Europe comprised option classes in Edinburgh University in 1976 or some optional undergraduate courses and an M.Sc. course in Spatial Data Analysis run in the University of Durham in the same year. Certainly the Edinburgh M.Sc. in GIS, run from 1984 to date, has had ramifications across Europe (and beyond). Whatever the origins of GIS teaching, the years from 1987 onwards have seen a huge growth in the GIS manpower supply side at all educational levels across Europe. Perhaps the greatest density of provision is in the Geography Department of the University of Utrecht. Taken together with other Dutch organisations at Delft, Wageningen, and Amsterdam, these centres provide teaching and research for 2,500 undergraduates, 150 Ph.D.s, and 400 academic staff at any one time, with GIS courses being obligatory in many degree courses (Burrough 1995). Former students from the universities in Europe have become a major force in the acceptance and use of GIS in many organisations across the continent. Moreover, institutions such as ITC (the International Training Centre in the Netherlands) have also helped to propagate GIS skills to many developing countries around the world.

At various times throughout this period, national governments in different parts of Europe have taken initiatives—sometimes almost by accident—which have fostered GIS development. The two best-known of these have been the British Committee of Inquiry into the Handling of

Geographic Information (DoE 1987) and the Dutch NexPri initiative. The British one, perhaps typically, was initiated by central government in 1985 and involved a study across all of the terrestrial application areas for which geographic information might be of use. The resulting Chorley Report (Rhind and Mounsey 1988) had wide influence within Britain. It led to the foundation of the Association for Geographic Information, the setting up of many courses in universities, two research programmes funded by the government's research councils and various changes within government itself. It succeeded not simply because it was timely but because the report was carefully written for the lay reader and considerable effort was expended outside the community in proselytising about GIS. The Dutch government followed up its own RAVI study by funding an interuniversity GIS Expertise Centre (though the funding was gained in competition with all other sciences) which has had considerable success.

As in the U.S.A., another key catalyst has been the success of major, general-purpose GIS conferences. Mainly, however, these developed in Europe long after they appeared in the U.S.A. though some relatively small national GIS meetings were held in the 1970s and early 1980s, and mapping and other conferences have long included papers on GIS matters. Ignoring smaller conferences like the EuroCarto ones initiated by Bickmore (see above) and held in different parts of Europe from 1982 onwards, there are good reasons for believing that the first European GIS conference was Auto Carto London, held in that city in 1986. The next was probably the first European GIS (EGIS) meeting held in Amsterdam in 1990 and continued annually thereafter in Brussels, Munich, Genoa, and Paris. A multiplicity of other conferences such as "GIS in Business" have grown up. There has been a boom in conferences in Eastern Europe, notably in the Czech Republic and Hungary. The success of these suggests that the market may be reducing so far as general purpose, continentwide conferences are concerned, with regional, national, and special theme ones becoming more viable.

Measuring the Spread of GIS

Comparison of the contributors to the 1986 Auto Carto London conference, the first EGIS one in 1990 and the 1994 EGIS/MARI one provide some indication of the way in which GIS has spread throughout Europe. Figure 15.1 shows the contributions by country of affiliation of the lead author of each paper. Any analysis is of course confounded by a number of other factors: the opening out of Eastern Europe post-1989 has certainly played a role in the changing balance of authorship. Equally, the location and provenance of the conference organisers also plays a role in where publicity is focused, the numbers of papers submitted, and who is selected to present papers. Thus, the Auto Carto London conference organisers deliberately set out to attract a good representation from North America as well as from Europe. The Netherlands-based organisers of EGIS '90 drew on their home base for many contributors to their first conference, and the 1994 one was held in Paris and served the dual role of European and French national conference (the number of French language papers and contributions from former French colonies was relatively high as a result). Nevertheless, the results in Figure 15.1 demonstrate the absolute increase in numbers of papers on GIS submitted, the continuing relative strength of the Nether-

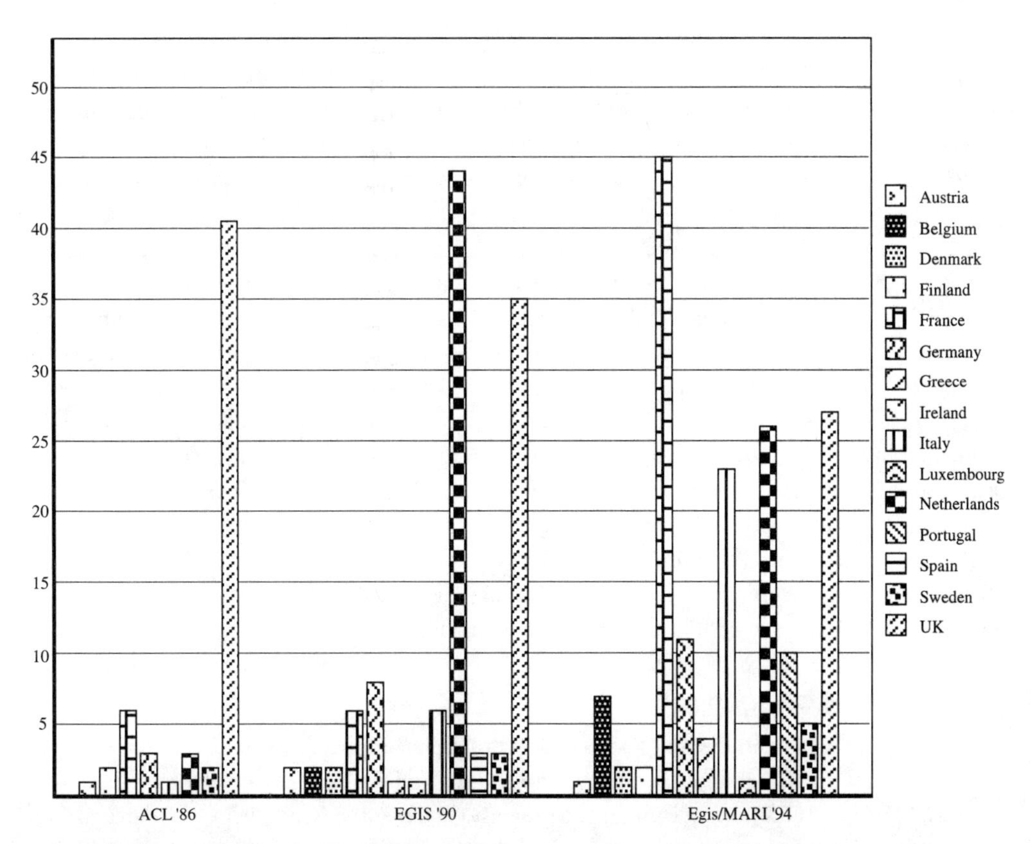

Figure 15.1 Geographical affiliations of first authors of papers given at three general-purpose GIS conferences held at intervals of four years.

lands and the U.K., the rapid growth in GIS in Italy and Portugal, a surprising lack of contributions by France until quite recently, and modest contribution by Sweden and Germany relative to their early work and size respectively. The German contributions are especially puzzling given such work as the ATKIS project and much environmental modeling. Rapid recent developments in GIS in Spain will probably raise that nation's contribution: it is playing host to both the European "Business in GIS" conference and the global International Cartographic Association Conference in 1995. The paucity of the contribution by the U.S. to the two more recent conferences is obvious: it is matched by that of Czechoslovakia.

Recent Pan-European Developments

Five further pan-national developments need to be outlined here. These show the extent to which Europe-focused GIS work is now commonplace. The first is the creation by national mapping agencies of a common geographical referencing system across Europe from Iceland to

Turkey and even into the Middle East, the EUREF system. The second is the GISDATA initiative funded by the European Science Foundation. This has consisted largely of a series of meetings mostly attended by academics, policy makers, and others. Modelled partly along the lines of those of the U.S. NCGIA, the topics covered have included generalisation, GIS and multimedia, the diffusion of GIS knowledge in local government, and conceptual models for geographic objects with undetermined boundaries. The third is EUROGI, an umbrella body with aspirations to be the European focus for Geographic Information (GI) issues. Its membership includes national GIS bodies plus various professional societies such as AM/FM, UDMS and OEEPE (a treaty-based experimental photogrammetric organisation). EUROGI's aims are to stimulate, encourage and support the development and use of GI at the European level. The European Committee for Standardisation (CEN) has initiated a working programme to define European standards in relation to all aspects of geographic data such as data quality, metadata and data transfer. Finally, the European Commission (effectively the public service of the European Union but with the role of initiating proposals for European Directives) has increasingly come to play a role in GIS. This is most obvious through its funding of studies, such as the GI2000 initiative centring on GI policy, base (or core) data, and metadata. In addition, however, other within-EC developments on regulation, procurement, and law in relation to environmental information influence the field of GIS within the continent.

International Linkages

It should be obvious from the text above that international connections, initially on an individual-to-individual basis but now also on a group-to-group basis, have played an important role in the development of GIS both in Europe and globally. Thus, individuals such as Waugh and Coppock spent early periods in Harvard, professional societies and SORSA and other organisations acted as catalysts in bringing pioneers together, and individuals such as Brassel and Burrough (and this author) migrated from university to university in the first period described above. A number of Europeans such as Michael Goodchild have made significant contributions to GIS development in North America yet retained strong links to Europe. Indeed, individuals such as Andrew Frank and Jean-Claude Müller have worked on both continents. And, in the last decade especially, individuals such as Baranowski, Csillag, Konecny, and Tikunov have helped to open up Eastern Europe to western GIS influences and tools through their myriad personal and institutional connections. Despite the numerous languages in normal use in Europe, English has—in research and development at least—become the *lingua franca* of GIS in Europe. A very large proportion of all conference papers is given in that language, and no paper in any other language has been submitted to the *International Journal of GIS*.

The Role of Personality

In a world-wide industry involving many thousands of employees in a multiplicity of organisations, the scope for any one individual effecting dramatic changes in the GIS world is now relatively small. This was not always so. This author has previously argued (Rhind 1988) that, in the early days of GIS, individual personalities made considerable and distinctive contributions to the development of the subject. Without some knowledge of their achievements, the evolution of GIS cannot be understood. Many of these people—Baxter, Bickmore, Bie, Bouille, Brassel, Burrough, Coppock, Hagerstrand, Nordbeck, Rase, Rystedt, Salmona, Salomonsson, Waugh, and others—have already been mentioned. Many other important players (e.g., Gardiner-Hill 1972 and Christ of IfAG in Germany) have not. No doubt many others deserve to be mentioned. The early history of GIS in Europe, as in the U.S.A., was one of times when great things (and great disasters) were achieved by the handful of people involved.

Conclusions

There can be little doubt that many early GIS developments occurred in Europe and that the strength of the GI and GIS industry in that continent is now considerable. The 15 countries and 350 million people within the European Union—plus those in surrounding countries waiting to join— constitute both a massive market and a highly innovative source of new ideas and developments in this field. The single most interesting research question in European GIS, however, is why, given the very early developments of GIS in various countries therein and the quality of national education and innovation, the continent did not prosper in the GIS software market in the critical decade from the late 1970s. This may well have profound policy significance for the European Union far beyond GIS.

Bibliography

Andersson, S. 1987. "The Swedish Land Data Bank." *International Journal of GIS,* 1 (3): 253–63.

Baigent. 1990. "Swedish Cadastral Mapping 1628–1700: A Neglected Legacy." *Geographical Journal,* 156 (1): 62–9.

Baxter, R. S. 1976. *Computer and Statistical Techniques for Planners.* London: Methuen.

Beckett, P. H. T., P. R. Tomlinson, and S. W. Bie. 1972. "Map Quality: The Second St. Cross Seminar." *Cartographic Journal,* 9 (2): 80–1.

Bickmore, D. P., and R. Boyle. 1964. "An Automated System of Cartography." *Proceedings, Technical Symposium of the International Cartographic Association,* Edinburgh, U.K.

Bouille, F. 1977. "Structuring Cartographic Data and Spatial Processes with the Hypergraph-Based Data Structure." In G. Dutton, ed., *Proceedings, First International Advanced Study Symposium on Topologic Data Structures in Geographic Information Systems.* Harvard Papers in Geographic Information Systems, vol. 1. Cambridge, MA: Harvard University, 22–35.

Bunge, W. 1962. *Theoretical Geography.* Lund, Sweden: Gleerup.

Burrough, P. A. 1980. "The Development of a Landscape Information System in the Netherlands, Based on a Turn-key Graphics System." *GeoProcessing,* 1 (3): 257–74.

Burrough, P. A. 1986. *Principles of Geographical Information Systems for Land Resources Assessment.* Oxford: Oxford University Press.

Burrough, P. A. 1995. "Academic Developments in GIS in the Netherlands." *GIS Europe* (March).

Coppock, J. T. 1962. Electronic data processing in geographical research. *Professional Geographer* 14, 1–4.

Department of the Environment (DoE). 1972. *General Information System for Planning (GISP).* Department of the Environment Research Report 1. London: DoE.

Department of the Environment (DoE). 1987. *Handling Geographic Information: The Report of the Government Committee of Inquiry.* London: Her Majesty's Stationery Office.

Döös, B. R., and M. A. Eaton. 1957. "Upper Air Analysis Over Ocean Areas." *Tellus,* 9: 184–94.
Experimental Cartographic Unit (ECU). 1971. *Automatic Cartography and Planning*. London: Architectural Press.
Geographical Algorithms Group (GAG). 1977. *Feasibility and Design Study for a Computer Algorithms Library*. London: Geographical Algorithms Group, University College.
Gaits, G. 1969. "Thematic Mapping by Computer." *Cartographic Journal,* 6: 50–8.
Gardiner-Hill, R. C. 1972. "The Development of Digital Maps." *Ordnance Survey Professional Paper.* New Series 23. Southampton: Ordnance Survey of Great Britain.
Hagerstrand, T. 1967. "The Computer and the Geographer." *Transactions Institute of British Geographers,* 42: 1–20.
Keisteri, T. 1986. "FINGIS: Software and Data Manipulation." In M. Blakemore, ed., *Proceedings, AUTO-CARTO London*, London: 69-75.
Koshkariov, A. V., V. S. Tikunov, and A. M. Tofimov. 1989. "The Current State and Main Trends in the Development of Geographical Information Systems in the USSR." *International Journal of GIS,* 3 (3): 257–72.
Milne, W. 1983. *GAG Library: Algorithms for Spatial Data*. Norwich, U.K.: University of East Anglia.
Nordbeck, S. 1962. "Location of Areal Data for Computer Processing." *Lund Studies in Geography,* Series C, General Mathematical and Regional Geography No. 2. Lund, Sweden: Lund University.
Nordbeck, S., and B. Rystedt. 1972. *Computer Cartography*. Lund, Sweden: Studentlitterateur.
Perring, F. H. and S. M. Walters. 1958. "Plant Atlas of Great Britain." *Times Science Review,* 13.
Rhind, D. W. 1971. "The Production of a Multicolour Geological Map by Automated Means." *Nachrichten aus den Karten und Vermessungeswesen* Heft 52: 47–52.
Rhind, D. W. 1988. "Personality as a Factor in the Development of a Discipline: The Example of Computer-assisted Cartography." *American Cartographer,* 15 (3): 277–89.
Rhind, D. W., and H. M. Mounsey. 1988. "The Chorley Committee and Handling Geographic Information." *Environment and Planning,* A (21): 571–85.
Rystedt, B. 1977. "The Swedish Land Data Bank." In Wastesson et al., eds., *Computer Cartography in Sweden*, 19–48.
Sawyer, J. S. 1960. "Graphical Output from Computers and the Production of Numerically Forecast or Analysed Synoptic Charts." *Meteorological Magazine,* 89: 187–90.
Simpson, S. N. 1954. "Least Squares Polynomial Fitting to Gravitation Data and Density Plotting by Digital Computers." *Geophysics,* 19: 250–7.
Smeds, B. 1973. "A 3-colour Ink Jet Plotter for Computer Graphics." *BIT,* 13 (2): 181–95.
Thompson, C. N. 1978. "Digital Mapping in the Ordnance Survey 1968–78." Paper given to International Society of Photogrammetry Commission IV Inter-Congress Symposium, Ottawa, Canada.
University of Stockholm Institute of Meteorology. 1954. "Results of Forecasting with the Barometric Model on an Electronic Computer (BESK)." *Tellus,* 6: 139–49.
Unwin, D. 1991. "The Academic Setting of GIS." In D. J. Maguire, M. F. Goodchild, and D. W. Rhind, eds., *Geographical Information Systems: Principles and Applications,* vol. 1. London: Longman, Harlow, 81–93.
Wastesson, O., B. Rystedt, and D. R. F. Taylor. 1977. "Computer Cartography in Sweden." *Cartographica Monograph* 20. Toronto, Canada: B.V. Gutsell.
Waugh, T. C. 1980. "The Development of the GIMMS Computer Mapping System." In D. R. F. Taylor, ed., *The Computer in Contemporary Cartography*. London: Wiley, 219–34.
Windsor, L. M. 1963. "The Directorate of Overseas Survey's Experience with Electronic Computers." *Proceedings, Conference of Commonwealth Survey Officers*, Part II. London: Her Majesty's Stationery Office, 99–108.

CHAPTER 16

The Development of GIS in Australia: From Infancy towards Maturity

Barry J. Garner and John F. O'Callaghan

Introduction

Australian applied and research scientists alike have been particularly responsive to recent developments in information technology, particularly in remote sensing and GIS—fields in which they have achieved international recognition. Several well-defined threads run through the history of GIS, principal among which are developments in land information systems, decision support systems for business, resource and environmental information systems, education and research, and the establishment of a small but significant national GIS industry (O'Callaghan and Garner 1991).

Rather than document the chronology of these developments individually, a more generalised approach is adopted based on the identification of critical periods in the development of GIS. The conceptual framework for this is the analogy with the stages in the growth of nations proposed many years ago by the economist Rostow (Rostow 1960). From empirical observation, Rostow argued that the process of economic growth could be divided into five stages, each characterised by a distinctive set of economic and social conditions: the traditional

Barry J. Garner is a professor of geography at the University of New South Wales, Sydney, Australia. His special interests are in the socioeconomic application of GIS.

John F. O'Callaghan is chief executive officer of the Cooperative Research Centre for Advance Computational Systems (ACSys), Canberra, Australia. He is a CSIRO and an adjunct professor at the Australian National University.

Acknowledgment: The authors are especially indebted to Mark Horn, CSIRO Mathematical and Information Sciences, for his constructive comments on drafts of the chapter and his valuable editorial assistance in producing the final manuscript.

society, the preconditions for takeoff, takeoff itself, the drive to maturity, and the age of high mass consumption.

The growth and development of GIS in Australia during the past 25 years closely parallels the stages in Rostow's schema, although obviously based on a different set of indicators. Any attempt to slice a continuous process into discrete time periods inevitably must be arbitrary, and this is certainly true of this particular interpretation. Even so, events indicate that the transition from a reliance on traditional analogue sources of geographic data commenced during the early years of the 1970s, and the *preconditions* for a substantial growth in GIS-related developments were established by the end of that decade. *Takeoff*—characterised by collaboration and consolidation—effectively occurred during the 1980s, by the end of which the *drive to maturity*—a period of integration and extension that dominated developments during the first half of the 1990s—was well under way. Indications now suggest that Australia is on the threshold of entering the final stage—the widespread use of GIS in business and society.

1970–1980: Preconditions for Takeoff

Events during the 1970s laid the foundations on which Australia's subsequent national and international GIS profile was built. The preconditions were established that made possible the consolidation of efforts made during the 1980s, a period of growing awareness, rapid diffusion, and ubiquitous adoption of GIS. The potential of the innovations in information technology was increasingly appreciated, and, by the mid 1970s, a keen interest in the design and development of computer-based information systems for administration, decision support, and particularly planning had developed (McCalden 1973). Developments also occurred on a number of other key fronts five of which may be singled out as being especially important in establishing the preconditions for later developments.

First were the early initiatives taken by state and territory governments to computerise land titles records and develop parcel-based land information systems (LIS). Australia has a sophisticated and tightly controlled system of recording land ownership based on the Torrens system of land registration. Land parcel records are complete and comprehensively maintained, the title to land is supported by government guarantees, and land and property taxes are a significant source of government revenues.

The legal and fiscal information embedded in the cadastre is thus central to land administration in the six states and two territories comprising Australia and is basic to the requirements of local governments as well as the utility authorities (Williamson 1982). It became increasingly obvious during the 1970s that the development of digital cadastral databases and computerisation of records attached to land parcels (2.5 million land titles on paper in New South Wales) could produce substantial economies and improved efficiency. Since land parcels are also the basic spatial unit for many planning-related activities, a multipurpose digital cadastre also became an important concept for the development of LIS.

Although the earliest attempt by a state government to develop a prototype LIS was in Western Australia in 1972, the first serious commitment to the development of a statewide computerised land administration system began in South Australia with the decision by the state government to build an integrated land records database in 1974 (Sedunary 1977). In the first stage, work commenced on the development of a central reference file based on land parcel and ownership data—the Land Ownership and Tenure System (LOTS) together with its associated coordinate graphics file (CLOTS). The system was implemented in 1979, and, based on its success, the government embarked on subsequent phases in the development of a statewide LIS that continued throughout the 1980s (Sedunary 1980).

Initiatives to develop LIS were also taken in other states and territories at this time. These included the land planning system in the Australian Capital Territory (ACT)—a system based on software for geophysical data analysis and mine planning capable of integrating terrain and cadastral data through the use of grid files to produce planning maps which was fully operational by 1974; the implementation of a computerised land title registration system and land parcel mapping system (the Torrens Title Automation Project) in New South Wales in 1977; and in Victoria, the recommendations of a Task Force set up to investigate the feasibility of developing an LIS adopted in 1979 as the basis for the development of the initial stage of a statewide LIS—the LANDATA project. By the end of the decade, the concept of a multipurpose digital cadastral database and computerised land records information systems had been embraced by all state governments.

Second were the technical developments in computer mapping and digital databases that continued throughout the decade. Attention quickly focussed on developing practical ways of using computers to map and graphically display digital data. Geocoding and geoprocessing became major concerns, and by middle of the decade it was widely considered that one of the most urgent needs in Australia was:

> ". . . to build up experience in geocoding applications and to provide administrators and policy makers with opportunities to learn what planning information systems are about, and to come to an appreciation of the options available. . . . The transfer of geocoding technology from overseas . . . could provide this type of experience, with considerable time and cost savings over in-house developments. . . . The initiation of a pilot and demonstration project in Australia is clearly a matter of urgency. . . ." (McCalden and Jarvie 1976).

Practitioners, especially those in the research community, were now fully aware of overseas developments in geocoding and computer mapping. By the end of the decade, these were being widely adopted or were the subject of subsequent research and development, for example, the mapping of agricultural statistics using SYMAP (Massey 1978) and the prototype application of ADMATCH (Davis and Paine 1979). Domestic developments also began during the early 1970s. The Australian Bureau of Statistics (ABS) released census data for the first time on magnetic tape and was exploring methods of geocoding census boundaries for mapping the 1976 Census of Population and Housing at the tract level (Byatt 1976), an initiative subsequently extended jointly with the Commonwealth Division of National Mapping (NATMAP, Australia's national mapping agency), which led to the development of the first fully automated census mapping system (King 1977). This was so sufficiently advanced by 1978–79 that it could be used to produce the first National Atlas of Population and Housing based on the results of the 1976 Census. The release by NATMAP of digital files describing boundaries of all local government areas in 1978 was to lead to a major expansion in the computer mapping of ABS census data during the 1980s. The Australian Bureau of Statistics was also taking a lead in the development of digital databases, an important early one of which was the Australian Municipal Information System (AMIS), which became available in 1972–73 and which contained summary financial data and population characteristics for all local government areas.

In the early 1970s, NATMAP also embarked on a program of automated cartography for small-scale mapping of Australia. The Australian Geodetic Datum for horizontal control was first computed in 1966, the generation of height data from stereoplotters began in 1969, and a revised continental Height Datum was computed in 1971 (Bomford 1971). The digital mapping program for the national topographic map series commenced in the 1970s, and there followed a steady production of a variety of map products. The national coverage of 1:250,000 scale topographic maps took until 1994 to complete. On another front, the army, which plays an important role in mapping Australia, began its AUTOMAP program with the installation of a Canadian computerised mapping system in 1977 for the production of 1:50,000 topographic maps of areas of defence significance.

A third important development during these formative years was the decision taken in 1977 by the Commonwealth (the federal government) to build a Landsat receiving station at Alice Springs. Australia's large and ancient land mass, the interior of which is relatively inaccessible, is rich in mineral resources which historically have underpinned the country's economic development. The environment is also a particularly fragile one; land degradation in its various forms has long been a serious problem. The importance of remote sensing as a source of environmental data was therefore quickly recognised, particularly by the geophysical and mining companies. The new station first received data for testing late in 1979 and commenced commercial operations at the beginning of 1980.

Remote sensing had become a significant focus for research by the mid 1970s, particularly in the minerals exploration industry and in the various divisions of the Commonwealth Scientific and Research Organisation (CSIRO), the national scientific research body. Early studies on the digital enhancement and display of Landsat data were conducted in CSIRO by a team led by McCracken, Green, and Huntington at the Division of Mineral Physics. The team's sustained research over a long period was recognised many years later in 1995 with the award of the Australia Science Prize for its contribution to remote sensing, particularly in the field of mineral exploration. Many other studies demonstrated the usefulness of the information in Landsat imagery for digital mapping, land assessment, and environmental monitoring during this period (Graetz et al. 1980).

Substantial links with North America were established during the 1970s. In 1978, a U.S.–Australia Workshop on Image Processing Techniques for Remote Sensing, jointly organised by John O'Callaghan and Azriel Rosenfeld, was held. The period of experimentation with digital processing techniques culminated in the organisation of the first Australasian remote sensing conference in May 1979 with more than 400 attendees, just six months before the receiving station made Landsat imagery publicly available, and in 1980, the late Professor David Simonnet, from the University of California–Santa Barbara, was invited to present the keynote address at a national GIS conference on "The integration of Landsat and ancillary remotely sensed data with GIS."

A fourth and especially significant contribution that established the preconditions for developments in GIS was the emergence of an active research community. Important advances were made during the 1970s in understanding the conceptual and technical problems associated with computerised spatial data handling and problems in the practical application of GIS. Although the universities became increasingly active in this study, the lead was quickly taken by researchers in the CSIRO, particularly in the Division of Land Use Research, as it was then called, where Bruce Cook and Peter Laut were among the most innovative individuals at the time.

Cook's fundamental research into the representation of plane region boundaries, a topic he had investigated while spending a period of leave with the organization developing Canadian GIS in the late 1960s, was published in a seminal paper in 1967 (Cook 1967). After 1970, he went on to make a number of other significant contributions to the understanding of conceptual and practical issues in the representation and handling of spatial data through his involvement in a major land use project in southern New South Wales. That was the first serious effort made in Australia to develop an operational, spatially referenced computer database with primitive GIS functionality for regional and environmental planning (Cook 1975; Austin and Cook 1977).

Laut's research was directed more at methodological issues. A particular concern was the development of flexible, multipurpose ecological information systems which he pioneered in a major project that began in South Australia on behalf of the Commonwealth in 1975 (Laut 1977). One of the important goals of the project, which incorporated the use of remote sensing, was the development of a framework for ecological surveys to provide critical environmental data for land use planning and policy evaluation. It was envisaged that this could form the basis of a national land resource database, although this did not become a reality until the 1990s.

The work of Laut influenced the development of the resource and environmental information systems which the Commonwealth, state, and territory governments actively promoted during the 1980s. Several early initiatives were taken by the Commonwealth during this period, including national soil degradation studies, wetland surveys, a national resources data bank, and the establishment of a national biological resources database, although GIS had not advanced to the stage where they could make much of a contribution to these. General purpose land resources inventories were being undertaken in most states and territories, along with investigations into the development of computerised information systems of forest resources. By 1977, pilot soil erosion and land use planning systems were being tried out in New South Wales, and a land availability data bank and an urban land capability information system were being established.

A fifth development of importance during the decade was the formation in 1975 of the Australian (later called Australasian to include New Zealand) Urban and Regional Information

Systems Association (AURISA), the founding father of which was Gerald McCalden, a geographer by training working at the time for a private research foundation. In 1973, McCalden organised the first of what has since become an annual national event—the Urban and Regional Planning Information Systems conference (URPIS, changed to the AURISA conference in 1990). Its success and that of URPIS 2, held the following year, resulted in the formal constitution of AURISA at URPIS 3 in 1975. McCalden's pioneering contributions to the development of GIS in Australia were recognised much later (in 1989) when he was made AURISA's first life member.

Since its inception, AURISA has been recognised as the premier interdisciplinary body for GIS in Australia, and the annual URPIS conferences have become the national forum for the discussion, promotion, and dissemination of information about LIS-GIS and other aspects of spatial data handling. The proceedings of those conferences mirror in large extent the growth and changing focus of developments in GIS in Australia, and they have been extensively used as a source of references for this chapter.

During the 1970s, the preconditions had been established for the substantial development of GIS in the following decade. The potential of the new information technology was widely appreciated, and opportunities for its application were being exploited on a broad front. Organisational frameworks had been progressively put in place, especially by governments, a relatively small but important research community was well established, and the institutional frameworks for information exchange and the dissemination of ideas were in place as a result of the formation of national associations.

By the end of the period, Australia was quite advanced in the application of digital techniques to capture various kinds of geographic data, notably topographic features, remotely sensed data, land tenure, and administrative boundaries. With the evolution of database technologies during the decade and the appearance of commercial relational database systems, Australia was well positioned to move beyond automated cartography, computer mapping, and computer aided design and enter an era of operational Geographical Information Systems.

1980–1990: Takeoff

GIS developments in Australia took off rapidly during the 1980s accompanying the acceptance of PC-based systems and communications networks. By the end of the 1980s, many GIS systems had matured to the point where they could be used in an increasing number of projects addressing problems in environmental and resource management and land administration. Many of these demonstrated the value of combining different kinds of geographic and aspatial data, and their success was reflected in the theme chosen for the URPIS 16 conference in 1988, Information Systems in Action.

By the beginning of the 1980s, it was clear, however, that applications projects typically involved considerable effort in data collection, digitising, and the registering of different kinds of data sets. The development of digital data on a "project by project" basis was clearly not cost effective which reinforced the need for better consolidation and coordination in the development of geographical data infrastructure in Australia. A number of important initiatives were taken to achieve this, and a series of important organisational changes were made. The Commonwealth was a leading player in many of these initiatives to consolidate the development of digital data sets and information systems by government agencies during the second half of the 1980s.

Land Information Systems

From a national perspective, the most visible and well coordinated activities continued to be those in LIS by the state and territory governments. Under Australia's federal structure, these have jurisdiction over all of the major areas for which land and geographic data are important. The role of the Commonwealth is, by and large, one of regulation and coordination in the implementation of national policies. By the end of the decade, it had assumed a key role in coordinating and consolidating developments involving land and geographic information and had become a major provider of information and infrastructure to support these.

An important early initiative taken by the Commonwealth was the organisation of a national conference of government agencies at the end of 1984 to address *Better Land Related Information for Policy Decisions.* A significant outcome of this was the decision to establish a national body—the Australian Land Information Council (ALIC, renamed ANZLIC in 1991 when New Zealand was included)—to better coordinate resources and share common issues confronting member jurisdictions in the development of land information systems. ALIC, first convened in 1986, became the peak intergovernmental body responsible for coordinating the collection and transfer of land related information between different levels of government in Australia and the promotion of the use of this information for decision making.

ALIC's first significant action was the endorsement early in 1987 of a set of policies and procedures that form the basis of the National Strategy for Land Information Management (Eddington et al. 1988). The strategy was subsequently revised in 1990 with the aim of establishing a national approach to land information management issues, the identification of procedural and policy principles for more effective land administration, the implementation of mechanisms for data access and transfer, and the coordination of land information management issues (ALIC 1990a). It is also of interest to note that ALIC announced in 1988 that LIS rather than GIS was the preferred term for government activities in the field.

Implementation of the National Strategy—the responsibility of ALIC and its participating jurisdictions—was based on eight objectives, key ones among which were to ensure the provision of appropriate education and training in land information management, the establishment of a national directory of land information, the development of data standards, the investigation of national priorities for the acquisition of basic land related data, the promotion of Australasian expertise in land information overseas, and support of research and development of technology relevant to land information management.

The Commonwealth also recognised the need for a higher level of coordination in the activities of its own departments and agencies involved with geographic information and environmental management, and undertook a number of important initiatives to bring this about during the second half of the 1980s. As part of a program of rationalisation of government agencies, a decision was taken in 1987 to amalgamate the Australian Survey Office with NATMAP to form the Australian Surveying and Land Information Group (AUSLIG). The Australian Centre for Remote Sensing was later attached to AUSLIG. In 1989, the Commonwealth Land Information Forum (CLIF) was established to promote interagency coordination of activities in the collection, storage, and dissemination of land information.

AUSLIG is now the Commonwealth agency responsible for geodesy, surveying, mapping (separate from responsibilities related to defence), and remote sensing in Australia. It provides a range of administrative, technical, and support services and undertakes selected application projects for the Commonwealth (Bell and Hobson 1990). During the 1980s, however, AUSLIG's involvement in GIS was limited mainly to exploratory demonstration projects, an important one at the time being the development of a prototype GIS for the Jervis Bay region in New South Wales (NSW) (Phillips and Blackburn 1989). In its role as an information provider, an early Commonwealth initiative taken through AUSLIG was the production of a directory of Commonwealth land information holdings, LANDSEARCH. The first edition of LANDSEARCH was published in 1985 (AUSLIG 1985). An enlarged edition, made available on-line over a national computer network in 1989, is still used.

State and territory governments progressively put in place the administrative and management structures necessary to develop and implement statewide LIS. A major review of these indicated that by the mid 1980s there were, however, still considerable differences between jurisdictions in the extent of their coverage, structure, and implementation (AURISA 1985). One of the most significant developments was the REGIS program started in northern Queensland in 1985. This consisted of seven individual projects including land use planning, evaluation of the location of tourist facilities, and river pollution and soil erosion studies. A major emphasis in the REGIS program was evaluation of the problems of networking computers and remote databases which included the pioneering use of satellite communications for voice, image, and data transfer between remote participants (Stanton 1987).

Another initiative at the state government level was the creation of the State Land Information Council (SLIC) in New South Wales in 1985 to coordinate the building of the statewide LIS. The design of the LIS was based from the outset on networking discrete databases, using the concept of an LIS Hub to facilitate the efficient management and transfer of land-related data between government agencies and other authorities (Hart 1988). Important among the latter were the large utility companies and the Water Board and Prospect County Council (an electricity provider) in Sydney, the operations of which relied increasingly on digital land parcel data (Hurle 1989). An important first step in coordinating databases was taken in 1980 with the integration of the computer systems of the NSW Valuer General's rolls with the Water Board's rating and billing systems.

In contrast to the formative developments taking place in most states, LIS developments had advanced to the point in South Australia and the Northern Territories where they were being used by the end of the period for a range of practical applications, and it was already becoming possible to assess the benefits of the considerable investments that had been made (Stephens 1988; Ralph et al. 1988). During this period, there were also significant developments in the application of GIS in public utilities for automated mapping and facilities management.

Compared with the LIS initiatives taken by the state and territory governments, the adoption of LIS-GIS and the implementation of systems for local planning by municipal governments was much slower (Garner 1990). A survey of 400 local governments in 1982 indicated that computers were still being used predominantly for engineering and accounting; only 15 per cent of those responding indicated that computers were being used for planning-related activities (Earle et al. 1984). Five years later, another survey revealed that this had doubled to 32 per cent and that an increasing number of municipalities were in the process of installing multipurpose land and property information systems, particularly those in the heart of the major metropolitan areas and their suburbs. Genasys, a leading GIS company headquartered in Sydney, played an instrumental role in the diffusion of GIS at the municipal level through the installation of its GENAMAP and other software systems.

Many of the new municipal developments were modelled on the pioneering initiative of the Council of the City of Sydney which has jurisdiction over Sydney's Central Business District. It began implementing its first system in 1972 when it commissioned the UK firm, ICL, to develop a corporate-wide LIS to service the day-to-day and strategic needs of the council, the driving force behind which was Kerry Nash, then the chief planner. The system was subsequently extended in 1976 to include a mapping facility and various subsystems for planning applications (Nash 1986, 1988). Developments at other major urban municipalities included the City of Adelaide which had an operational LIS by 1984; Brisbane City was able to produce its town plan by computer in 1986 and had a fully operational planning system by the end of the decade; and, by 1988, the City of Melbourne had embarked on the development of a GIS.

Environment and Natural Resources

Throughout the 1980s, increasing concern about the state of the environment became a national issue, and environmental policies and their implementation was high on the political agendas of all governments, especially those of the Commonwealth. In Australia, responsibility for implementing environmental policies is primarily the concern of the state and territory governments. Although these were largely preoccupied with developing parcel-based LIS, it was envisaged that this would eventually be extended to include environmental and resources data as well. In particular, the various agencies realised that this information would play a critical part in monitoring compliance with the requirements of state and Commonwealth environmental impact assessment legislation.

By and large, however, the state and territory governments have yet to incorporate environmental coverages in their LIS to any significant extent although they became increasingly active throughout the decade in developing regional resource inventories and using GIS for environmental projects (AURISA 1985). Notable examples include the Queensland rainforest

GIS (Stanton and Bundock 1988), the South Australian environmental database (Stubbs 1985), and the natural resources data management system developed for Victoria (Alexander 1987). To accelerate these activities, ALIC organised workshops in 1989 and again in 1990 to address a range of issues relating to intergovernmental coordination of natural resources data to serve the needs of national environmental policies (ALIC 1990b).

Building on initiatives taken during the 1970s, the Commonwealth introduced a wide range of new environmental legislation, policies, and programmes during the decade of the 1980s. Among the more significant of these from the viewpoint of GIS applications were the National Conservation Strategy (1983), National Forest Strategy (1986), the Australian Wilderness Inventory (1986), National Land Capability and Land Degradation Surveys (1986-87), and the National Soil Conservation Strategy (1988), all of which reinforced the need for better coordination of national resources data. The need for better coordination in the provision of environmental information also became important in another context—Australia's defence needs. The 1987 Policy Information Paper *Defence of Australia* which stressed the need for adequate geographic information for military purposes resulted in a major effort by the Australian army to develop GIS capability.

The Commonwealth continued to strengthen its role as a coordinator and provider of information and at the end of the period undertook two significant initiatives. The first of these was the creation in May 1988 of the National Resources Information Centre (NRIC) within the Department of the Environment. NRIC aimed to develop systems and procedures for the rapid identification, access to, and integration of resource data held by the Commonwealth to support national environmental policies and programs. From the outset, NRIC was heavily committed to GIS, and it has since become a major proponent of the application of the technology (Johnson et al. 1989).

The second initiative was the Environmental Resources Information Network (ERIN) which was established in 1989 to provide the information, systems infrastructure resources, and technical advice required for environmental decision making by the Commonwealth. ERIN quickly became a major centre for the establishment and management of a diverse set of national environmental databases, ranging from those for endangered species to those required for monitoring drought and pollution. Later ERIN, together with MRIC, was responsible for a number of important initiatives in the development of resource databases and information systems during the 1990s.

Significant developments were also occurring during this period in the research community. The benefits of combining Landsat imagery with digital maps were being demonstrated for applications in environmental monitoring and management. An important early one of these was the Land Image-Based Resource Information System (LIBRIS) which applied image processing techniques to the quantitative analysis of land cover using "greenness" indices in the semi-arid rangelands (Graetz et al. 1983). Australia also began to produce its own image processing software during this period, for example, ER-Mapper.

Another significant research contribution was the Australian resources information system (ARIS), a continental-scale GIS for studies of biophysical and socioeconomic resources that was registered to local government area boundaries for mapping. By the mid 1980s, ARIS had been developed to the point where it could be used in a number of applications including the production of the Electoral Atlas of Australia, mapping rangeland environments, and assisting in locating the route for a high-speed ground transportation system between Sydney and Melbourne (Cocks et al. 1988).

Socioeconomic Applications

Compared with the rapid progress that was being made in LIS and resource and environmental information systems, comparatively little emphasis was given to socioeconomic applications of GIS despite obvious opportunities for this (Garner 1982). Although the ABS continued to issue the Census of Population and Housing in digital form for mainframes, an increasing range of data sets was becoming available on floppy diskettes. A major new development occurred in 1988 with the release of CDATA 86, a package designed specifically for desktop machines which contained items from the 1986 Census of Population and Housing and selected data from 1981 on CD-ROM. This was coupled to a commercial PC-based colour mapping system that enabled statistical profiles and maps to be produced for a variety of statistical units which the ABS had progressively standardised to its Australian Standard Geographical Classification.

By the mid 1980s, an increasing number of projects were exploring the use of GIS for demographic and economic applications. A major aim of many of these was to demonstrate the ways in which land parcel data could be aggregated, integrated with other kinds of socioeconomic data, and processed for planning applications using GIS (Zwart and Williamson 1988). Others focussed on the mapping of census data (O'Callaghan 1981) and the development of land use databases for planning, for example, the URBAN 2 project in Perth (Devereux 1985). Very few were concerned with spatial analysis; one exception was the use of ARC/INFO for mapping and analysing the distribution of customers of a major bank (Roberts 1988).

The most important developments relating to socioeconomic GIS applications during the 1980s, however, were occurring in the research community, especially at the CSIRO where substantial advances were made in the development of systems with GIS capabilities. Significant among these were a number of packages developed to bridge the gap between analysis and implementation in the planning context. These included LUPIS, SIRO-PLAN, and LUPLAN, decision support tools for land use planners which allocated preferred land uses to predefined mapping units on the basis of different weightings attached to alternative policy criteria (Ive and Cocks 1983; 1988). These packages were important because they extended GIS analytical capabilities to decision-making rules with traditional cartographic modeling capabilities. These decision support software packages foreshadowed the development of knowledge-based systems linked to a GIS (ARC-INFO), for example, ADAPT, a more advanced rule-based package subsequently developed for deriving zoning schemes (Davis and Grant 1987), and ONKA, a policy analysis decision support system (Davis et al. 1988).

One of the most innovative new packages developed during this period at CSIRO was designed specifically to solve territory assignment problems (ITA), a modeling system based on the incorporation of location-allocation algorithms with GIS (Horn et al. 1988). ITA was originally developed for marketing applications in conjunction with Tony Buxton (now managing director of TACTICIAN Corporation, Boston). It was subsequently applied to site location analysis for financial institutions and was selected by the Australian Electoral Commission to carry out electoral redistricting for elections to the Commonwealth House of Representatives. Another significant contribution in this period was built on the pioneering research into urban activity location modeling undertaken at the end of the 1970s under the leadership of John Brotchie at the former CSIRO Division of Building Research, that led to the development of a powerful planning software package called TOPAZ (Brotchie et al. 1980). By the end of the decade, many systems for desktop mapping and modeling had been developed (Newton et al. 1988), including LAIRD, a package for assessing impacts of retail developments (Roy and Anderson 1988), and MULATM, a traffic network planning package (Taylor 1988).

Education and Research

The rapid growth in LIS-GIS activities reinforced the need for the provision of an adequate level of education and training that had been identified in the *ALIC/ANZLIC National Strategy* of 1987. The Commonwealth tacitly recognized this in the Key Centre initiative which was introduced in the mid 1980s to enable the universities to respond to the need for high level training and applied research in areas considered important for national development. As part of this initiative, a fund was provided in 1985 to establish the Australian Key Centre for Land Information Systems (ACKLIS), based at Queensland University under the directorship of Professor Ken Lyons. ACKLIS's mission was to create and maintain a world-recognised centre of excellence in the areas of georeferenced data, land administration systems, natural resources data, and computer-aided mapping. It has been engaged in a wide range of teaching, training, and research programs in these areas and has played a key role in supporting another component of the National Strategy—promotion of Australian expertise in LIS-GIS throughout the Asia-Pacific region. In recognition of its accomplishments, the Institute for Land Information in the U.S.A. designated ACKLIS a Centre of Excellence in Land Information Studies in 1985.

A report commissioned by AURISA revealed that in 1989 some 35 different educational institutions were actively engaged in teaching and research in subjects relating to LIS-GIS principally through the disciplines of geography and surveying (AURISA 1989). Important among these was the University of New South Wales where a Centre for Remote Sensing had been established in 1981 (the first at an Australian university). Courses in LIS-GIS were well established in the Schools of Surveying and Geography; at Curtin University in Western Australia where the first full program in computing and GIS was introduced in 1988; and at Melbourne University where a Centre for GIS and Modelling was established jointly by Surveying and Landscape Architecture and where new programmes in LIS-GIS were introduced in 1989 (Garner and Zhou 1993). The Centre for Resources and Environmental Study at the Australian National University also emphasised GIS applications using an integrative systems approach following the appointment of Henry Nix as director in 1987.

Contributions to research at CSIRO continued to be especially significant throughout the decade. Following earlier research on databases for geographic data (McKenzie and Smith 1977), the potential for using relational data models to manage spatial and aspatial data was explored by Abel (Abel and Smith 1984). Rapid retrieval of the spatial data was achieved with a toolkit SIRO-DBMS (Spatial Indexing in a Relational Organisation-DBMS), and the new approach was successfully demonstrated in a zoning application in the Great Barrier Reef Marine Park (Abel 1988). The toolkit was further developed in the 1980s and subsequently licensed to ARC Systems, a leading GIS supplier in the utilities sector, which is now incorporating it into operational systems, with the Convergent Group Asia Pacific.

Recognition of the growing significance of research into spatial data handling led in 1987 to the creation of the Centre for Spatial Information Systems within the Division of Information Technology under the direction of John O'Callaghan. The Centre brought together the research efforts in spatial databases (Abel 1988; Abel and Smith 1984), remotely sensed data processing (O'Callaghan 1984), interactive visualisation and colour modeling (Robertson and O'Callaghan 1986; O'Callaghan and Robertson 1986), and computer mapping (O'Callaghan and Simons, 1983). Since its formation, the centre has underpinned a number of significant developments through collaborative projects in the Australian GIS industry and user community.

By the end of the 1980s, research into the application of GIS was well established in the private sector, particularly in the mining and exploration industry, and in various public agencies, including Telecom Australia (now Telstra) and the Australian Defence Forces. Following the successful installation of a GIS for cable plans records, Telecom began investigating the potential of GIS for network planning and forecasting demand for services, and, by the end of the decade, its efforts were increasingly focussed on the role telecommunications would play in the networking of spatial information systems (Cavill and Greener 1988; Edney and Cavill 1989). The Australian Defence Forces had also become increasingly aware of the importance of automated command and control systems, digital mapping, and computerised land information. By the arrival of the '90s, military applications of GIS were well established (Puniard 1988; Laing and Puniard 1989), and tactical and strategic systems based on the integration and analysis of image-based data for decision support roles were being developed (Nichol et al. 1987).

International recognition of Australia's contributions to GIS research and development continued during the '80s. In 1982, a U.S.–Australia Workshop on the Design and Utilisation of Computer-based Geographic Information Systems was organised by John O'Callaghan and Donna Peuquet, while Barry Garner and O'Callaghan organised the Third International Symposium on Spatial Data Handling which was held in Sydney in 1988.

1990–95: The Drive to Maturity

The range and technical capabilities of GIS products commercially available by the beginning of the 1990s provided the basis for a widening front of uptake and application in both the public and private sectors. In government agencies, attention was increasingly focussed on the important issue of integration, which provided the theme for AURISA 90: From Innovation to Integration: Bringing it all Together. GIS was extended into an important new range of application areas, a small but nationally significant domestic GIS industry had matured, and there was increased international recognition of Australia's contribution to developments in GIS.

The Move to Integration

By the 1990s, GIS was increasingly being viewed as a "data integration machine," and the move towards integration was gathering pace at various levels including the integration of spatial and textual data, system components, diverse data sets within individual systems as well as across systems, and in applications. Computing issues relating to distributed systems, user interfaces, and "open systems" architecture became prominent topics in the research literature (O'Callaghan 1990; Ackland et al. 1993). Managing databases of increasing size and complexity for a widening set of users had become a major issue, and unlocking the wealth of information they contained focussed efforts on data access, standards, interchange, and improving connectivity between systems (Abel et al. 1994).

The lack of standards for geographic data, which had been recognized as a priority for action in the National Strategy, was a major obstacle to the development of a national spatial data infrastructure (Clarke 1991). The existing standard (AS 2482—interchange of feature coded digital mapping data) was by now obsolete. Although Standards Australia had formed Committee IT/4 (GIS) in 1987 with responsibility for developing standards for land and geographic data and data transfer, no substantial progress was made until after 1992 when ANZLIC contracted the Australasian Spatial Data Exchange Centre (AUSDEC) to develop the new Standard (AS-NZ 4270) which was adopted in 1995 (AS-ANS, 1995). Subcommittees established under the IT/4 initiative were also investigating standards for land use classification (a draft had been issued in 1989), utilities, topographic and hydrographic data, and for street addressing (a draft was released in 1993). Standards were also being addressed at the time by the Inter-Government Advisory Committee on Surveying and Mapping, established in 1988, which was charged with maintaining and developing technical specifications and standards for national mapping.

Integration of a different kind—that between remote sensing and GIS—also became increasingly important for research and GIS applications. A major upgrading of the Australian remote sensing receiving station and processing facilities was completed at the end of 1989 to enable the direct reception of Thematic Mapper (TM), SPOT, and National Oceanic and Atmospheric Administration (NOAA) Advanced Very High Resolution Radiometer (AVHRR) data. A new range of products subsequently became available, geocoded and registered to the Australian Map Grid. This development essentially opened the way for greater integration of image with other data sets for resource and environmental applications. The pioneering research into the development of image-based GIS that started in the latter part of the 1980s intensified during the 1990s, and applications based on the integration of the two technologies became increasingly common (Zhou and Garner 1990).

The successful use of GIS for data integration across different jurisdictions by the Murray-Darling Basin Commission, an intergovernmental body established in 1988 to coordinate land use planning and management of Australia's largest catchment covering one-seventh of the continent (Nanninga and Tane 1990), was extended to integrate remote sensing for investigating land degradation, salinity and vegetation change, and the modeling of interconnections between surface and groundwater processes (Jupp et al. 1990). Similar applications were occurring elsewhere, particularly in arid grazing lands, for example, the assessment by the NSW Environment Protection Agency of the impacts of land use activities on the environment in western New South Wales (Turner and Ruffio 1992).

A concern with standards and integration also underpinned the development of data dictionaries and directories by the Commonwealth. Following an initiative by NRIC, the initial stages of a National Directory of Australian Resources (NDAR) were established to provide information (largely metadata) on geographic data sets held by government agencies throughout Australia (Pahl 1991). By 1993, it contained information on more than 3,500 data sets. An important feature of NDAR was the use of consistent and accurate terminology for indexing and searching the database, including organisational tables containing standard terminologies and codes. Access to the directory was provided by a software system called FINDAR developed by NRIC, and access is now also being provided over the Internet, through the World Wide Web (WWW) (http://www.agso.gov.au).

ERIN also became increasingly involved in the development of standards and national databases after 1990, including standards for bioenvironmental spatial information systems (flora, fauna, vegetation, ecosystems, and landscapes) and the development of the national database on plant distribution based on recorded observations of more than 7,000 species. In 1992, it embarked on a significant joint venture with CSIRO, NRIC, and the Australian National Parks and Wildlife Service to develop the national maritime information system which was designed to underpin the Ocean Rescue 2000 program—a decade-long program to protect Australia's marine environment. Another of ERIN's new ventures was the development of the Environment Resource Information System (ERIS), a distributed network providing access to environmental information through a number of nationally dispersed nodes. In 1993, ERIN received two major awards for ERIS—a gold award at the Australian Government Technology Event and the prestigious Computerworld-Smithsonian award for excellence—and in 1994, access to ERIN's computerised information system became available through the Internet WWW (http://kaos.erin.gov.au/erin.html).

The Commonwealth was also now involved in broader interagency integration issues and in 1992 created the Commonwealth Spatial Data Committee (CSDC) to address all issues relating to spatial data of concern for the national government. The principal task of CSDC was the maximisation of benefits from the increasingly widespread application of GIS by Commonwealth agencies through the introduction of policies to better integrate activities and to develop a national spatial data infrastructure. A major concern of CSDC has been the development of a policy for the transfer of spatial data between Commonwealth agencies.

The National Strategy was further revised by ANZLIC in 1994 to focus more directly on the development of a national structure for the collection, classification, management, and transfer of land and geographic data. In the ANZLIC *Strategic Plan 1994–97*, new objectives were presented aimed at fostering standardisation in the natural resources area, the development of a national information infrastructure and standards for custodianship to enable good stewardship of fundamental data (ANZLIC 1994). Recent efforts have focussed on establishing a Spatial Data Infrastructure for Australia and New Zealand and implementation of the Spatial Data Transfer Standard (see http://www.auslig.gov.au/pipc/anzlic/anzlicma.html).

The emphasis on integration was also reflected in important new directions in LIS developments by the state and territories. New South Wales could claim that it had established a truly integrated system; databases from more than 350 different agencies had been successfully integrated into a logical database so that inconsistencies in entries could be checked and removed. Significant advances had also been made in data integration and distribution and administrative arrangements (Bullock 1993). Developments in South Australia and Queensland have addressed the integration of land information databases to support the delivery of information services to clients and the public. These developments have received awards for innovative uses of Information Technology in Government. Integration has also underpinned LIS developments in Western Australia, and a start had been made on the development of state geographic data standards (Burke 1993). In 1992, Victoria embarked on one of the most comprehensive LIS initiatives so far undertaken by any state government, the development of a statewide multiagency GIS (Alexander 1993). This was based on the Strategy for Geographical Data Management developed by Roger Tomlinson and Associates which quantified the benefits obtainable by rationalising the generation of geographic data and by supporting effective use of these data sets throughout government agencies. For the first time, LIS-GIS was beginning to be integrated within "whole of government" architectures for information systems.

The concern with integration also led to new directions for research (AURISA 1993). Recognising that user and corporate needs were now changing, decision support became an increasingly important focus, and exploration began of a new generation of GIS designed to handle distributed, multimedia data and heterogeneous systems. A CSIRO team led by David Abel continued to make significant contributions by developing prototype environmental decision support systems based on the integration of diverse data (digital maps, taxonomic data, satellite imagery, and photographic records) with the facility for exploratory data analysis (Abel et al. 1992) and the development of software to integrate specialised predictive models with GIS for water modeling (Abel et al. 1993). These innovative contributions which incorporated spatial data access, analysis, modeling, and interactive visualisation into GIS architectures were the recipient of four best-paper awards at AURISA conferences during the 1990s and the prestigious CSIRO Medal in 1996.

Extending Applications

An important feature of the developments in GIS in Australia during the early 1990s has been the diffusion of the technology into new areas of application. One of the most significant of these has been hazard management. A response information system (RISC) for monitoring chemical spills in Moreton Bay (Brisbane) was operational in 1990 (Barker 1990), and other important extensions related to the two most serious natural hazards in Australia—floods (Buckle 1994) and bushfires. The latter had been the subject of some isolated interest in earlier years—for example, the development of PREPLAN, a system in which dynamic models were linked with GIS for monitoring fire behaviour in National Parks (Kessel and Good 1981; Kessel 1988)—but no significant GIS applications occurred in this area until the 1990s. Since then, applications have been developed by many authorities. A number of them, as well as the CSIRO Division of Forestry, have developed GIS-based bushfire modeling systems since the 1980s (Garvey et al. 1992).

Applications for GIS are occurring in several areas of intelligent transport systems, notably navigation, fleet management and dispatching, vehicle routing and scheduling, and traffic management. These applications call for much finer granularity of data than traditional road centreline data sets, and a new draft Australian geographic data standard (GDF), promulgated in 1995, has been developed to meet the needs of professionals and organisations involved in the creation, update, supply, and application of referenced and structured road network data. The SCATS traffic management system developed by the New South Wales Roads and Traffic Authority employs a spatial model of traffic lanes and the connectivity of intersections and has been installed in many overseas cities as well as in most Australian capitals.

Emergency services provision and management also emerged as a significant new area of application. The Queensland government embarked on a major project (ESMAP) to apply GIS to enhance the capabilities of the police and emergency services to access and use a comprehensive range of geographical information based on the state's DCDB (Granger 1991). In New South Wales, GIS was being developed by the fire services for use in monitoring levels of fire protection and incident management as well as planning service delivery based on computer-aided dispatch systems (Thompson 1994).

Socioeconomic applications of GIS continued to expand during this period. More sophisticated, model-based systems were being applied to land use planning at various scales. Examples range from the use of GIS in conjunction with small area population forecasting and allocation modeling to match the demand for housing with the supply of land at the urban scale (McDougal et al. 1992) to the CYPLUS project, a regional scale application to identify land use options for sustainable resource development in the Cape York Peninsula, northern Queensland (McNaught 1994). Other new application areas that were emerging at this time included the health sector (Hennicke et al. 1993; Gaheegan and Gilchrist 1994), recreational resource management (Robinson and Colless 1992), and assets management by the public sector.

Many of these applications depended on data provided by the ABS, which continued to provide an increasing range of digital products that now included CMAP91, a CD-ROM containing all digitised boundaries used for the 1991 Census as well as digital topographic data for the whole of the country coupled to a reduced functionality version of Mapinfo for mapping. Another CD-ROM product, CDATA91, complements CMAP91. CDATA91 is designed for desktop analysis of 1991 Census small area census data. Subsequently ABS announced the release of the Integrated Regional Data Base (IRGB), aimed at broadening the availability of small area statistical data, which contained key economic and social indicators for non-metropolitan Australia to assist regional policy initiatives. An expanded version of IRGB which included data from various Commonwealth departments was released in October 1994. The 1996 Census is based on a new integrated national digital topographic map base.

AUSLIG was also engaged in new ventures aimed more directly at servicing the needs of the GIS user community. A major shift in the production of cartographic products occurred in the early 1990s with the launch of the Australian Geographic Data Base (GEODATA), a program aimed at providing high quality digital geographic data specifically designed for use in GIS (Bell 1991). The first of the new products—TOPO-10M and COAST-100K (coastline and state boundary files)—were released in 1992, and national coverage of TOPO-250K (hydrography, infrastructure, and relief files) had been completed by 1994. Another significant product was the digital postcode boundaries released at the end of 1992, and, in 1993, work commenced jointly with the Australian Geological Survey Office on the production of a medium level DEM of the Australian continent. Together with the digital products from the ABS, an expanding spatial data infrastructure was put in place on which future research, development, and application could be based.

Conclusion

The development and application of GIS reflects Australia's ability to adopt and extend many facets of information technology. The size of the country and the distinctive features of its physical and economic geography have made the application of GIS cost effective. Unlike most other countries, governments have played a key role in developing and providing the organisational and data infrastructure necessary for effective applications. Several innovative projects have been accorded national and international recognition. Overseas, URISA awards for Exemplary Systems in Government have been given to five Australian initiatives, including the REGIS project, the Port Adelaide urban LIS-GIS, and most recently ESMAP, the Queensland emergency services system.

A significant contribution to LIS-GIS research and development has been made by Australian research groups, particularly within the CSIRO. Australian authors have made a significant contribution to the *International Journal of Geographical Information Systems* in which, on a per capita basis, a larger number of Australia's papers has been published than from any other single country. Australia has enjoyed a high profile through attendance and presentations at international conferences, representation on the editorial boards of major publications in the field, and as collaborators in the organisation of major GIS conferences in Australia and southeast Asia.

Developments in research and applications also underpinned the development of a small-scale but highly successful and innovative domestic GIS industry. Major players included ESRI Australia, Genasys, ERMS, and ARC Systems (now Convergent Group Asia Pacific), and MAPTEK which, together with public agencies, are now increasingly involved in the export of

GIS services, technological expertise, and applications "know how," particularly to the Asia-Pacific region. In 1992, the total annual expenditure on spatial information systems and desktop mapping in Australia was estimated to be of the order of AUD400-500 million (Price Waterhouse 1994).

The strength of research, development, and application in LIS-GIS developed in Australia during the past quarter century has provided a very solid basis on which to expand during the next decade. Important new emphases in research and development are already emerging, notably research into the impacts of the convergence of information and communication technologies for applications based on geographically referenced data and the organisational and social impacts of the use of GIS.

Education and training for GIS has increased rapidly and now 49 tertiary education institutions offer courses in GIS and related areas (AURISA 1995). Funds were allocated by the Commonwealth in 1995 to establish a new key centre based at Adelaide University, South Australia, to undertake teaching and research into the social applications of GIS. This is a significant initiative in that it will be the first key centre in the social sciences.

The use of spatial information systems technology is also expected to become widespread in the provision of consumer services in the future, particularly in tourism and travel services (Australia's fastest growing source of overseas earnings), transport services, and the social services and health sectors. At the same time, it is anticipated that the importance of Australia's industrial and commercial GIS base will continue to expand to provide an increasing range of GIS services in the Asia-Pacific region. All of these developments are consistent with the aspiration of Australia to play a critical role in the effective application of information and communications technology in the future for the benefit of Australians and the international competitiveness of Australian industry.

Bibliography

Abel, D. J. 1988. "SIRO-DBMS: A Database Tool-kit for Geographical Information Systems." *International Journal of Geographical Information Systems*, 3: 103–16.

Abel, D. J., P. J. Kilby, J. R. Davis, and A. Deen. 1993. "The Spatial Water Modelling Program: A Case Study in Integrated Spatial Information Systems." *Proceedings, AURISA 93*. Canberra, Australia: AURISA, 406–15.

Abel, D. J., J. R. Davis, and P. J. Kilby. 1994. "The Systems Integration Problem." *International Journal of Geographical Information Systems*, 8: 1–12.

Abel, D. J., and J. L. Smith. 1984. "A Data Structure and Retrieval Algorithm for a Database of Areal Entities." *The Australian Computer Journal,* 16: 147–54.

Abel, D. J., S. K. Yap, M. A. Ackland, M. A. Cameron, D. F. Smith, and G. Walker. 1992. "The Environment Decision Support System Project: An Exploration of Alternative Architectures for Geographical Information Systems." *International Journal of Geographical Information Systems*, 6: 193–204.

Ackland, R. G., D. Abel, D. Campbell, P. Lamb, H. Lei, T. Tran, T. Collins, and T. Mittiga. 1993. "The LISA Project: An Open Systems Land Information Access System." *Proceedings, AURISA 93*. Canberra, Australia: AURISA, 362–72.

Alexander, D. 1987. "Using a Statewide Natural Resources Geographic Information System to Support Management, Planning and Policy Development in a Large Multidisciplinary Government Department." *Proceedings, URPIS 15,* vol. 4. Canberra, Australia: AURISA, 2–21.

Alexander, D. 1993. "Victoria's Comprehensive GIS Planning Program." *Proceedings, AURISA 93.* Canberra, Australia: AURISA, 241–49.

AS-ANS. 1995. *Geographic Information Systems—Spatial Data Transfer Standard.* Sydney, Australia.

Australia–New Zealand Land Information Council (ANZLIC). 1994. *Strategic Plan 1994–1997.* Canberra, Australia: ANZLIC.

Australian Land Information Council (ALIC). 1990a. *National Strategy on Land Information Management.* Canberra, Australia: ALIC.

Australian Land Information Council (ALIC). 1990b. *National Coordination of Natural Resources Data.* Canberra, Australia: ALIC.

Australian Surveying and Land Information Group (AUSLIG). 1985. *LANDSEARCH1: Directory of Commonwealth Land Related Data.* Canberra, Australia: Commonwealth Department of Local Government and Administrative Services.

Australian Urban and Regional Information Systems Association (AURISA). 1985. *Report of the Working Group on Statewide Parcel-Based Land Information Systems in Australasia.* Technical Monograph No 1. Canberra, Australia: AURISA.

Australian Urban and Regional Information Systems Association (AURISA). 1989. *Report of the National Working Party on Education and Research in Land and Geographic Information Systems.* Technical Monograph No 3. Canberra, Australia: AURISA.

Australian Urban and Regional Information Systems Association (AURISA). 1993. *AURISA GIS Research Inventory 1993.* Canberra, Australia: AURISA.

Australian Urban and Regional Information Systems Association (AURISA). 1995. *Education and Training in Urban Regional Information: The 1995–96 Australasian Inventory.* Monograph No. 10. Canberra, Australia: AURISA.

Austin, M. P., and K. D. Cocks, eds. 1977. *Land Use on the South Coast of New South Wales: A Study of Methods of Acquiring and Using Information to Analyse Regional Land Use Options.* 4 volumes. Canberra, Australia: CSIRO.

Barker, T. 1990. "A Response Information System for Chemical Spills (RISC) in Moreton Bay." *Proceedings, URPIS 18.* Canberra, Australia: AURISA, 105–14.

Bell, K. C. 1991. "Data Quality and Spatial Data Integration: The Australian Geographic Data Base Program." *Proceedings, AURISA 91.* Canberra, Australia: AURISA, 596–75.

Bell, C., and D. Hobson. 1990. "Coordination of Land Information: AUSLIG Supporting the National Interest." *Proceedings, URPIS 18.* Canberra, Australia: AURISA, 359–67.

Bomford, A. G. 1971. "Automated Cartography in the Division of National Mapping." *Cartography,* 7: 119–25.

Brotchie, J. F., J. W. Dickey, and R. Sharpe. 1980. *TOPAZ—General Planning Model and its Applications at the Urban and Facility Planning Levels.* Heidelburg, Germany: Springer-Verlag.

Buckle, P. 1993. "Applying Geographic Information Systems to Reduce Losses Caused by Natural Disasters." *Proceedings, AURISA 94.* Canberra, Australia: AURISA, 71–80.

Bullock, K. 1993. "Integration: The Key to Successful Development in NSW." *Proceedings, AURISA 93.* Canberra, Australia: AURISA, 262–72.

Burke, A. 1993. "The West Australian Land Information System (WALIS)." *Proceedings, AURISA 93.* Canberra, Australia: AURISA, 250–62.

Byatt, P. 1976. "Investigations into the Geocoding of Census Data." *Proceedings, URPIS 4.* Canberra, Australia: AURISA, 1.1–1.10.

Cavill, M. V., and S. Greener. 1988. "Introducing Geographic Information Systems Technology: Concepts, Approval and Implementation." *Proceedings, URPIS 16.* Canberra, Australia: AURISA, 323–30.

Clarke, A. L. 1991. "Spatial Data Base Standards—Technical and Management Issues." *Proceedings, AURISA 91.* Canberra, Australia: AURISA, 444–455.

Cocks, K. D., P. A. Walker, and C. A. Parvey. 1988. "Evolution of a Continental-scale Geographical Information System." *International Journal of Geographical Information Systems*, 2: 263–280.

Cook, B. G. 1967. "A Computer Representation of Plane Region Boundaries." *The Australian Computer Journal,* 1: 44–50.

Cook, B. G. 1975. "A Computer Data Bank in a Regional Land Use Study." *Proceedings, URPIS 3.* Canberra, Australia: AURISA, 2.01–2.14.

Davis, J. R., and I. W. Grant. 1987. "ADAPT: A Knowledge-based Decision Support System for Producing Zoning Plans." *Environment and Planning B,* 14: 53–66.

Davis, J. R., P. M. Nanninga, and R. D. S. Clark. 1988. "A Decision Support System for Evaluating Catchment Policies." *Proceedings of the Conference on Computing in the Water Industry.* Melbourne, Australia, 205–9.

Davis, J. R., and T. A. Paine. 1979. "Address Matching of Australian Spatial Data." *Proceedings, URPIS 7.* Canberra, Australia: AURISA, 5.4–5.6.

Devereux, D. 1985. "The URBAN 2 Project." *Proceedings, URPIS 13.* Canberra, Australia: AURISA, 225–34.

Earle, T. R., E. P. Fitzgerald, and R. Learmonth. 1984. "Computer Utilisation in Australian Local Government." *Proceedings, URPIS 12,* vol. 6. Canberra, Australia: AURISA, E7–E15.

Eddington, B., M. Phillips, and K. Bell. 1988. "National Strategy on Land Information Management: A Blue-print for National Coordination." *Proceedings, URPIS 16.* Canberra, Australia: AURISA, 130–38.

Edney, P., and M. Cavill. 1989. "The Melbourne Knowledge Precinct GIS Pilot Project." *Proceedings, URPIS 17.* Canberra, Australia: AURISA, 325–31.

Gaheegan, M., and J. Gilchrist. 1994. "GIS and Healthcare—Design of a Suitable Data Model." *Proceedings, AURISA 94.* Canberra, Australia: AURISA, 407–18.

Garner, B. J. 1982. "Towards More People-oriented Geographic Information Systems." *Proceedings, URPIS 10.* Canberra, Australia: AURISA, 200–211.

Garner, B. J. 1990. "GIS for Urban and Regional Planning and Analysis in Australia." In L. Worrall, ed., *Geographic Information Systems: Developments and Applications.* London: Bellhaven Press, 41–64.

Garner, B. J., and Q. Zhou. 1993. "GIS Education and Training: An Australian Perspective." *Computers, Environment, and Urban Systems,* 17: 61–71.

Garvey, M., K. Stephenson, and M. Whelan. 1992. "The Use of Geographic Information Systems at the Country Fire Authority." *Proceedings, AURISA 92.* Canberra, Australia: AURISA, 220–28.

Graetz, R. D., W. R. Gentle, and J. F. O'Callaghan. 1980. "The Application of Landsat Image Data for Land Resources Assessment in the Arid Lands of Australia." *Bulletin Remote Sensing Association of Australia,* 4: 5–26.

Graetz, R. D., R. P. Pech, M. R. Gentle, and J. F. O'Callaghan. 1983. "The Application of Landsat Image Data to Rangeland Assessment and Monitoring: The Development and Demonstration of a Land Image Based Resource Information System (LIBRIS)." *Journal of Arid Environments,* 4: 236–45.

Granger, K. 1991. "Information for Public Safety and Emergency Management: The Queensland ESMAP Project." *Proceedings, AURISA 91.* Canberra, Australia: AURISA, 118–30.

Hart, A. L. 1988. "The NSW LIS in Action." *Proceedings, URPIS 16.* Canberra, Australia: AURISA, 25–29.

Hennicke, G., I. Chukwodozie-Ezigbalike, and I. D. Bishop. 1993. "Implementing a GIS for Analysing Accident and Emergency Services." *Proceedings, AURISA 93.* Canberra, Australia: AURISA, 448–57.

Horn, M., J. F. O'Callaghan, and B. J. Garner. 1988. "Design of Integrated Systems for Spatial Planning Tasks." *Proceedings of the 3rd International Symposium on Spatial Data Handling.* Columbus, OH: International Geographical Union, 107–16.

Hurle, G. 1989. "The Status of Development of Facility Management Systems within the Australian Electricity Supply Industry." *Proceedings, URPIS 17.* Canberra, Australia: AURISA, 350–8.

Ive, J. R., and K. D. Cocks. 1983. "SIRO-PLAN and LUPLAN: An Australian Approach to Land Use Planning: 2—The LUPLAN Land Use Planning Package." *Environment and Planning B,* 10: 346–359.

Ive, J. R., and K. D. Cocks. 1988. "LUPIS: A Decision Support System for Land Planners and Managers." In P. W. Newton, M. A. P. Taylor, and R. Sharpe, eds., *Desktop Planning: Microcomputer Applications for Infrastructure and Services Planning and Management.* Melbourne, Australia: Hargreen, 129–39.

Johnson, B. D., J. Mott, and T. Robey. 1989. "Providing Effective Access to Resource Information—Progress Towards a National Directory of Australian Resource Data." *Proceedings, URPIS 17.* Canberra, Australia: AURISA, 260–65.

Jupp, D. L. B., T. R. McVicar, J. Walker, and T. I. Dowling. 1990. "Using AVHRR Data, Image Processing, and GIS Methods to Monitor the Spatial Changes in Land Degradation and Salinisation in the Murray Darling Basin of Australia." *Proceedings, URPIS 18.* Canberra, Australia: AURISA, 95–103.

Kessel, S. R. 1988. "Fire Hazard Modelling: The PREPLAN and FIREPLAN Systems." In P. W. Newton, M. A. P. Taylor, and R. Sharpe, eds., *Desktop Planning: Microcomputer Applications for Infrastructure and Services Planning and Management.* Melbourne, Australia: Hargreen, 222–31.

Kessel, S. R., and R. B. Good. 1981. *PREPLAN—the Pristine Environmental Planning Language and Simulator for Kosciusko National Park.* Canberra: Australian National Parks and Wildlife Service.

King, C. W. 1977. "The Automated Census Mapping Program." *Proceedings, URPIS 5.* Canberra, Australia: AURISA, 6.1–6.10.

Laing, A. W., and D. J. Puniard. 1989. "The Australian Defence Force Requirements for Land-related Information." In D. Ball and R. Babbage, eds., *Geographic Information Systems: Defence Applications.* Sydney, Australia: Pergamon Press, 61–79.

Laut, P. 1977. "The South Australian Ecological Survey—A Base for a State-wide Land Information System?" *Proceedings, URPIS 5.* Canberra, Australia: AURISA, 3.1–3.12.

Massey, J. S. 1978. "The Australian Small-area Agricultural and Pastoral Land Use Information System: 1900 to the Present." *Proceedings, URPIS 6.* Canberra, Australia: AURISA, 4.39–4.80.

McCalden, G. 1973. "Design for an Urban and Regional Information System." *Proceedings, URPIS 1.* Canberra, Australia: AURISA, 35–46.

McCalden, G., and W. Jarvie. 1976. "How Transferable is Geocoding Technology?" *Proceedings, URPIS 4.* Canberra, Australia: AURISA, 3.1–3.10.

McDougall, K., M. Bell, and I. McQueen. 1992. "The Development of a GIS Based Decision Support System for the Forecasting of Urban Growth." *Proceedings, AURISA 92.* Canberra, Australia: AURISA, 365–71.

McNaught, I. 1994. "CYPLUS GIS—Serving the Land Use Strategy for Cape York Peninsula." *Proceedings, AURISA 94.* Canberra, Australia: AURISA, 89–95.

McKenzie, H. G., and J. L. Smith. 1977. "The Implementation of a Data Base Management System." *The Australian Computer Journal,* 9: 138–44.

Newton. P. W., M. A. P. Taylor, and R. Sharpe, eds. 1988. *Desktop Planning: Microcomputer Applications for Infrastructure and Services Planning and Management*. Melbourne, Australia: Hargreen.

Nanninga, P., and H. Tane. 1990. "Integration through Coordination: Designing a GIS for the Murray-Darling Basin." *Proceedings, URPIS 18*. Canberra, Australia: AURISA, 281–92.

Nash, K. R. 1986. "The Application of Computers to the Planning Tasks in the City of Sydney." *Australian Planner*, 24: 36–49.

Nash, K. R. 1988. "The Sydney City Council LIS a Decade On: The Dream and the Reality." *Proceedings, URPIS 16*. Canberra, Australia: AURISA, 1–13.

Nichol, D. G., M. J. Fiebig, R. J. Whatmough, and P. J. Whitbread. 1987. "Some Image Processing Aspects of a Military Geographic Information System." *The Australian Computer Journal*, 19: 154–60.

O'Callaghan, J. F. 1981. "Evaluation of Interactive Census-type Data Analysis and Display Systems." *Proceedings, URPIS 9*. Canberra, Australia: AURISA.

O'Callaghan, J. F. 1984. "The Analysis of Remotely Sensed Data in an Image-based Geographic Information System." *Proceedings, Landsat 84*, 332–38.

O'Callaghan, J. F. 1990. "Trends in Geographical Information Systems: An Information Technology Perspective." *Proceedings, URPIS 18*. Canberra, Australia: AURISA, 397–400.

O'Callaghan, J. F., and B. J. Garner. 1991. "Land and Geographical Information Systems in Australia." In D. J. Maguire, M. F. Goodchild, and D. W. Rhind, eds., *Geographical Information Systems: Principles and Applications*. London: Longman Scientific, 57–70.

O'Callaghan, J. F., and P. K. Robertson. 1986. "Colour Image Display of Geographical Data Sets Using Uniform Colour Spaces." *Proceedings, International Symposium on Spatial Data Handling*, 322–25.

O'Callaghan, J. F., and L. Simons. 1983. "COLOURMAP: An Interactive Colour Mapping System." *Proceedings, 1st Australasian Conference on Computer Graphics*. Sydney: Australian Computer Graphics Association, 190–195.

Pahl, W. J. 1991. "Maximising the Use and Sharing of Natural Resource Information between Commonwealth, State, and Territory Agencies within Australia." *Proceedings, AURISA 91*. Canberra, Australia: AURISA, 467–73.

Phillips, M., and J. Blackburn. 1989. "The Chrysalis Project: A Regional GIS over Jervis Bay." In D. Ball and R. Babbage, eds., *Geographical Information Systems; Defence Applications*. Sydney, Australia: Pergamon Press, 204–31.

Price Waterhouse. 1994. *The Australian Spatial Information Systems Industry*. Sydney, Australia: Price Waterhouse.

Puniard, D. J. 1988. "Some Defence Applications of GIS." *Proceedings, URPIS 16*. Canberra, Australia: AURISA, 172–180.

Ralph, G., F. Bryant, P. Brooke-Smith, I. Pedler, and T. Mittiga. 1988. "Land Information Systems in Action—Realising the Benefits, Land Information South Australia." *Proceedings, URPIS 16*. Canberra, Australia: AURISA, 139–150.

Roberts, N. 1988. "Using GIS to Target the Market for Financial Institutions." *Proceedings, URPIS 16*. Canberra, Australia: AURISA, 247–60.

Robertson, P. K., and J. F. O'Callaghan. 1986. "The Generation of Colour Sequences for Univariate and Bivariate Mapping." *IEEE Transactions on Computer Graphics and Applications*, Feb.: 24–32.

Robinson, P., and R. Colless. 1992. "MSB Waterways: We're with You on the Water Integrated Information." *Proceedings, AURISA 92*. Canberra, Australia: AURISA, 395–402.

Rostow, W. 1960. *The Stages of Economic Growth: A Non-communist Manifesto*. Cambridge, U.K.: Cambridge University Press.

Roy, J. R., and M. Anderson. 1988. "Assessing Impacts of Retail Development and Redevelopment." In P. W. Newton, M. A. P. Taylor, and R. Sharpe, eds., *Desktop Planning: Microcomputer Applications for Infrastructure and Services Planning and Management*. Melbourne, Australia: Hargreen, 172–79.

Sedunary, M. E. 1977. "The Development and Extension of the South Australian Land Ownership and Tenure System (LOTS)." *Proceedings, URPIS 5.* Canberra, Australia: AURISA, 2.1–2.13.

Sedunary, M. E. 1980. "LOTS—An Operational Land Information System for South Australia." *Proceedings, URPIS 8.* Canberra, Australia: AURISA, 21.1–21.12.

Stanton, G. 1987. "The REGIS Program—Building a Tool for Policy Making in Northern Queensland." *Proceedings, URPIS 15.* Canberra, Australia: AURISA, 4.1–4.13.

Stanton, G., and B. Bundock. 1988. "Development of the Queensland Government Rainforest Geographic Information System." *Proceedings, URPIS 16.* Canberra, Australia: AURISA, 261–68.

Stephens, V. 1988. "Getting It Together—The Northern Territory Land Information System." *Proceedings, URPIS 16.* Canberra, Australia: AURISA, 52–65.

Stubbs, T. J. 1985. "The Application of Geographic Information Systems to Environmental Planning." *Proceedings, URPIS 13.* Canberra, Australia: AURISA, 165–73.

Taylor, M. A. P. 1988. "Computer Models for Traffic Systems Applications." In P. W. Newton, M. A. P. Taylor, and R. Sharpe, eds., *Desktop Planning: Microcomputer Applications for Infrastructure and Services Planning and Management*. Melbourne, Australia: Hargreen, 264–98.

Thompson, K. 1994. "GIS and Indicators of Fire Service Performance." *Proceedings, AURISA 94.* Canberra, Australia: AURISA, 53–57.

Turner, G. W., and R. M. C. Ruffio. 1992. "Environmental Auditing Using GIS." *Proceedings, AURISA 92.* Canberra, Australia: AURISA, 350–58.

Williamson, I. P. 1982. "The Role of the Cadastre in a Statewide Land Information System." *Proceedings, URPIS 10.* Canberra, Australia: AURISA, 19–26.

Zhou, Q., and B. J. Garner. 1990. "GIS and Remote Sensing: Towards the Better Integration of Data for Land Resource Management." *Proceedings, URPIS 18.* Canberra, Australia: AURISA, 185–94.

Zwart, P. R., and I. P. Williamson. 1988. "Parcel-based Land Information Systems in Planning." In P. W. Newton, M. A. P. Taylor, and R. Sharpe, eds., *Desktop Planning: Microcomputer Applications for Infrastructure and Services Planning and Management*. Melbourne, Australia: Hargreen, 44–53.

CHAPTER 17

Canada Today

David Forrest

Introduction

With the challenge of managing the world's second largest national land mass, it is not surprising that mapping has played an important part in Canada's history. The country's surveying and mapping tradition can be traced back to the explorers of the sixteenth century. The Geological Survey of Canada, founded in 1842, was one of the first national geological surveys in the world.

The foundation for today's surveying and mapping industry was created when military pilots, surveyors, and cartographers returned after World War II and found employment in government and private companies. Air photo coverage, available for only 25% of the country in 1945, was completed by 1957. Canada was quick to use computers in mapping and benefited from the seminal contributions of Roger Tomlinson, Raymond Boyle, David Douglas, Thomas Peucker, and other pioneers in automated cartography and GIS.

Since the 1960s, a dynamic industry has evolved in Canada, based on the collection, management, and application of geographic information. The country has created a world-class GIS software industry. It has also developed an international reputation in remote sensing, with customers in more than 100 countries. Approximately 25% of the image processing systems, 50% of the satellite ground receiving stations, and 90% of the high-resolution airborne radar systems

---

David Forrest is president of Global Vision Consulting Ltd., a management consulting firm located in Victoria, British Columbia, Canada, that provides services internationally in market analysis, strategic planning, requirements definition, and application design to the geomatics industry and its clients. A long-time analyst of the geographic information industry, he has written more than a hundred magazine and newspaper articles on GIS.

Author's Address: David Forrest, 4478 Annette Place, Victoria, B.C., Canada V8N 3J7. E-mail: dforrest@gvcl.com.

used worldwide have a Canadian origin. Canada's Long Term Space Plan forecasts continuing rapid growth in products and services related to Earth observation. While most Canadian companies tended in the past to specialize either in remote sensing or GIS, the geomatics industry has benefited from synergies between the two fields and a community of skilled professionals that moves easily between them. The term geomatics is widely used in Canada as defined by the Geomatics Industry Association of Canada (GIAC). See under Industry and Professional Associations below.

Canadian firms are active in the Middle East, Africa, Asia, South and Central America, and Eastern Europe. Canada generates approximately $800 million in geomatics revenues annually and exports close to $200 million worth of geomatics products and services every year. Its pioneering role in geomatics, its ability to create innovative technology and institutions, and a demanding, increasingly diverse, and sophisticated home market have been key factors in this international success.

Geographic information processing evolved rapidly in Canada in little more than three decades, going from a pioneering period of innovation in the 1960s and 1970s through a period of data and technology integration in the 1980s, when the infrastructure was created for subsequent growth and diversification. The 1990s have begun a new era of growth characterized by wider use of geographic information and the emergence of a new mass market. These three periods of development (summarized in Figure 17.1) provide a helpful framework for understanding the history and future direction of GIS in Canada.

Building the Foundation

Innovation (1960–1980)

Canada is credited by many with developing the world's first GIS, the Canada Geographic Information System (CGIS), created in the 1960s by the federal government to support the Canada Land Inventory (Chapter by Tomlinson). Design of the system started in 1963, and it moved into production in 1971. Land capability information was stored in five coverages describing an area of approximately 2.7 million square kilometers. Access to the system was supported through computer terminals located across the country.

Canada's federal mapping agency—the Department of Energy, Mines and Resources—introduced computer cartography in 1968. The Canada Centre for Remote Sensing (CCRS) was created in 1972 to lead the country's national remote sensing program, and it quickly earned an international reputation as a world-class research and development centre. Agriculture Canada created the Canadian Soil Information System (CanSIS) in the early 1970s to maintain soil survey information and produce interpretive maps.

Computers were first used to handle geodemographic information at about the same time. Statistics Canada developed the Area Master File—a geographic database describing urban street networks—for the 1971 Census. The Postal Code Conversion File was added in 1978, allowing census data to be accessed by postal code.

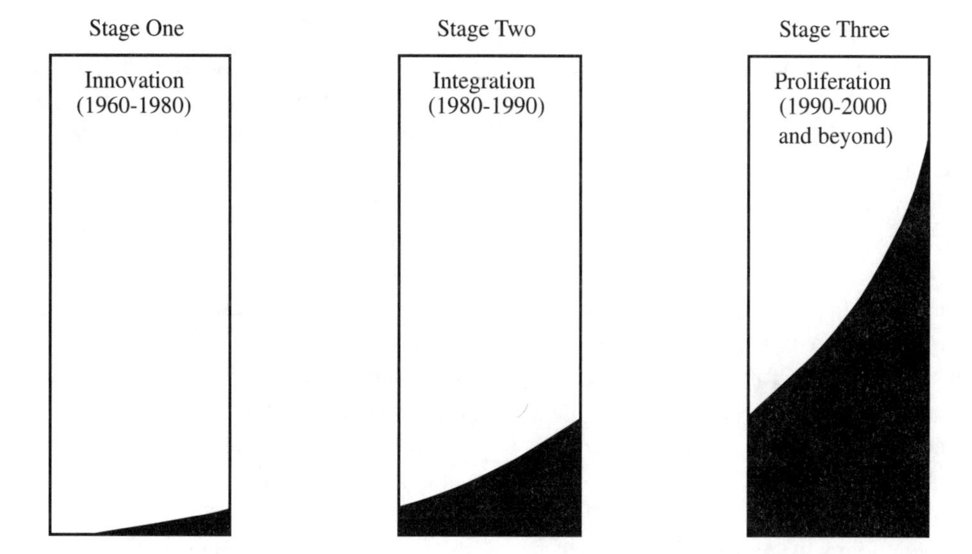

Innovation

Trends

technology development, isolated initiatives, champions in a few diciplines, standalone systems, narrow applications, informal committees and working groups, high cost, low data availability, limited growth

Dominant Products

maps and charts

Integration

Trends

data integration, technology integration, corporate/enterprise strategies, multi-participant initiatives, new applications, reduced cost, improved data availability, moderate growth

Dominant Product

geographically referenced information

Proliferation

Trends

system integration, creation of mass market, universal application, universal access, industry standards, new applications, low cost, high data availability, exponential growth

Dominant Products

value-added products and services, decision support systems, executive information systems, geographic enabled applications, geographic appliances

Figure 17.1 Trends in geographic information processing.

Pioneering activity was not limited to the federal government. Canada's Maritime Provinces were also quick to get involved. The Department of Surveying Engineering at the University of New Brunswick helped NASA with reconnaissance mapping to identify lunar landing sites for the Apollo program in the early 1960s. The university hosted the first international conference on land information systems in 1968 and conducted the first study of land information costs and benefits in 1971.

The first multiparticipant project in North America was undertaken by the Council of Maritime Premiers, when it created the Land Registration and Information Service (LRIS) in 1973. The mandate of the new organization was to coordinate and cost-share land surveying and mapping programs in the provinces of New Brunswick, Nova Scotia, and Prince Edward Island.

Other provinces followed shortly thereafter. Ontario created a task force on geographic referencing in the early 1970s. In Alberta, the Task Force Report on Urbanization and the Future recommended in 1974 that the province create a new coordinating agency for land-related information systems. A project was begun in 1978 to develop a plan for interagency coordination. The Province of British Columbia bought an Intergraph system in 1978 and began to produce forest inventory maps, pioneering computer mapping for resource management applications in government. GIS activities in Quebec started in the late 1970s in the Secteur Terres of the Ministère de l'Energie et des Ressources.

There were a number of early initiatives by municipalities. The City of Burnaby, British Columbia, pioneered GIS development in the mid 1970s for infrastructure, municipal, and utility management applications. Metropolitan Toronto's Central Mapping Agency was created in 1974 to provide six local municipalities with shared facilities to develop and exchange survey and mapping data—a forerunner of later multiparticipant initiatives. The City of Edmonton, Alberta, launched its Geographic Base Information System project in 1978. The City of Calgary, Alberta, started geocoding in the 1970s and began planning for a multipurpose cadastre.

By the end of the decade, computer mapping was in widespread use in the oil and gas industry, and many telephone and electric utilities had started to use the technology. The fledgling geomatics industry began producing digital data in the late 1970s in response to demands from government agencies. Canada was a dominant player in international surveying and mapping until the end of the 1970s, with its companies winning two out of every three foreign aid contracts.

Integration (1980–1990)

GIS and land-related information systems evolved rapidly during the 1980s from standalone, special-purpose digital mapping systems to multipurpose, multiuser, and multidisciplinary decision support systems. Whereas they initially served mainly the needs of a single department, these systems eventually became more critical to corporate information management in many organizations.

This change was accompanied by many difficult issues. Greater emphasis was placed on the ability to share geographically referenced information, and data and systems integration became of paramount concern.

By the end of the decade, most governments recognized the value of geographic data and understood the benefits of corporate planning and implementation. Many formed coordinating agencies to help integrate departmental systems and to ensure that maximum benefits were obtained from investments in geographic technology. The 1980s were a time of consolidation and integration, characterized by database, infrastructure, and institutional development. The foundations established during this period prepared GIS users and suppliers alike for the growth and diversification that lay ahead.

The Federal Government

The most active federal government departments in the 1980s were: Energy, Mines and Resources Canada (responsible for land mapping, some water mapping, aeronautical charts, and remote sensing data acquisition); Fisheries and Oceans Canada (responsible for producing and maintaining nautical charts); the Department of National Defence (responsible for providing maps, charts, and geographic support for the Canadian Forces); Environment Canada (responsible for producing weather maps and environmental information); Statistics Canada (responsible for conducting and reporting on the census); Agriculture Canada (responsible for soil mapping); and Forestry Canada (responsible for maintaining forest inventory in cooperation with other agencies).

The Surveys, Mapping and Remote Sensing Sector (SMRSS) was formed in the Department of Energy, Mines and Resources when the Canada Centre for Remote Sensing and the former Surveys and Mapping Branch merged in 1987. The sector was given the mandate to develop a digital framework for national mapping and GIS applications in Canada. It was also directed, with some restrictions, to make its resources and facilities available to private sector companies interested in exporting Canadian expertise and technology.

A national coordinating organization—the Interagency Committee on Geomatics (IACG)—was formed in 1988 to promote data sharing among federal government agencies. Its terms of reference included:

- creating institutional and cooperative arrangements to facilitate the development and use of shared databases;
- establishing a mechanism to develop national GIS interchange and geographic referencing standards;
- defining requirements for effective linkage between government GIS; and
- promoting and supporting an interagency GIS network.

Located in Energy, Mines and Resources Canada, the committee created the following technical subcommittees: GIS Data Models; Data Communications and Networking; Government Databases; Standards, Research and Education; and User Needs.

A Canadian General Standards Board committee was created and made responsible for developing national geomatics standards. The Data Communications and Networking subcommittee conducted a Federal Network Requirements Study to analyze requirements for data sharing among departments, define data communications needs, develop implementation scenarios and benefits, identify communications services within government that could form the basis for an eventual network, and recommend demonstration projects.

The Provinces

Natural resource applications were the catalyst for early GIS implementations in many provinces. In 1982, New Brunswick's Timber Management Branch was the first client in the world to install ESRI's (Environmental Systems Research Institute) ARC/INFO software. Geographic information systems for forestry multiplied in other provinces in the next five years, encouraged by funding support from the Canadian Forestry Service. The objective was improved long-term planning for timber management.

Government departments quickly adopted GIS technology, and data sharing and coordination was a major preoccupation from the middle to the end of the decade.

British Columbia British Columbia created the Ministry of Crown Lands in 1988 and made it responsible for consolidating mapping services and developing a government-wide strategy for land information. The ministry's goals included planning for future information needs, improving access to land information, facilitating information exchange, improving the quality of land information, pursuing cost-sharing opportunities, promoting private sector involvement, and supporting the marketing of innovations.

The government developed a Corporate Land Information Strategic Plan that would allow land-related information systems to evolve within individual agencies, while maximizing opportunities for data sharing and exchange. The strategic plan defined a global information systems architecture—the Land Information Infrastructure—to coordinate all land-related data, applications, and technologies. The Government Land Information Data Exchange group was created to manage these initiatives. A Land Information Strategic Committee was established in 1989 to involve ministries and crown corporations in defining and implementing corporate strategy.

The province contracted with a consortium of private mapping companies in 1986 to begin work on a state-of-the-art three-dimensional digital topographic base. Terrain Resource Information Management (TRIM) maps will eventually cover the entire province, at a total cost of approximately $80 million.

Alberta Alberta committed to one of Canada's earliest land information systems initiatives when it created the Land-Related Information Systems Steering Committee in 1980—providing a forum for interdepartmental discussion and cooperation. A report delivered in 1981 recommended the creation of a province-wide communications network. The recommendation was approved by the cabinet in 1982, and the LRIS Group was formed in 1983 to support systems development and coordinate implementation.

A strategic plan for the network was developed in 1986. A business plan was produced in 1987, evaluating products and services, markets, expected benefits, and project economics.

Approximately 120 manual and automated land-related information systems were identified as being maintained by the government at that time. The objective was to support integrated online access to land information through terminals, microcomputers, and other computer systems distributed throughout the province.

The Electronic Access to Government Land Data Policy, announced in 1987, cleared the way for the provincial government to provide land information to private-sector organizations and the public in digital form. A technical study in 1989 evaluated alternatives for network implementation. In September 1991, Land Information Alberta, a business organization in Alberta Forestry, Lands and Wildlife, was created to operate the network.

Saskatchewan Saskatchewan created a Land Information Users Committee in the mid 1980s, representing approximately 16 agencies and crown corporations already active in GIS or contemplating GIS projects. The committee was chaired by the Central Survey and Mapping Agency of the Saskatchewan Property Management Corporation.

The committee defined provincial GIS strategy, project requirements, standards, and funding arrangements, conducted demonstration projects, and specified technologies required to support a shared approach. Working groups were established to set overall project objectives, develop strategic plans and policies for GIS implementation, define architectures for a shared provincial GIS, provide information to the user community, define education and training requirements, and investigate market, industry, and export development opportunities.

In 1988, Saskatchewan committed to developing an integrated, multiuser GIS funded by the user community. The federal government, provincial government departments, municipalities, crown corporations, and the private sector all participated. The undertaking was structured as a capital project with a total cost of approximately $31 million over six years, to be completed by 1994. Cost recovery was expected over a period of nine years through cost-sharing arrangements with major participants and the sale of digital data products.

Manitoba Manitoba began looking at opportunities for data sharing in the mid 1980s. Early in 1987, a number of organizations cooperated in a study that evaluated the costs and benefits of an integrated land-related information system. Participants included Manitoba Natural Resources, Manitoba Telephone Systems, Manitoba Hydro, the City of Winnipeg, Treasury Board, and Manitoba Data Services. The study identified potential savings of more than $30 million from the shared use of land-related information.

Participants in the study planned to create a shared data facility that would include 14 categories of information: geodetic, cadastral, utility, transportation, administrative/zoning, buildings, structures, hydrography, hypsography, environment, natural resources, demographic, remote sensing, and digital orthophotography. An information utility would be created to package and market this information, paying royalties to data providers.

The multiparticipant initiative was led by Linnet Graphics International Inc. of Winnipeg (since renamed Linnet Geomatics International Inc.), a company established specifically to promote and develop the Manitoba Land Related Information System. Approval to begin implementation was received from the provincial cabinet in 1990, based on results of the cost-benefit analysis and anticipated economic development opportunities.

Ontario Ontario created a prototype of a new automated land registration system in Oxford County starting in November 1984. Database development was completed in 1987. Based on the prototype's success, the cabinet granted approval to implement an enhanced version of the system—the Province of Ontario Land Registration Information System (POLARIS)—in 65 Land Registry offices province-wide.

A strategic alliance between government and the private sector to implement POLARIS was contemplated early in 1988, when a joint government-industry committee made recommendations on the private sector's role. The objective was to develop an internationally competitive land-related information systems industry in Ontario.

The government published a Notice of Intent in September 1988, indicating that it would rely on a competitive bidding process to choose an industry partner. Two acceptable proposals were received by the February 28, 1989, deadline, from consortia representing most of the 21 firms invited to respond. After a protracted evaluation process and detailed negotiations, the government finally selected one of these consortia as a partner. The strategic alliance was approved by the cabinet in 1991.

Quebec Quebec completed a study in 1988 that identified 30 government programs, with budgets totaling approximately $1.5 billion, that depended on geographically-referenced information. This data was being managed by 100 information systems, costing approximately $70 million a year. Enhancing these systems, the government estimated, would cost approximately $300 million over ten years.

Concerned with the need to improve data compatibility, reduce costs, and coordinate technology acquisition, the province committed to a ten-year plan to create a governmentwide network of land information systems. An initial three-year plan was defined for 1988 to 1991, and three coordinating groups were established: a management committee, a users forum, and a committee of deputy ministers.

The management committee was given the mandate to prepare a strategic plan for geomatics; develop design concepts for the network; identify costs, benefits, cost-sharing alternatives, and funding sources; involve the private sector; investigate the potential of geomatics as a tool for economic development; provide guidance for experimental projects; and support government ministries in accelerated geomatics development. Quebec's Ministry of Communications developed a detailed methodology for government agencies to use in estimating the return on investment in geomatics projects.

New Brunswick Within the first decade after the Land Registration and Information Service was formed, the Province of New Brunswick moved quickly to develop land-related databases—implementing an on-line property assessment database, completing conventional base and property maps, and initiating two major digital mapping programs.

A post-graduate research project at the University of New Brunswick in 1984 assessed progress and proposed that a network be created to integrate the province's spatial information systems. In 1985, the provincial Surveys and Mapping Advisory Committee approved funding for a cooperative research and development initiative with the university—the LANDNET program—to further investigate the feasibility of such a network. The final report, submitted in 1988, summarized the results and presented a framework for phased implementation.

New Brunswick created a crown corporation in 1989—the New Brunswick Geographic Information Corporation—to be responsible for all geographically referenced information in government. Geomatics was identified as a strategic opportunity for economic development. NBGIC was mandated to define government strategies for land information systems integration and to provide geographic information to users at a reasonable cost.

Municipalities

Cities in most provinces began implementing enterprise-wide GIS in the 1980s, addressing the needs of multiple departments and integrating municipal information management. Digital landbase development created new corporate entities in many cases, plus special partnering and consortium arrangements.

Oxford County, Ontario, participated in a user-needs study conducted by the Ontario Ministry of Natural Resources in 1983 and began implementing a multipurpose GIS in 1986. The system was designed to integrate the county's own data with data provided by federal, provincial, and municipal governments. Local municipalities and public utilities commissions were given access to the system.

Edmonton, Alberta, committed to a corporate system when it appointed a full-time project manager and established the Geographic Base Information System office in 1985. By the end of the decade, the city had completed all of the primary geographic databases (land survey control, property maps, tax parcel ownership, and registry maps, an integrated utility information system, and the single-line street network), and its GIS was serving more than 40 user groups.

Winnipeg, Manitoba, undertook an initial study of land-based business functions in 1987. A more extensive user-needs study was completed in 1989, developing the requirements and strategy for a corporate initiative. The city made a parallel investment in database development, signing a five-year Memorandum of Understanding with the Province of Manitoba in 1988 to develop a digital property base, the foundation for its integrated Land Based Information System.

Halifax, Nova Scotia, received council approval to develop a corporate GIS in 1989. Unlike many North American municipalities, usable base maps were available from the start. Halifax was able to use existing topographic and cadastral maps developed by the Land Registration and Information Service of the Council of Maritime Premiers.

The Geomatics Industry

The Canadian geomatics industry grew significantly during the 1980s. By the end of the decade, it included firms in many disciplines: surveying, mapping, remote sensing, engineering, geography, forestry, geology, and computer science. Companies invested heavily in hiring, training, research, and development during this period and acquired digital technology to enhance their capabilities in a rapidly changing field.

Surveying and mapping firms diversified from data compilation to data integration and data analysis. Remote sensing firms offered new services based on the integration of image analysis and GIS technologies.

A world-class software industry was created that was successful in marketing its products worldwide. Companies active in this sector (and their products) included GeoVision Corporation (VISION), Universal Systems Limited (CARIS), TYDAC Technologies Inc. (SPANS), PAMAP Graphics Ltd. (PAMAP), Digital Resource Systems Ltd. (TerraSoft), Generation 5 Technology Limited (Geo/SQL, Mun-MAP), EngHouse Systems Ltd. (CableCAD), PCI Enterprises (EASI/PACE), and many others. In hardware technology, companies like Gregory GeoScience (PROCOM), Dipix Technologies Inc. (ARIES), ITRES Research Limited (CASI), Gentian Electronics, and others acquired international reputations.

By 1990, Canada had more than 1,300 geomatics companies, employing 12,000 people. Larger and more sophisticated firms were beginning to emerge, and large consortia were being created to undertake major GIS/LIS projects at home and abroad.

Industry and Professional Associations

Reflecting changes in the industry, the Canadian Association of Aerial Surveyors redefined its mandate in 1988 to include digital mapping, remote sensing, geographic information processing, and GIS. The new organization—renamed the Geomatics Industry Association of Canada (GIAC)—further broadened its charter in 1989 to include equipment and software suppliers.

GIAC defined geomatics as "the field of scientific and engineering activities involved in the application of computer and communication technologies to the capture, storage, analysis, presentation, distribution, and management of spatial information to support decision making." With such a broad charter, membership in the association grew and became increasingly diverse.

Geomatics practitioners in Canada were represented by national professional and technical associations that included the Canadian Institute of Surveying and Mapping (CISM), the Canadian Council of Land Surveyors (CCLS), the Association of Canada Lands Surveyors (ACLS), the Canadian Hydrographic Association (CHA), the Canadian Cartographic Association (CCA), and the Canadian Remote Sensing Society (CRSS).

CCLS represented members of Canada's ten provincial land survey associations at the national level. CISM represented the country internationally through its association membership in world surveying and mapping organizations. (It has since been renamed the Canadian Institute of Geomatics.)

Canadian survey technicians and technologists were represented by the Canadian Association of Certified Survey Technicians and Technologists in Alberta, Saskatchewan, Manitoba, Ontario, New Brunswick, and Nova Scotia and by other organizations in British Columbia, Ontario, Quebec, and Newfoundland.

Education and Training

Many universities offered geomatics-related programs. These included four surveying engineering schools: the University of New Brunswick, the University of Toronto at Erindale College, the University of Calgary, and Université Laval. There were approximately 40 degree programs in geography. Other departments, in disciplines such as forestry and environmental sciences, offered GIS-related courses.

The University of New Brunswick's Department of Surveying Engineering first offered a graduate degree program in Land Information Management in 1979. It founded the Canadian Laboratory for Integrated Spatial Information Research (CanLab INSPIRE) in 1988 to provide a national focus for research in spatial analysis and spatial information systems. An Ocean Mapping Group was established in the same year. The university received international recognition for these initiatives and for its research activities in land information applications, management, and policy.

The University of Toronto's Institute for Land Information Management—a multidisciplinary institute with faculty drawn from other departments and universities, including the National Center for Geographic Information and Analysis in the United States—worked with the Geographic Information Standards Laboratory of the U.S. Department of Commerce in the late 1980s on research in object-oriented GIS, geographic database design, expert systems applications, and geomatics standards.

The University of Calgary's Centre of Expertise in Land Information Systems conducted research in such diverse fields as the revision of positional data in spatial databases, phenomena-based terrain data modeling, design of knowledge-based route guidance systems, and intelligent image interpretation. Université Laval led the way for other North American universities when it implemented a formal undergraduate program in geomatics, following an extensive review of its surveying and mapping curriculum in 1984–85.

By the end of the decade, approximately 20 technical schools offered geomatics-related training programs.

Graduates of the College of Geographic Sciences in Lawrencetown, Nova Scotia, were in high demand for government and industry positions across North America. Founded in 1949 to train surveyors and known in the late 1950s as the Nova Scotia Land Survey Institute, the college took its present name in 1986, providing post-secondary training programs in surveying, mapping, land-use planning, remote sensing, GIS, and software development for the geographic sciences.

Sir Sanford Fleming College in Lindsay, Ontario, began offering a two-year diploma program in GIS in 1981. The B.C. Institute of Technology in Burnaby, British Columbia, implemented a one-year post diploma program in GIS in the fall of 1987 and began offering part-time courses in 1989. Geomatics programs were also developed by the Southern Alberta Institute of Technology in Calgary; the Northern Alberta Institute of Technology in Edmonton; Georgian College in Barry, Ontario; Algonquin College in Ottawa; Seneca College in Toronto; Collège Ahuntsic in Montreal; and the Cabot Institute in St. John's, Newfoundland.

The Banff Centre for Management in Banff, Alberta (one of three centres in the Banff Centre for Continuing Education), began providing specialized workshops and seminars for managers and professional administrators in February 1988, based on lectures, group discussions, case studies, and tutorial exercises.

Proliferation: The 1990s and Beyond

Canada's geomatics industry has grown and diversified rapidly in the last three decades, as GIS has converged with other technologies and new user communities have emerged with new application needs. Markets are now developing that are further and further removed from traditional surveying and mapping. Faced with fierce global competition and a changing client base, many of the traditional technology and service firms may eventually disappear. There have already been significant casualties. However, strong competitors are emerging in their place that have already adapted to the realities of the new market.

The decade of the 1990s will be a period of major adjustment. Whereas in the past governments at all levels have tended to set the direction of the industry, recession and budgetary restraint are having a major impact on their geomatics programs. Reduced revenues from government clients will force suppliers of geomatics products and services to look elsewhere to sustain their business. The Canadian geomatics industry will be influenced much more heavily in the future by developments in the commercial and consumer marketplace and by export opportunities.

Some governments have sought to raise additional revenues through the sale of geographic information and the provision of geomatics services. They have faced vehement opposition from the geomatics industry, which views them as unfair competitors that are subsidized by the taxpayer. Such initiatives are unlikely to survive in the long term, given the concern of Canadians about spiraling public debt and uncontrolled growth in government. The public is demanding a return to basics that could not only eliminate these entrepreneurial initiatives but constrain government in its traditional role as a mentor for industry development as well.

The future is already becoming clear: opportunities will be larger, joint ventures and strategic partnerships will be more common, and competition within and outside Canada will be much more intense. Canadian companies will have to develop products and services from the outset with an eye on international markets.

A Cooperative Approach

In recent years, multisector partnerships have become much more critical to winning international business. Large geomatics projects require the combined expertise of the private sector, government, and educational institutions. Clients want to see all of these elements in an integrated proposal. Canada has developed a model for international projects today that is based on cooperation between these three sectors.

A project in Mexico provides a recent example. The Mexican government selected a Canadian consortium in 1992, in open competition with 22 institutions and companies from other countries, to help implement a modernized national mapping program in the National Institute for Statistics, Geography and Informatics. The prime contractor for the $22-million project was Photosur Geomat Inc. of Montreal, Quebec, a geomatics subsidiary of SNC-Lavalin Inc., the world's third largest engineering services company. The Canada Centre for Geomatics in Sherbrooke, Quebec, was included in the project team to transfer Canadian government management expertise. Less than a year after the bid was published, Mexico was ready to begin producing topographic and thematic products using the new system, for a national database of Mexican territory.

New Technologies

Large markets are emerging for geographic information-based products and services today that are based on new technologies. Electronic Chart Display and Information Systems and spaceborne radar are two major opportunities that Canada is developing.

Electronic Chart Display and Information Systems

The Canadian Hydrographic Service has participated in activities of the International Hydrographic Organization and the International Maritime Organization and has played a lead role in preparing international draft standards for Electronic Chart Display and Information Systems.

Offshore Systems Ltd. of Vancouver, British Columbia, has developed a system that very closely meets these standards. The system integrates information from Loran, GPS receivers, depth sounders, radar reflectors, microwave beacons, gyrocompasses, and other sensors to show a ship's position, course, and track in real time on a high-resolution electronic chart display. Visual and audible alarms warn of hazards and deviations from the prescribed course.

Founded in 1977 to provide positioning services for offshore oil exploration in the Beaufort Sea, OSL began selling electronic chart technology to the Canadian navy in 1985. Its systems are now widely installed in passenger ferries, oil tankers, oceanographic ships, submarine rescue vehicles, icebreakers, buoy tenders, and coast guard vessels.

OSL, Matrix Technologies of St. John's, Newfoundland, and the Canadian Hydrographic Service formed a joint venture company—Nautical Data International—to distribute electronic chart data. The first objective of the company was to provide data to vessels operating in Canadian waters. It will then look for opportunities to partner, joint venture, and play a role in international chart distribution.

RADARSAT

NASA launched RADARSAT—Canada's first remote sensing satellite and the world's first operational radar satellite—in November 1995. RADARSAT International, a consortium of

Canadian companies in the space and remote sensing technology industries, processes and distributes the satellite's data worldwide.

RADARSAT significantly improves the collection of ice and weather information in high latitudes. It also supplies information on ocean wave spectra and ocean features for applications in wave forecasting and climatology.

Other applications include ship detection and monitoring of offshore economic zones and the detection and monitoring of oil pollution. The satellite is expected to improve operating efficiency and safety for oil and gas operations, marine transportation, and naval activities in northern regions. Resource-based companies and government agencies with reconnaissance mandates will be major customers. Radar data will improve capabilities for land-use mapping and will enhance the development of global stereo data sets for geomorphologic and topographic mapping and petroleum and mineral exploration. Data from the satellite are also expected to be useful in agricultural monitoring.

Provincial Initiatives

Many provincial governments have continued with major initiatives to implement land information systems, networks, and information utilities.

British Columbia

British Columbia awarded a $2.8-million contract to MacDonald Dettwiler Associates in 1990 to develop an operational prototype of a land information utility. The prototype of the service—called LandData BC—was completed in 1993, providing on-line access to cadastral and topographic information, plus the capability to order more than 30 off-line databases. Funding for the production system was approved by cabinet in June 1994. Development will be completed by the end of 1997. The utility is eventually expected to support between 3,000 and 5,000 public and private sector users.

Alberta

By the spring of 1994, Alberta's LRIS Network had signed up 200 customers and had passed the $1-million mark in revenues. True to its original vision, the network has attracted customers from a cross-section of the Alberta economy: the surveying industry, the oil and gas industry, the real estate industry, lawyers, banks, consulting engineers, agricultural organizations, the federal government, provincial agencies, and municipalities.

In a period of downsizing and fiscal restraint, the government moved Land Information Alberta to the registries division of Alberta Municipal Affairs. By late 1994, it began evaluating future options for service delivery, including privatization, outsourcing, or an agency arrangement.

Manitoba

The Government of Manitoba is the largest partner in the Manitoba Land Related Information System project, a cooperative initiative with the private sector. Participants include provincial government departments, the City of Winnipeg, Manitoba Hydro, Manitoba Telephone System, Inter-City Gas, and Linnet Geomatics International Inc., a Winnipeg-based company formed to promote and develop the project.

Ontario

The Province of Ontario partnered in 1991 with a consortium of 25 Ontario-based firms called Real/Data Ontario Inc. to implement POLARIS. TERANET Land Information Services Inc. was created—jointly owned by the Government of Ontario and the consortium—to convert the province's land registration records, create a digital cadastral mapping base, and market geographic data through an information utility. The government estimated it would save more than $50 million by implementing the land registration system in partnership with the private sector. The project will develop expertise that can be marketed worldwide.

Quebec

Quebec created the legal foundation for cadastral reform in 1985, assigning responsibility to the Ministère de l'Energie et des Ressources. The reform program was accelerated in 1992, with a projected expenditure of $500 million over 14 years. The project—affecting approximately four million parcels—generated more than 1,500 surveying contracts.

DMR Group Inc. of Montreal, Quebec, was awarded a five-year, $27-million contract to develop and operate four cadastral reform information systems in the largest civilian geomatics contract ever undertaken in Canada. The company is integrating hardware, software, services, and training programs delivered by a project team that includes land surveyors, service providers, and hardware and software suppliers.

A New Information Infrastructure

Spatial data is more widely available in Canada than ever before. The number of GIS installations is increasing rapidly, particularly with the growing affordability of desktop technology. Delivering data to end-users in a way that is timely and affordable will require new infrastructure for data sharing. The concept of a National Spatial Data Infrastructure is receiving growing support, and many new networks are being planned.

The Inland Waters, Coastal and Ocean Information Network

Support has grown in the last decade for a new national network providing access to coastal, ocean, and inland waters information. This initiative is known as the Inland Waters, Coastal and Ocean Information Network (ICOIN).

ICOIN applications will include coastal zone management, emergency preparedness, pollution monitoring and control, weather forecasting, site-specific oceanography and meteorology, marine transportation routing, habitat evaluation, fish distribution, catch analysis, and tourism development. The Canada Ocean Mapping System developed by the Canadian Hydrographic Service will be a fundamental component.

A New Brunswick-based consortium of companies—ICOIN Industries Inc.—began the first phase of the project late in 1992, designing the ICOIN information management system, demonstrating a prototype to illustrate the system's functionality, and conducting a market study to confirm its commercial potential.

Land Information Network for Canada

The Surveys, Mapping and Remote Sensing Sector (in the federal department renamed Natural Resources Canada) committed in November 1992 to implementing a new infrastructure to manage and distribute geographic information describing the Canadian landmass. The proposed Land Information Network for Canada (LINC) will provide electronic access to geographic data and metadata, query and integration services, and data visualization and manipulation capabilities through a standard user interface. Facilities will also be provided for constructing business applications.

The target system is intended to be operational in 1998. Many SMRSS databases will then be accessible on-line Canada wide. The government plans to work with industry partners who will share the risks and benefits of development.

Canadian Earth Observation Network

The Canadian Earth Observation Network (CEONet)—a component of Canada's Long Term Space Plan—has received initial funding from the Canadian Space Agency, the Canada Centre for Remote Sensing, the Department of Fisheries and Oceans, and Environment Canada. The system will link users and providers of Earth observation data within Canada and will eventually allow access to international Earth observation networks. The system will use Internet initially, and services will eventually be provided over the CANARIE network.

Human Resources Development

Geomatics education and training programs have come under considerable pressure in recent years. Between 1985 and 1991, the number of private geomatics firms increased by approximately 18%, and the number of jobs grew by about a third. By the early 1990s, many firms were experiencing at least some difficulty in hiring personnel. Many companies were concerned that the educational system would be unable to produce the graduates needed to meet rising business demands. Succession planning was becoming a growing problem for managers faced with replacing senior personnel.

Confronting an impending crisis in the geomatics labor force, the Geomatics Industry Association of Canada and the Canadian Institute of Surveying and Mapping (now the Canadian Institute of Geomatics) joined forces to conduct a Human Resources Planning Study that would address these issues. A Geomatics Industry Adjustment Committee was created to oversee the study.

The objectives of the study—funded by Employment and Immigration Canada—were to examine industry demographics and identify human resource shortages, evaluate the impact of technological change in the workplace, examine the effectiveness of government policies and programs in helping the geomatics industry adjust to change, assess the capabilities of the Canadian education and training system to meet future industry demands, define new approaches to training and retraining, and develop strategies that would help the industry in human resources planning.

The Industry Adjustment Committee published its report and recommendations in November 1991. Three panels were formed under the direction of a Committee on Human Resources Planning to continue its work: a Technical Skill Set Panel to develop an evolutionary model of the geomatics business and identify the essential skill sets required by the industry, a Geomatics Management Panel to help develop and implement a national management training program in geomatics, and a Structures and Organizations Panel to design national mechanisms for program delivery.

The Committee on Human Resources Planning published its final report in August 1994, recommending regularly scheduled reviews to identify ongoing requirements for skills development in an industry that is experiencing revolutionary change.

Summary

Within 30 years, computer mapping has entered the Canadian mainstream. From a period of innovation in the 1960s and 1970s through a decade of integration in the 1980s, geomatics is now moving into a period of proliferation. New applications in market analysis, retail site selection, advertising, direct marketing, real estate, and others are emerging. A new infrastructure is being developed that will integrate digital geographic databases, geographic information processing, and data communications networks.

While the future of this integration is still unknown and the rate of progress may be constrained by government cutbacks, there are early indications of the potential benefits. The New Brunswick Telephone Company and the New Brunswick Emergency Measures Organization implemented a geographically-enabled emergency warning system in 1994 that links subscriber telephone numbers to the provincial geographic base. When areas endangered by flooding are identified through an interactive graphic interface, NBAlert™ automatically calls affected subscribers to relay emergency messages and evacuation instructions. We can expect to see many similar applications in the future.

Geomatics will play a major role in Canada's society and economy. Spatial information will become widely accessible in schools, homes, and offices. Most Canadians will be touched by geographic information processing—some directly, through the use of GIS and related technologies, many others indirectly, through systems and appliances that incorporate geographic intelligence. Such an outcome could scarcely have been visualized only three short decades ago.

CHAPTER 18

International Applications of GIS

D. Wayne Mooneyhan

Introduction

It is, of course, not possible to document even a small percentage of the international applications of GIS in one chapter of one book. Neither could it be done by any one person or small group of authors. Such an endeavor would require numerous books and hundreds of writers. Indeed many of the preceding chapters of this book are truly about international applications of GIS.

This chapter will instead attempt to document the efforts of the United Nations Environment Programme (UNEP) and the United Nations Institute for Training and Research (UNITAR) to build the capacity in developing countries for use of GIS in local and national environmental and resource management models. An attempt also is made to document the efforts within UNEP/GRID (Global Resource Information Database) to capture and archive appropriate spatial datasets developed by national and international sources both for use in

---

D. Wayne Mooneyhan began his career with NASA as a chief engineer for the Saturn-Apollo rocket program. He served as director of NASA's Earth Resources Laboratory in Mississippi from 1970 to 1985. He then served as director of the United Nations Environmental Programmes Global Resource Information Database (UNEP/GRID) program in Geneva, Bangkok, and Sioux Falls until 1992. He currently serves with the Universities Space Research Association in a consulting capacity, developing programs in remote sensing and GIS for core data needs in global and sustainable development strategies. *Author's Address:* D. Wayne Mooneyhan, P. O. Box 1785, Picayune, MS 39466.

Acknowledgments: The material for this chapter was contributed by the following UNEP personnel: Ole Hebin, Facility Manager for GRID operations in Geneva, Switzerland, 1989–1994; Dr. Ashbindu Singh, Facility Manager for GRID operations in Sioux Falls, South Dakota, U.S.A., 1992–present; Surendra Shrestha, Liaison Officer for the Environmental Assessment Program for Asia and the Pacific in Bangkok, Thailand, 1993–present; Ron Witt, Facility Manager for GRID operations in Geneva, Switzerland, 1994–present; and Stephen Gold, Programme Officer for Environmental Training Programmes of UNITAR in Geneva, Switzerland. The material was edited by D. Wayne Mooneyhan.

UNEP's own regional and international assessment programs and for dissemination to users worldwide upon request.

Background

During the 1972 United Nations Conference on the Human Environment held in Stockholm, Sweden, the concept of a coordinated environmental assessment and monitoring program for the Earth was developed. One of the significant results of this conference was the establishment of the United Nations Environment Programme with a mission "to provide leadership and encourage partnership in caring for the environment by inspiring, informing, and enabling nations and peoples to improve their quality of life without compromising that of future generations." UNEP was to have the primary responsibility, through cooperation and coordination with other United Nations organizations, to assess the state, trends, and problems of the environment on a global basis, and to provide early warning of impending environmental dangers. At the Stockholm meeting, "Earthwatch" was established as the coordinating mechanism. Through Earthwatch, U.N. bodies, in collaboration with governments and scientists, would gather data for comprehensive assessments of environment issues. These assessments were to provide early warning of environmental threats and also to afford decision-makers and planners a foundation to devise effective responses and policies. It was clear that this responsibility would require a well-coordinated program of high-quality data collection and a significant analysis effort. The original organization of UNEP established the Global Environment Monitoring System Programme Activity Center (GEMS/PAC) to coordinate the monitoring and assessment activities.

Also in 1972, the U.S. National Aeronautics and Space Administration (NASA) launched the Earth Resources Technology Satellite (ERTS 1, later known as Landsat) as a civil space science program with an "open skies" data release policy. High-quality data for the entire surface of the Earth became available to all requesters for the first time. The system was operated by the U.S. Government during the period from 1972 to 1981, and data were provided to users at the nominal cost of reproduction. [In 1981, the Landsat satellites were turned over to private industry. This resulted in significant changes in acquisition and pricing policies that remained in effect until the system was returned to government operation in 1992.] Landsat broke the "open skies" myth and was soon joined by other high resolution Earth-looking civilian systems from France, Japan, and other countries. High resolution satellite data were now available on a global basis, and the Global Environmental Monitoring System (GEMS) program of UNEP soon recognized the value of these data in global monitoring and assessment activities.

The advent of satellite data had two significant impacts on the history of GIS development. First, it provided a source of data for the entire Earth's surface that could be readily georeferenced and renewed over time, thus vastly increasing the applications of GIS that relate to monitoring and change detection. Second, and perhaps more significant, it provided a huge quantity of digital data that required unaffordable amounts of computer time to process. This, in turn, resulted in a demand for research and development efforts to make everyday applications practical. Many new software programs for data handling and management and many new hardware approaches to data storage and high speed processing were developed in the decade that followed. The cost-effective data sources and cost-effective processing procedures that were forced by the satellite data supply provided a turning point in the history of GIS development and applications.

The NASA Earth Resources Laboratory, located in Mississippi, was one of the organizations involved in the development of both hardware and software improvements and applications of satellite data for environmental assessment and resource management models during the 1970s. By 1983, the laboratory had developed, documented, and released a comprehensive set of software that included a high speed image processing capability and a raster-based GIS capability. The software package was known as ELAS (Earth Resources Laboratory Applications Software). The management of the UNEP/GEMS/PAC, which had for two years studied the need for a global geographic database within GEMS/UNEP/Earthwatch, reviewed ELAS in 1983 and determined that it would fill the needs of UNEP for processing and analyzing spatial data.

In late 1984, the executive director of UNEP signed a Memorandum of Understanding (MOU) with NASA in which NASA agreed to provide hardware, software, and technical personnel for a period of three years to support the transfer of image processing and GIS technology to UNEP and its member countries. NASA also agreed to provide all NASA-held regional and global datasets to UNEP for distribution to all requesting them on a no-cost basis. UNEP named the project the Global Resource Information Database (GRID), and centers were opened in 1985 in Geneva, Switzerland, in a facility provided by the University of Geneva and in 1986 at UNEP headquarters in Nairobi, Kenya. In addition to NASA, there were other important contributors to the initial phase of GRID. Hardware was donated by both the Concurrent Computer Company and the Prime Computer Company, and vector-based GIS software was donated by the Environmental Systems Research Institute (ESRI). The Swiss Directorate for Development Cooperation also funded a training program for developing country participants for five years starting in 1986.

Progress Through Time

Progress can be judged from many aspects and perhaps it would be appropriate to review the progress of GIS applications by examining the progress of GRID. This section will review the first decade of GRID, including organizational status, projects, assignments, number of participating centers, and the nature and long-term affects of training activities. Contributions by others to the progress of GIS in this time frame are also reviewed.

Organizational Status

The status of GRID within the UNEP organizational structure is one appropriate way to measure progress. In mid 1985, the management of GEMS/PAC recognized the importance of spatial data and GIS technology to the environmental assessment process and established GRID within an existing line item of the GEMS budget. Although GRID had the approval of upper management, it had no funds of its own. In 1986 and 1987, GRID was elevated to a line item in the GEMS budget with its own funds and was recognized as a pilot project within GEMS. During 1988 and 1989, GRID moved into an implementation phase—with specific approval of the UNEP Governing Council—to expand the network to other regions. GRID reached the operational phase in 1991 and was elevated to a Programme Activity Center (GRID/PAC) with a specific role to develop spatial databases and GIS applications in the UNEP Earthwatch mission.

In the context of Earthwatch, GRID continues to provide a link between data collection and monitoring activities and the experts who use these data for environmental assessments, policy making, and resource management. The Plan for Action that emerged from the 1992 United Nations Conference on Environment and Development (UNCED) underscored the need for: 1) enhanced accessibility of integrated environment and development information, and 2) enhanced national capacity to deal with such information in decision making and policy setting. The existing GRID system became the base for UNEP's revised Environmental Assessment Programme (EAP). The new EAP mission is to provide the world community with improved access to meaningful environmental data and information and to help increase the capacity of governments to use environmental information for decision making and action planning for sustainable human development (UNEP 1994). The new programme emphasized: 1) an information base to support policy formulation and raise awareness of environmental issues, 2) sectoral information and production of integrated environmental information for sustainable development, 3) scientific assessments and products for a wide range of users, 4) descriptions of current situations and focus on emerging issues and early warning, and 5) reorganization of the existing set of disparate data systems into an integrated UNEP information delivery system.

The rapid elevation in the organizational status of GRID was due almost entirely to the visibility provided by products derived through the use of spatial data models and GIS technology in local, national, and regional level demonstration projects.

Character of Projects

Progress also can be measured by the changing character of GIS projects that have been and are being pursued by the various centers in the GRID system. During the early phases of development (1985–1989), numerous case studies and demonstration projects were conducted with strategic national and international partners as "proof-of-concept" for GIS applications. The purpose of these projects was to confirm to UNEP's governing bodies, national governments, and the international community that spatial data models and GIS systems were necessary tools for complex environmental assessments and resource management.

National and subnational projects were designed to demonstrate that GIS has multiple utility on the local-to-national scale. Scientists and resource managers from the participating country defined the application projects and provided the needed data as available. The remainder of the data (most often the satellite-based data) was provided by GRID. The application model and the analysis were developed jointly by GRID scientists and the country participants. GRID provided the equipment, software, and training and produced the final project results, including GIS databases, map products, and technical reports. Countries in which GRID has helped to carry out national or subnational case studies are: Argentina (Patagonia), China (Three Gorges area), Costa Rica (Gulf of Nicoya), Indonesia (western Java), Kenya (national), Nepal (Bagmati zone), Panama (national), Peru (Chumbivilcas Province), Saudi Arabia (various), Thailand (Chiang Mai Province/Mae Klang watershed), and Uganda (national).

It is not practical to discuss each of the above case studies, but perhaps two of the more successful should be briefly mentioned. In the Bagmati zone study in Nepal (which includes the capital of Kathmandu), the purposes were to demonstrate the use of GIS technology for the planning process, to create a digital environmental database from data of differing scales and resolutions, and to help establish a GIS capability at the International Center for Integrated Mountain Development (ICIMOD) in Kathmandu. As noted later, ICIMOD has since become an active GIS capability in the Himalayan Region and is a cooperating center in the GRID system. A second example, the GRID-Uganda/Ministry of Environment Protection joint national case study, was to develop a nationwide natural resource database to aid Uganda's reconstruction efforts that began in 1987. The case study was specifically oriented towards an analysis of the current distribution of forest and wetlands and changes in these patterns over a ten-year period. It included a retrospective analysis, based on satellite data, by a team of Ugandan resource managers and ultimately concluded with the creation of a nationwide thematic atlas in digital format (UNEP 1987). This digital atlas later formed the base of an expanded and continuing nationwide database activity, the National Environmental Information Center (NEIC) of Uganda.

Several regional level studies also were conducted by GRID in conjunction with sponsoring partners. For a number of years, GRID has participated in a joint UNEP/FAO (Food and Agricultural Organization of the United Nations) project to map and monitor the tropical forests of Amazonia and West Africa. The primary data source is one kilometer data from the U.S. weather satellite AVHRR (Advanced Very High Resolution Radiometer). The processed satellite data were combined with other ancillary spatial information to form a multilayered geographic database. Typically satellite imagery with high spatial resolution, along with existing (often outdated) local and national forestry maps, were used to "guide" the forest/non-forest categorization of the coarser AVHRR one kilometer data. Initial mapping results were field checked and reprocessed to improve accuracy of the classifications. GRID's GIS capabilities and ancillary datasets from the global database were used for three purposes: 1) selecting "non-cloud" pixels from among many satellite scenes for the final mosaics; 2) interpreting the forest/non-forest information content of the satellite data (e.g., based on boundaries, elevation, distance from major roads, etc.); and 3) updating the results using multitemporal datasets from later years.

More recently, GRID has completed a study known as the Global Assessment of Human-Induced Soil Degradation (GLASOD), in conjunction with UNEP's Desertification Control unit (DC/PAC) and the International Soil Reference and Information Centre (ISRIC) in the Netherlands. An atlas of world soil conditions and degradation (UNEP 1994) and a series of digital soils datasets are two of the results of this project.

In recognition of the need to provide improved and more timely datasets to national, regional, and global users, UNEP opened a GRID office in Sioux Falls, SD, in January 1991. The center is located at the EROS Data Center (EDC) and is a cooperative effort of UNEP, the U.S. Geological Survey (USGS), and NASA. Through joint efforts with many North American scientists and the advanced technology available at the EDC, GRID/North America has access to numerous datasets that are valuable to both developed and developing countries. The datasets are mostly derivatives of ongoing scientific studies that are further processed for international distribution, thus taking maximum advantage of the joint location with EDC. This center has become a major contributor of regional and global datasets and therefore plays a significant role in making environmental information available to the world community through the GRID network. Although GRID/North America has other mission objectives in technology transfer and training, it is the first GRID institution devoted primarily to data capture.

The projects pursued by the GRID program have progressed from the initial GIS demonstration projects to the intermediate applications projects associated with specific training programs to today's efforts that have more emphasis on providing easy access to more and better spatial data and information for use in the GIS systems of the UN organizations and their member countries.

Mission Assignments

Another approach to judging progress might be to examine the changes in the GRID mission. The initial role for GRID was to assemble, archive, and disseminate data in geographic form. The move to operational status resulted in significant changes in those goals.

The expanded mission of GRID is to provide:

- timely and reliable georeferenced information; and,
- access to a unique international geographic information system (GIS) service to address environmental issues at global, regional, and national levels in order to bridge the gap between scientific understanding of Earth processes and sound management of the environment.

The tasks and objectives of the expanded GRID mission can be summarized as follows:

1. Database management, including the acquisition, verification, and dissemination of georeferenced environmental datasets and development of methodologies to handle global datasets. This involves close contacts with international scientific, conservation, and development communities to improve access to relevant data.
2. GIS and Image Processing (IP) applications to provide GIS/IP capabilities and expertise to support environmental assessments and problem solving at local to global scales.

3. GIS/IP technology transfer and capacity building in developing countries through formal training programs and on-the-job case study applications in conjunction with technical assistance programs.
4. GRID system development through the establishment of additional regional and national operating GRID centers to ensure that support capabilities for UNEP and Earthwatch continue to grow.

Number of Participating Centers

Another way to judge progress of the GRID system might be through the growth in the number of participating centers. There are four centers that have regional responsibilities. The centers are funded partially or wholly by UNEP and are managed by UNEP personnel. These centers are:

1. Geneva, Switzerland, opened in 1985, responsible for Europe plus global data set maintenance and distribution;
2. Nairobi, Kenya, opened in 1986, responsible for Africa plus UNEP internal;
3. Bangkok, Thailand, opened in 1989, responsible for Asia and the Pacific; and
4. Sioux Falls, SD, opened in 1991, responsible for North America plus the generation of new global and regional datasets.

Two other UNEP regions, Latin America and Western Asia, do not have regional centers as yet; however, there are cooperating national organizations within those regions.

There are presently seven national or regional centers that receive some funding from UNEP but are staffed and managed entirely by the national governments or regional organizations. These centers are:

1. Arendal, Norway, opened in 1989, responsible for Norway and polar region datasets;
2. Tsukuba, Japan (National Institute for Environmental Studies, NIES), opened in 1990, responsible for Japan and supercomputing processing requirements;
3. Warsaw, Poland, opened in 1991, responsible for Poland and selected regional environmental studies;
4. São Jose dos Campos, Brazil (Instituto Nacional de Pesquisas Espacias, INPE), opened in 1992, responsible for Brazil and datasets of the Amazon region;
5. Kathmandu, Nepal (International Center for Integrated Mountain Development, ICIMOD), opened in 1992, responsible for datasets of the Himalayan region and GIS training for eight member countries;
6. Apia, Western Samoa (South Pacific Regional Environmental Programme, SPREP), opened in 1993, responsible for 22 island countries in the region; and
7. Copenhagen, Denmark, opened in 1994, responsible for Denmark and selected regional environmental studies.

Other regional organizations are participating in various ways. In Asia, these are:

1. Association of South East Asian Nations, ASEAN;
2. Committee for Investigations of the Lower Mekong Basin; and
3. South Asia Cooperative Environment Programme, SACEP.

In Africa, the participating regional organizations are:

1. Economic Community of Central African States, CEEAC;
2. Economic Community of West African States, ECOWAS;
3. Permanent Interstate Committee for Drought Control in the Sahel, CILSS;
4. Inter-governmental Authority on Drought and Development, IGADD; and
5. Southern Africa Development Coordination Conference, SADCC.

There also are an increasing number of national centers with GIS capabilities that are labeled "GRID Compatible Centers." Many of these centers were developed as a result of UNEP/UNITAR project and training activities. One such project in Africa that provided hardware, software, data, and training resulted in the following centers:

1. Botswana—established in the Department of Town and Regional Planning, Ministry of Local Government, Lands, and Housing;
2. Burkina Faso—established in the Department of Forestry, Ministry of Environment, and Tourism;
3. Côte d'Ivoire—established in the Service Autonome de Teledetection, Direction Generale et Controlle de Grandes Traveaux;
4. Ghana—established in the Environment Protection Council;
5. Kenya—established in the National Environment Secretariat;
6. Lesotho—established in the Environment Division, Lesotho Highlands Development Authority;
7. Mali—established in the Projet d'Inventaire des Ressources Terrestres;
8. Mozambique—established in the Environment Division, National Institute for Physical Planning;
9. Niger—established in the Department of Hydraulics, Ministry of Environment and Hydraulics;
10. Tanzania—established in the National Environment Management Council; and
11. Uganda—established in the Ministry of Environment Protection (this unit has evolved into a National Environment Information Center for Uganda).

More than 100 persons were trained in the use of GIS software in this one African Program through the UNITAR training unit. Similar programs in the Asia/Pacific region involve the countries of Bangladesh, Bhutan, Cambodia, People's Republic of China, Fiji, India, Indonesia, Lao P.D.R., Maldives, Myanmar, Nepal, Pakistan, Sri Lanka, Thailand, Viet Nam, and Western Samoa. While a successful start in technology transfer to a developing country does not assure long-term survival, there are enough signs of acceptance of GIS results by decision makers to indicate that progress is being made.

Changes in Approaches to GIS Training Activities

An assessment of progress of any new technology must be examined in the context of training activities. The GIS training activities for essentially all of the above projects were conducted by UNITAR, which is responsible for training and research programs within the United Nations system. In the beginning of the UNEP/UNITAR joint effort, the training was developed and conducted by the GRID professional staff in Geneva. Because the technology was very new to developing countries, the training periods ranged from three to six months with the shorter time frame for midlevel managers and the longer for beginning professionals. Funding for the living expenses of the trainees was provided by the Swiss Directorate for Development Cooperation, and the program was administered by UNITAR. During the first year of the program, UNITAR developed a cooperative effort with a nearby Swiss technical university, École Polytechnique Federale de Lausanne (EPFL), to add the basic sciences to the curriculum. In the second year, the six-month training periods consisted of three months of instruction in science and applications and three months of training in GIS. The entire training effort was shifted to EPFL in the third year, with UNITAR administering and coordinating the activities on behalf of UNEP and the Swiss Government.

Each trainee was required to develop an application project around an existing environmental or resource management problem in his or her country. The data for the applications projects were provided and processed by the trainees so that they could apply each step to a real life situation. Assessment models were developed by the trainees, with support from the instructors. The project results were published in a technical report and presented to the student's national sponsor to demonstrate the utility of GIS-derived products to specific country problems. All of the projects involved the use and digital interpretation of satellite imagery, and the results were integrated into a GIS database covering the study areas. As GIS technology became more common and hardware and software became more user friendly, the length of time required to train new users was reduced from the original six months to two to four weeks.

UNITAR now provides GIS training to support several United Nations organizations. The UNITAR European office in Geneva has developed its own training program in managing environmental data related to natural resources, primarily using methods and models employing GIS technology. Training in GIS, which is only one of UNITAR's training activities, is usually supported by associate institutions. Although there are numerous such institutions, those most often involved are Ecole Polytechnique Federale de Lausanne (EPFL) in Lausanne, Switzerland; Clark University in Boston, MA; and the International Center for Integrated Mountain Development (ICIMOD) in Kathmandu, Nepal.

UNITAR also has developed a training workbook series entitled "Explorations in Geographic Information Systems." There are at present seven volumes:

1. *Change and Time Series Analysis;*
2. *Applications in Forestry;*
3. *Applications in Coastal Zone Research and Management;*
4. *GIS and the Decision Making Process;*
5. *Applications in Mountain Environments;*
6. *GIS and Hazard and Risk Assessment;* and
7. *GIS and Urban Management in Developing Countries.*

These materials are produced to meet the growing demand for discipline specific exercises for training courses and to assist training program alumni in addressing environment, resource management, and human development issues in developing countries.

It will not be practical to cover all the UNITAR training activities. However, a contrast of the first year, 1986, with 1993 demonstrates significant progress.

In 1986, there was one three-month training course for middle management officials from developing countries. The training was conducted by GRID scientific staff using the operational equipment at the GRID facility in Geneva. All of the training was on the job, and there was no standard training material. Each of the four trainees worked his or her project, with the assistance of the GRID staff, and prepared final reports to sponsors.

In contrast, 1993 consisted of the following training activities:

In March, UNITAR conducted the second Latin American Regional Advanced Workshop on GIS for Environmental Management in Chile. The three-week workshop, conducted in cooperation with EPFL and the University of Chile, was attended by 20 alumni and scientists and planners from eight countries. June, Africa GIS93 was conducted in Tunisia. This one-week workshop, which was attended by more than 100 professionals from Africa, Europe, and North America, helped establish the framework for a comprehensive training program in geoinformation, communications technology, negotiations, and decision-support systems in the region for 1993 and beyond. August, the Introductory Workshop on GIS for Health Care and Urban Management was conducted in conjunction with the Italian Cooperation and the São Paulo Municipality, Brazil, for 20 professionals of the São Paulo Municipality.

Also during August, a follow-up Advanced São Paulo Municipality Workshop on GIS for Urban Management (with an emphasis on specific vector-based GIS tools and the decision-making process) was conducted for 23 professionals. The workshop was conducted in cooperation with the São Paulo Municipality, São Paulo State University and the Environmental Systems Research Institute of Redlands, CA. October, the Advanced Slovakian Workshop on GIS for Ecological Planning was conducted in Banska Bystrica, Slovakia, in cooperation with the Slovak Ministry of Environment, the Landscape Ecology Center, and Clark University.

Some 20 professionals from the ministry's field offices participated in the workshop. Also in October, UNITAR conducted the first Training of Trainers workshop in cooperation with the EPFL, the University of Freiberg, Longman GeoInformatics, and Clark University. The two-week workshop at the EPFL trained ten of the most active GIS alumni from Asia, Africa, and Latin America. The workshop explored the tools and techniques in technology transfer most appropriate for developing countries. November, the Mediterranean Regional Advanced Workshop on GIS for Coastal Zone Management, in cooperation with the PAP/RAC (Priority Action Program/Regional Activities Center, an element of the UNEP Regional Seas Program for the Mediterranean), Split, Yugoslavia, the University of Alexandria, Egypt, and ESRI, USA. The workshop trained 16 alumni and their colleagues from six countries in coastal zone planning and monitoring.

Also in November, UNITAR conducted the UNEP/UNITAR/ICIMOD Cycle II Bangladeshi National Workshop on the use of Environmental Information Systems for Integrated Environmental and Urban Planning in cooperation with GRID office for Asia and the Pacific in Bangkok, Thailand; ICIMOD in Kathmandu, Nepal; the Local Engineering Department in Dhaka, Bangladesh; and Clark University, U.S.A. This workshop was a follow-up to a 1992 introductory workshop and was but one activity within a larger five-year cycle of regional training for the ICIMOD member countries of Afghanistan, Bangladesh, Bhutan, China, India, Myanmar, Nepal, and Pakistan. Also in November, the Advanced Asian Regional Workshop on Environmental Information Systems for Environmental Management was conducted in Thailand. The workshop focused on multicriteria/multiobjective land use planning in Thailand and was attended by ten alumni and national representatives from environmental institutions in Asia.

In addition, a three-month feasibility study was conducted in 1993 to develop an International Networking Programme. The study, financed in part by the National Center for Geographic Information Analysis (NCGIA), Santa Barbara, CA, examines the possible mechanisms for matching scientists and planners from universities in developing countries with their counterparts in developed countries to facilitate international cooperation in research and education. Another activity was the translation of the manuals for the GIS software package (Idrisi) to French. Plans call for the translation of both the UNITAR Workbook Series and the Idrisi software manuals to Spanish, Portuguese, and Arabic.

The statistics on the total number of personnel trained in GIS technology by the combined efforts of UNEP/GRID, UNITAR, and their associate institutions are not available, although the number would be several hundred. However, it may be more meaningful to discuss one of the mechanisms employed by this program in order to leverage the impact of the training efforts.

In 1990, the UNEP/GRID/UNITAR joint program provided in-depth training to 14 scientists and engineers from ICIMOD and its member countries through the GRID/Asia offices located on the campus of the Asian Institute of Technology in Bangkok, Thailand. The effort was an early training of trainers designed to prepare the participants to conduct regular training courses. Since 1990, ICIMOD, with some support from UNITAR and GRID/Asia, has trained member country personnel as follows:

- Bangladesh, 20
- Bhutan, 5
- China, 15
- India, 35
- Myanmar, 45
- Nepal, 60
- Pakistan, 15

Thus, the training of 14 professionals in 1990 has resulted in the training of 195 others four years later. The success is, of course, not happenstance. It is the result of selecting a stable institution with a staff of well-qualified professionals and an appropriate mandate to carry out the training effort.

Long-term Effects of Training

It would not be accurate to cite the apparent success of training by only presenting the numbers of persons reported as trained. There have been continuing efforts by hundreds of organizations and thousands of well-meaning trainers over the past several decades to help developing country officials take advantage of modern-day information management tools for resource management and policy making. The list of training providers is almost endless and includes all United Nations and intergovernmental organizations, aid-to-development agencies from most developed countries, international lending institutions, industrial organizations, non-governmental organizations, universities, private foundations, and volunteer groups.

However, many, if not most, of these efforts have failed. The failures often result not from the quality of the training or the trainers but because of unstable institutional conditions or changing political situations. While some of the failures could have been predicted during the selection process, often it is only in the follow-up that the cause of failure becomes apparent. Some of the problems and lessons learned in the UNEP/GRID/UNITAR joint training program are significant and are discussed in the following pages.

One of the most obvious requirements for success is that the trainee-graduates have working access to appropriate computer equipment and software in order to carry on the GIS database development and related work in their institutions. The newly-learned GIS skills are quickly lost if there is no opportunity for application or practice. At one time in the GRID program, a large equipment vendor provided 20 sets of PC level computer equipment for training under the condition that the equipment be transferred to the home institutions of the trainees. The success rate resulting from this approach was considerably higher, at least in the short term, but was by no means perfect.

Not so obvious in the beginning was the fact that home-going trainees would not automatically be accorded carte blanche assignments to carry on new activities within their home institutions, which had their own priorities, methods of work, and operating rules. In the early phase, the trainees were nominated by their institutions and selected by the country for six months of intensive training to establish and maintain a national database. Still, many of the trainees returned home to find a changed political situation or, in some cases, no support for the intended follow-up activities. Whenever serious training efforts are provided, the providers should be certain that the selected institution has a clear mandate from the appropriate governing body and that there is a commitment to institutionalize the new capability.

Another lesson not immediately evident, and one that had to be learned by all, was the extent of follow-up required once the trainees returned to their home institutions. There were hundreds of practical issues to be dealt with from questions on how to carry out a certain GIS procedure to software updates and even the general feeling of needing moral support as well as technical dialogue with a larger group. UNITAR developed a comprehensive newsletter, and there were periodic follow-up workshops by region to keep the trainees current. However, this was not enough because the trainees' problems could not be predicted or scheduled. The trainee should have access to a help desk for the particular system selected for operational use. New networks such as the Internet could begin to solve this problem if, in fact, they ever become affordable in developing countries.

A problem well known in other technology transfer programs, and informally known as the brain drain, occurred at a later stage. Essentially, the better-trained—and thus more marketable—GIS graduates become more mobile both within their own institutions and government agencies and as candidates to move to other jobs or to higher education opportunities at home or abroad. These trainees were less likely to stay in post to apply the acquired skills, thus resulting in a higher-than-average rate of attrition among the graduates. This turnover results in either accepting the inevitable loss or developing ways to train replacements. Even very rapid efforts toward such training sometimes find that the discontinuity brings great difficulty in regaining political support. This problem was addressed in later stages of the GRID/UNITAR program by training two or more candidates from the same country institution. If one can afford to consider the long term, a sometimes unrecognized and less-than-tangible benefit of this problem of mobility may be to sensitize a bureaucracy to the newer technology. In practice, this is very hard to measure and is likely only to be the case if the ex-trainee stays in his or her home institution or moves to one very similar. Otherwise, the knowledge in a given institution is likely to be proportional to the level of responsibility to which trainees rise and may indeed be negligible unless the trainees have significant decision-making or fund-disbursement power that is directly related to the technology.

There are several objectives of any training activity. These range from increasing the number of persons who have an awareness of a new technology to setting a new curriculum in the educational system to institutionalizing the new technology into the management/decision-making process. If the latter is the purpose, the provider must be sure that: 1) the receiving institution has the appropriate mandate from the central governing body; 2) the central governing body is committed to implementation of the new technology as an operational system; 3) funding provisions have been made for equipment and software in the operational phase; and 4) the funding sources for training are prepared to stay the course through follow-on support to the trainee until the new system is fully accepted as an operational entity.

International Progress in GIS

Finally, this discussion could not be concluded without acknowledging a critical group of participants. There are many persons and organizations around the world that design, develop, expand, or improve spatial data management systems. They range from individual investigators of specific applications to university laboratories concerned with tools-for-teaching to value-added companies marketing new products to industrial organizations developing hardware and/or software turn-key systems to market. They exist in almost every developed country and in some developing countries. The dedication to provide GIS tools that are less complex and expensive and are easier to use, teach, operate, and purchase is no doubt the most significant contribution to progress. A decade ago, there were very few system options to choose from, and it has been said that to learn to master any one of them completely would require approximately three lifetimes.

Today, the system options are numerous. Most of them are relatively easy to use, relatively inexpensive to purchase and maintain, and operating on less expensive hardware. According to a recent survey, there are more than 93,000 GIS installations worldwide. Analysis of the distribution by continent reveals that approximately 65% are in North America, 22% in Europe, 5% in Australia, and 8% combined in South America, Africa, and Asia (Singh 1994). The dedication of this diverse group to development, improvement, cost reduction, and aggressive international marketing strategies is responsible for most of the progress that has been made in the past decade. Many times, it is the generosity in pricing policies that permits GIS capabilities to be implemented in developing countries and educational institutions.

Current Landscape of International Activities

In 1987, the World Commission on Environment and Development introduced the concept of "sustainable development," which promotes an approach to economic development that preserves and/or enhances the environment for future generations (Anonymous 1987). The concept of sustainable development was the centerpiece for the United Nations Conference on Environment and Development (UNCED) held in Rio de Janeiro, Brazil, in June 1992. Leaders from almost every country attended the conference, which appeared to have a positive effect on the views of individuals and institutions around the world. It raised the awareness of the relationship between development and environment activities and produced Agenda 21, a framework for action for the twenty-first century. The conference also laid the groundwork for the establishment of the United Nations Commission on Sustainable Development. In addition, almost every organization of the United Nations and international lending institutions have since developed programs in sustainable development, and there appears to be an improved spirit of cooperation among the organizations involved in environment and development.

One aspect of sustainable development that became apparent almost immediately is that the projects now required new and different information. Most planners, from global to regional to national and local, find that they are "information limited" in their post-Rio activities. Chapter 40 of Agenda 21 (United Nations 1992) deals with the potential role of information as a major cross-cutting issue that must be dealt with in future development. Most of the organizations with significant development responsibilities have attempted some action related to information and information networks. However, two years after Rio, a much-needed central forum to coordinate or harmonize activities had not been established. There have been a number of informal meetings and/or conferences to discuss common interests in environment, development, and information. One such consultation, attended by representatives of 24 organizations with significant development roles, was held at the International Development Research Centre (IDRC) in Ottawa, Canada (IDRC 1994). Part IV of the report discusses Potential Areas for Collaborative Action and deals with two conditions that are destined to influence the future of GIS development. In some respects, the report says, there is too much "information" floating around post-Rio in the sustainable development field, particularly in the form of tracts, books, organizational reports, research reports, seminars, and conferences. The busy practitioner, the report continues, does one of three things:

- creates a pile in the office saying, One day I will look at that;
- has a budget (or time) for extracting from it or at least flicking through it (lucky person); or
- junks it all (or perhaps recycles it) after a delay, to soothe his or her conscience.

So much of what is available is valuable, very valuable, the IDRC report says, but sadly, few have time to match the resource to their day-to-day needs. The result is waste, not only in terms of reinvention and duplication but also in terms of missed opportunity and networking.

Set against a glut in some places is a huge deficit in others. Many sectors of importance in the post-Rio world never even receive what is out there and thus lose out on the multiple efforts made to follow Rio up. Solving the dual problem of both information deficit and glut is where collaborative effort could be targeted, the report concludes.

To begin with, the information required to pursue economic development anywhere in the world is very complex. When complicated with the idea of sustainability, the required information becomes even more complex. Even where there appears to be too much sustainable development information, it is sometimes wrong or incomplete information. Most often, however, the method and format of presentation requires too much time and study on the part of the decision maker. In truth, the "more data than is needed" situation is too often a "more data than can be analyzed with existing methods" situation. Decision makers will not have more time in the future, and therefore there is an increasing need to institutionalize the capability to analyze data, model the options, and present the results in graphic and geographic formats. GIS will play a major role in the future to solve the interface problems that exist between data and decisions. It is the perfect tool for both analysis and presentation and lends itself to standards that make information networking more practical.

The UNDP, a major development organization, has focused its post-Rio efforts on Sustainable Human Development (SHD) that adds even more complex social issues to the information equation. To help developing nations access more information and make better use of it, UNDP has implemented two pilot projects. Sustainable Development Networks (SDN) have been developed in a number of countries to provide all sectors of society with access to information sources both in-country and abroad. Presently a pilot project is planned to develop SHD-Reference Centers (SHD-RC) that provide additional information sources from abroad but, more importantly, additional tools for analysis, modeling, and presentation. GIS capability is a core requirement for these reference centers.

A deficit of appropriate information is more often the case in developing countries. While there are numerous projects that require site specific information gathering, there are few efforts to provide national scale datasets to support sustainable development. To address this issue, the UNDP and the UNEP jointly sponsored the International Symposium on Core Data Needs for Environmental Assessment and Sustainable Development Strategies in Bangkok, Thailand, in November 1994. Some 65 individuals from 28 nations participated in the symposium, including policy makers, scientists, and researchers from developing and industrialized countries. Also present were representatives from the UN, industry, aid-to-development agencies, and data suppliers. One result of the deliberations was the identification of ten high-priority core data sets necessary to support sustainable human development initiatives at any point on Earth. While there was much discussion as to the makeup of the list, there was total consensus that these data sets were most useful in geographic format (Estes et al. 1994).

The current and most assuredly the future requirements for information and information networking to support sustainable development will for many years press the GIS and network research and development communities for more and better ways to analyze, model, and present complex development scenarios to decision makers.

Summary

An increasing number of world leaders believe that if the Earth as we know it is to survive, then policy makers must begin to consider sustainability to be of prime importance in all decisions, whether related to economic, social, or environmental issues. These issues are examined by thousands of scientists who in the past have limited their studies to the individual science sectors, although there are recent attempts to develop interdisciplinary models. Still, when the issue is complex, it has too often been said that "the policy maker needed the answer and the scientists gave him science." If policy makers are to be held accountable for the increasingly complex decisions needed to maintain sustainable human development, then they must have the tools to deal with large quantities of information in short periods of time. Policy must now consider the national and global economy, needs of society, and impact on the physical environment in almost every decision. In most countries today, these sectors have their own methods of gathering data, their own disciplinary models, and their own traditional methods of presenting results. The quantity and quality of information to balance the equations among economic development, preservation of the environment, and enhancing the quality of life will most certainly demand that new institutional capabilities be developed to analyze, model, and present integrated results for decision makers.

There appears to be an emerging need for a technical/scientific capability between research scientists and policy makers to apply the results of scientific research directly to complex political decision models. Many of these models require spatial information that will demand new methods of acquiring data, especially in the social and economic sectors. While economic development for a country may be represented statistically by GNP per capita, human development programs will require much more detailed, location-specific information in order to deal with poverty, medical needs, education requirements, and social inequities. Many countries are now implementing new data gathering programs that will force the development of models to integrate multisectoral information. These models will of necessity be spatial models, and continued research and development is required to align GIS technology with the new perspective of human development.

It is not likely that any spatial model capable of such complex analysis will ever be simple or easily manipulated by policy level decision makers. It is therefore critical that they be supported by well-trained technical and scientific staff. While industrialized countries generally produce an ample supply of scientists and technologists through their normal educational system, this is not often the case in developing countries. For many years, well meaning aid-to-development organizations have provided training in the form of short-term programs that are usually presented by visiting instructors. This approach is sometimes successful with a more simple technology, but even then the numbers trained are often too few to sustain a critical mass. With the complexity of decision models related to sustainable human development, it is unlikely that one-shot short-term training will ever develop the capability to cope with the rapidly expanding science and technology. It is critical that aid-to-development agencies support long-term programs that create educational systems in developing countries to train their students in the sciences of sustainability and the technologies of the new information culture.

Bibliography

Anonymous. 1987. *Our Common Future*. Oxford, U.K.: Oxford University Press.

Estes, J., J. Lawless, D. Mooneyhan, et al. 1994. *Report of the International Symposium on Core Data Needs for Environmental Assessments and Sustainable Development Strategies*. 2 volumes.

Hebin, O. and R. Witt. 1993. "Global Information Moving to Full-Service Database." *GIS World,* 6 (4): 50–55.

IDRC. 1994. *Information and Agenda 21*. Report of an Informal Consultation on Environment and Development, and Information, IDRC, Ottawa, Canada.
Singh, A. 1994. "Global Installation Base of Geographic Information Systems." *International Journal of Remote Sensing*.
United Nations. 1992. "Information for Decision-Making." Chapter 40, *Agenda 21*.
United Nations Environment Programme (UNEP). 1994. "Environment Assessment Programme Progress Report. September 30, 1994." Unpublished report.
United Nations Environment Programme (UNEP). 1987. "UNEP/GEMS/GRID." Unpublished report.

PART VII

Reflections

Don Hemenway, Vice President, GIS World, Inc.

I probably first heard of GIS in 1984 or 1985, just after joining the staff of the American Society for Photogrammetry and Remote Sensing (ASPRS). I got so involved in it, I started the "GIS Observer" column in 1988 in ASPRS's journal, *PE & RS.* I wondered where the term came from and ran a contest in my column, asking for the earliest citation of the term GIS. Roger Tomlinson won by citing his own coining of the term and its publication in the mid to early 1960s.

For the last decade, Don Hemenway has been promoting GIS technology through national associations and journals.

John Townshend, Department of Geography, University of Maryland College Park

Like so many people I learnt about GIS from the work of Roger Tomlinson and specifically from the two volumes he edited entitled *Geographical Data Handling,* published by the IGU in 1972. Looking at these two volumes again one is struck by how far-seeing were Tomlinson and his collaborators all those years ago.

John Townshend is former chair of the geography department at UMCP and a leading researcher with the NASA Pathfinder program.

CHAPTER 19

What Next? Reflections from the Middle of the Growth Curve

Michael F. Goodchild

Introduction

It is now almost exactly 30 years since the heady early days of the Canada Geographic Information System (CGIS) (Chapter by Tomlinson), when the creative energy of a group of Canadian geographers and computer scientists developed the first GIS design and coined the term. Much has happened in the world of GIS since then: much more research has been done; crucial breakthroughs have been made in the key algorithms and data structures; and GIS has emerged as a strong and still-growing application of electronic data processing. The field now has its own journals, magazines, and conferences, its own programs in institutions of higher education, and its own societies. Together, the varied chapters of this book document that history and identify the key events along the way, marked by releases of new software, moves of individuals, decisions of government, and a host of other more or less significant events that together have marked and assisted the emergence of GIS.

Dr. Michael F. Goodchild is a professor of geography at the University of California at Santa Barbara, director of the National Center for Geographic Information and Analysis (NCGIA), and associate director of the Alexandria Digital Library project. He has a broad international experience in GIS and was the director of the NCGIA's GIS Core Curriculum project. *Author's Address:* Michael F. Goodchild, National Center for Geographic Information and Analysis, and Department of Geography, University of California, Santa Barbara, CA 93106-4060. E-mail: good@ncgia.ucsb.edu. 805/893-8049 (voice); 805/893-7095 (FAX)

Acknowledgment: The National Center for Geographic Information and Analysis is supported by the National Science Foundation. The Alexandria Digital Library project is supported by the National Science Foundation, NASA, and the Advanced Research Projects Agency.

It falls to this final chapter and this author to try to bring the story to some kind of conclusion. Of course, the story is far from concluded—GIS is still growing strongly, perhaps more strongly than at any point in its past. Students of growth patterns might claim that we are still on the upwardly accelerating part of the growth curve and have not yet reached the point of inflection and the decelerating growth which can be expected when it passes 50% of potential. But the skeptical might be more demanding. When all is said and done, what exactly is GIS? Clearly it is very different from the analytic machine for geographic information envisioned by Tomlinson and others in 1966 (Chapter by Tomlinson). No one, however prescient, could have anticipated the impacts of digital technology that have occurred in the interim, particularly in the past five years. The wider world of digital technology that made GIS possible continues on its own growth path and continues to influence both current GIS and our understanding of its future.

Accordingly, the first part of this chapter attempts a summary and synthesis of the current state of GIS. The following two sections first review changes that have occurred in the larger environment of digital computing and then link them to the development of GIS. This leads to a section on prospects for continued growth. Current limitations to growth are reviewed, with a look at the potential dangers facing GIS through misuse and failure to understand the technology's social implications and its setting within broader societal concerns. Finally, the chapter ends with some comments on growth directions and the role current research may play in making them feasible.

GIS: Where Are We Now?

Despite 30 years of development and growth, information on the actual size of the GIS phenomenon is surprisingly difficult to come by. The largest GIS conferences attract many thousands of participants (approximately 5,000 at the 1995 ESRI User Conference), suggesting that the population of GIS users in the U.S. is 100,000, to within an order of magnitude. Figures on numbers of sites from the major GIS vendors suggest that there are a similar number of installed systems worldwide (GIS World Inc. 1996); and the total circulation of the major GIS magazines and total sales of major GIS textbooks are also of the same order of magnitude. Recent figures indicate annual sales by the GIS industry of $500 million (GIS World Inc. 1996). Some 300 vendors list their products in software directories (GIS World Inc. 1996), though not all would meet more rigorous definitions of GIS.

Over the years, GIS has expanded to include significant areas of related disciplines. It is now almost inseparable from cartography, and software for automating cartographic production is often identified as GIS. It overlaps substantially with geodesy, photogrammetry, image processing, remote sensing, land records management, and computer-assisted design. It is now difficult to draw the line between packages for image processing and GIS and similarly between CAD (computer-aided design) and GIS, as vendors have expanded functionality in the gray areas between them.

Today, the term "GIS" means much more than it did 30 years ago, almost defying definition. At its most broad, GIS now refers to any activity involving geographic information in digital form. The convenience of the three-letter acronym has allowed us to give it a meaning of its own that may have little to do with the three words represented. We often hear software referred to as a "GIS system," because GIS now means more than a software system; and we often hear of "GIS data." In fact, GIS can now refer to any combination of software, hardware, data, and communications, and similarly, "doing GIS" can mean building a database, analyzing its contents, making decisions, or plotting maps.

The multiple roots of GIS have been well documented in the earlier chapters of this book. Many threads have converged, from the need to analyze large volumes of natural resource data (Chapter by Tomlinson) and to manage the collection and dissemination of the census (Chapter by Cooke) to the addition of geographic features to systems built for design activities (Chapter by Moyer). GIS remains largely a technology of two-dimensional, static, deterministic data at a single level of resolution. Attempts to build time, the vertical dimension, uncertainty, and hierarchies of generalization into GIS remain largely confined to the research community, although there are excellent products available to link GIS with capabilities in these areas in other systems, and many GIS vendors have made progress in representing hierarchies of objects and in temporal change. But despite these developments, GIS remains largely dominated by the metaphors of maps and images, and many might argue that it should remain so. Whether we will move beyond these limitations is discussed at length in a subsequent section.

Within the limited context of map and image data, GIS is an enormously powerful tool for inventory, query, analysis, and decision-making; the larger vendor systems do almost everything one would ever want to do with geographic information in a digital computer. The number of alternative ways of capturing map and image contents into a digital database provides much of the richness of current GIS but also makes it difficult to transfer data between systems that have taken different approaches or to achieve interoperability. Again, these are topics of active research and are discussed at length later.

The Broader Context

The Computer as Calculator

Although the concept is much older, the first machine we would recognize as a digital computer appeared on the scene almost exactly 50 years ago. ENIAC (Electronic Numerical Integrator and Computer) and its mechanical precursors addressed a particularly simple and well-defined human problem—the need to make great numbers of arithmetic calculations quickly. Although it would have been possible to do so by hand, the economic and social costs of doing so were clearly too severe, and the challenge of building a massive calculating machine was too challenging. Many of the early motivations for massive calculation were digital. In the immediate postwar period, there was heavy demand for predictions of trajectories of assorted ballistic weapons and for solution of the equations governing nuclear fission and fusion (Stern 1981). Consequently, ENIAC was quickly followed by others and by the emergence of a commercial manufacturing industry.

Civilian applications followed. Massive arithmetic computations are needed for weather forecasting to find numerical solutions to the equations governing the atmosphere. They are needed in statistics, in the inversion of large matrices for factor analysis, and in optimization of models in traffic analysis. By the late 1950s, commercially available computers were being installed in major universities to serve an increasingly diverse range of needs. All, however, were driven by the need for a fast calculating machine.

The Computer as Information System

By the advent of GIS, it was becoming apparent to far-sighted followers of the digital computer that something much more significant than a calculating machine was beginning to emerge. Although FORTRAN included the ability to store alphabetic characters in addition to numbers, its main use was in making numerical output easier to read and understand. But if a computer could process text, it could be used to search for particular words or to store messages. By the 1960s, visions of a digital world had begun to emerge (Negroponte 1995). If the computer could store other kinds of information besides numbers, process them, and communicate them to others, then one could begin to imagine a future in which computers allowed a much more comprehensive and adaptable approach to all of society's information needs. Methods to store text, images, and even sound began to emerge in the 1960s, along with terms like "artificial intelligence" that reflect this vastly expanded view of the power of digital computers using metaphors of human reasoning. Early experiments were made in digital communication technology, which offered advantages in much lower error rates and more powerful switching.

Today, we are much more likely to think of computers as information systems than as calculators. The byte became the basis on which to store a full range of alphanumeric characters. Images are stored by allocating a fixed number of bits or bytes to each picture element or pixel. Maps, however, remain more problematic, as there are many ways to capture the contents of a topographic map using concepts of raster or vector storage. Modern database management systems freely incorporate numbers, text, and images but rarely maps except in primitive scanned raster format or in specialized GIS.

Digital Worlds

Today, we encounter digital technology in almost all aspects of our lives. It is used to compose text and print newspapers, to reproduce music, to make maps, to transmit voice messages by wire or radio, and it is about to become the basis of television as well. Digital chips are installed in automobiles, microwave ovens, and even some credit cards. It is difficult now to find any kind of information of value to modern society that is not represented in digital form at some point in its life. Digital computers are now found in a significant proportion of households in the industrial world and are beginning to replace many traditional information resources such as encyclopedias and atlases. Yet while many people are now familiar with digital information systems and regularly use digital word processors and electronic mail, few make use of the computer's power to store, process, and retrieve the contents of maps. In the case of geographic information, digital technology has only just begun to have an impact on our daily lives.

New Societies

With the recent explosion of interest in digital communication using tools such as electronic mail, the Internet, and the World Wide Web, what was once an exotic academic specialty has now become a major topic of public interest. Predictions of a pervasive impact of digital technology on society that were once distributed only among a select handful of futurists are now openly debated, even by politicians (Gore 1993; Gingrich 1995). Open communication between individuals using technologies that are almost impossible for governments to control has been heralded as the essential ingredient of a new democratic age, in which the individual is empowered as never before. Whatever one's reaction to the hype, it is clear that digital information technology is beginning to have an impact on society. It permits a massive redistribution in the geography of employment by supporting telecommuting, and it exacerbates inequities by offering vastly greater opportunity to those who can afford to invest in it.

The previous sections have suggested that the history of digital computing can be organized into four distinct stages. The next sections turn to GIS and trace its development within the same basic framework.

GIS within the Broader Context

The Analytic Engine

As Roger Tomlinson shows in Chapter 2, the design of CGIS was oriented primarily to the analysis of the Canada Land Inventory, a vast collection of data on land resources gathered over a significant fraction of the surface of Canada. The results needed by the provincial and federal sponsors were essentially numeric—totals of land area in various categories—and the computer as calculating machine seemed the only means available to perform the necessary analysis at reasonable cost. Early designs included no facility for map output for two main reasons—the computer-driven plotters of the time were far from reliable and high quality cartography was available as a simple by-product of the input process.

Other early applications of computing to geographic data similarly reflected the thinking of the time (Coppock and Rhind 1991). Computers were introduced to the cartographic production process first as calculators to ease the numerical calculations involved in producing maps as projections of the Earth's surface. The idea of using computers to edit maps came somewhat later after the perfection of coordinate input and output devices, including the digitizer and plotter.

By the 1970s, a dominant concern in the developing field of GIS was how to achieve an efficient and effective representation of the contents of a map other than by primitive scanning. CGIS had exploited the idea of arcs to represent the contents of a natural resource map by creating a digital representation of each common boundary between a pair of adjacent areas. But, in other cases, particularly when areas overlap or do not exhaust the space of the map, it is clearly more efficient to represent areas as separate, independent entities. Many more choices had begun to emerge by the 1970s, prompting the Harvard lab to see format conversion between these alternatives as an important and legitimate function of a geographic information system (Chapter by Chrisman). POLYVRT was the first of these exchange modules and reflected the emergence of a new idea—that GIS would have to be more than automated cartography and analysis if it were to exploit the full power of the digital computer to handle geographic information.

The General-purpose Tool for Geographic Information

Although the term had been coined a decade earlier, it was not until the late 1970s that the idea of a general-purpose processor of geographic information emerged. By then, it was clear that a GIS could: (1) use the digital computer to support a wide range of alternative methods of representation or data models, including raster and vector options; (2) be used to structure such data in a rigorous way to make it easy to share it and communicate it to others; (3) perform a full range of editing, retrieval, and analysis functions; and (4) provide the means to make high quality maps. As we have seen, much of the development work that implemented this vision was performed at a small number of institutions in the 1970s, notably the Harvard lab. By 1980, the vision had become sufficiently clear to justify private sector investment, and a range of new commercial products began to appear.

Database management systems offer generic capabilities for storing, retrieving, and presenting data, So too does a GIS but specifically for geographic data. A user skilled in the use of a DBMS can quickly gain the skills needed to work with a well-structured set of data, using standard methods of access such as SQL (Standard Query Language). As GIS evolved as a general-purpose tool, it began to look more and more like a special kind of database management system for geographic data. Of course, a GIS must have many functions concerned with the registration of its data on the surface of the Earth as well as many specialized functions for visualization and analysis, but nevertheless there are many areas in which the application of principles of database management to GIS leads to greater effectiveness. In the 1980s, strong links developed between GIS and DBMS that continue to evolve today.

GIS as an Intrusive Technology

Today, the number of installed GIS is perhaps 100,000 or at least two orders of magnitude less than the number of home computers in the United States. While digital technology already touches our lives in many ways, the intrusion of GIS is much more limited. If advertising in airline magazines is anything to go by, many of us are about to acquire laptop mapping systems to help us find our way in strange cities, and in-car navigation systems are beginning to appear in rental vehicles. But despite this, GIS remains largely the domain of the professional, in specialized departments of local governments, utilities, resource industries, and resource agencies. GIS has yet to penetrate everyday life in the way that word processing or spreadsheets have.

Yet the consumer marketplace is in many ways the current frontier of GIS. In-car navigation systems have begun to appear in consumer electronics stores, and global positioning satellite (GPS) receivers are widely used in amateur boating. Mapping software is now available in spreadsheet packages, and digital atlases have begun to appear on CD for home use. Digital maps for the media are now available for sale over the Internet. Perhaps we will see the emergence of a TV shopping channel for digital imagery in the next few years or a digital library of geographic data for use in schools (for an example of a prototype digital spatial data library, see the Alexandria Digital Library project: http://alexandria.sdc.ucsb.edu).

The GIS Society

If there has been much speculation about the social and political consequences of digital technology, the same is probably not true of GIS and the use of digital technology for geographic information. Yet the social implications of GIS are potentially very significant. GIS is now used to support a wide range of decisions, particularly in local government. The computer still intimidates many, and output from a computer is often given far more credibility than it deserves. Undoubtedly, there is in GIS technology the potential for abuse, when results are presented as truth but based on bad data or bad analyses. The computer empowers, and if it is available to only one side in an argument it almost inevitably biases the outcome. As GIS technology becomes more widely available, it is increasingly used by both sides—but the potential for abuse is still there.

GIS is expensive, particularly when the costs of assembling data are taken into account. To date, it tends to have been more widely available to those already in power in society and to have served to strengthen that power rather than diminish or share it. Continued decreases in the costs of hardware and software and vastly greater access to data provide grounds for hope that the eventual effect will be to increase equity in society rather than decrease it, but a continued supply of new innovations available only to those able to afford them casts doubt on this argument. These and other concerns are explored at length in the volume edited by Pickles (1995).

In summary, while four stages have been identified in the growth of digital technology in general, it seems that the current state of GIS is somewhere between the second and third—a general-purpose information technology that has not yet penetrated everyday life to a significant degree. The average person might encounter some aspect of GIS and read about it, but access to its full capabilities is currently limited to professional settings.

The Future of GIS

The previous section identified four stages in the development of digital computing and four associated stages in the development of GIS. It was argued that GIS currently lies somewhere between the second and third stages; the view of GIS as a comprehensive information system for geographic data is being gradually supplanted by a sense that GIS is about to intrude on many aspects of everyday life. The fourth stage, of digital technology driving fundamental changes in society may lie in the future for GIS, but there is little sign of it at present, and many problems remain to be overcome before geographic information is as easy to handle as other kinds of information.

One conclusion from this analysis is that in terms of installed capacity and numbers of users of GIS, we remain at an early point on the growth curve. The mass market for geographic information technologies is probably two orders of magnitude larger than the current largely professional market, although the average investment per installation will clearly be much lower. It may be represented by GPS transponders or navigation systems in every car; by road and street map access through interactive television; by a mass market for digital imagery of local neighborhoods; by map-making capabilities in spreadsheet packages; or by digital atlases in the household. All of these hold great promise for helping to develop a mass market for geographic information technologies.

While the acronym "GIS" has become the central rallying-point in a coalescence of related disciplines and activities, it seems unlikely to persist in the mass-market era, where users are less likely to see a commonality between a digital atlas on CD and an in-car navigation system. In that sense, it has been argued that GIS as an identifiable computer application may disappear, as geographic information is integrated into other forms of data handling. Similar arguments are made about GIS as an information system application. Extensions to standard query languages such as SQL may allow its users to make many of the types of queries now restricted to specialized GIS. Recently, there has been a spate of new collaborative ventures between GIS and DBMS vendors, perhaps in anticipation of a greater blurring between the two areas.

There are other views, however. GIS is now identified with much more than software. It includes data, journals, conferences, educational programs—in short, an entire community. The concerns of the community over training, terminology, concepts, and methods of decision-making will not go away as GIS becomes increasingly integrated with other software. The term "geographic information science" has been coined to describe a discipline concerned with the deeper common issues arising from the digital handling of geographic information (Goodchild 1992a), and a University Consortium for Geographic Information Science has been formed recently in the U.S.

In the early days of GIS, the lack of sufficiently powerful computer hardware was critical. CGIS developed in an era when map scanners were unknown, and the cost to the project of developing one on an experimental basis was far higher than the cost of a scanner today, even after allowing for inflation (Chapter by Tomlinson). By the 1970s, the appropriate peripherals had become available along with vastly increased computing resources, and attention switched to the development of necessary software and data structures (Chapter by Chrisman). The 1980s were an era of great improvement in communication technology and infrastructure, allowing distributed GIS databases to function together as if located in one place. Client-server architectures developed in the 1980s, also in response to vastly improved communications, exemplified by the Internet.

Today, technology allows us to consider the possibility of sharing digital geographic information over integrated, worldwide networks, using such Internet services as the World Wide Web. But while sharing is possible in principle, in practice it is severely impeded by the lack of appropriate methods and tools for describing data. Thus, issues of data access may be the most important issue for the 1990s. They are discussed at length below.

Directions in GIS

This section examines emerging development directions in GIS from three perspectives: data models, data access, and the data life cycle. All three focus on data rather than on analysis or modeling, reflecting a tendency in the field to see data issues as of particular significance at this point in the history of the technology.

Data Models

A data model is defined as the set of entities and relationships used to build a representation of some real-world phenomenon in a digital computer (Goodchild 1992b). Geographic data modeling is particularly complex because of the large number of options available. Any GIS implements some combination of the available models, but variations between systems, particularly in terminology and in the particular combination implemented, create major difficulties for users wishing to transfer data from one to another. At the simplest level, the models fall into two categories—raster and vector—though the story is actually much more complex.

Three broad classes of geographic data models are implemented in current GIS (Goodchild 1992b). The first conceptualizes the world as an empty space littered with discrete point, line, or area objects. Any point in space can be occupied by any number of objects, since objects are allowed to overlap and do not normally fill the space. Objects fall into different classes depending on their topological dimensions (zero, one, or two for points, lines, and areas respectively) and on their semantic meaning. These discrete object data models are often found in brands of GIS that have evolved from CAD systems.

In the second class are the field models. Here, the world is conceived as describable by a finite number of variables or fields, each of which has a single value at every point in space. There are five ways of representing fields implemented in GIS: as rasters of cells; as grids of regularly spaced points; as triangular meshes; as digitized contours; and as randomly located points. Examples of each can be found in traditional methods of sampling fields, but not all are implemented in every GIS. A raster GIS, for example, typically implements only the first two.

The third class of data models is made up of those used to represent continuous or discrete variation over networks in applications such as transportation or hydrology. In such cases, the network forms a connected one-dimensional space embedded in two dimensions, and geographic phenomena are confined to the network itself.

Current research in GIS seeks to generalize these basic data models in five ways, each reflecting the need to overcome a constraint inherited from the world of maps and images. The first is time, which may affect the contents of a database through changes in attributes, changes in the positions and shapes of objects, or changes in entire fields (Langran 1992). Temporal information is crucial for many GIS applications, and yet most GIS provide no explicit tools for dealing with it. The second is the vertical dimension and the need to generalize from two-dimensional data models in order to handle applications in geology, atmospheric science, and oceanography (Turner 1992). Although many GIS allow a third coordinate to be added, much more comprehensive solutions are needed to represent three-dimensional fields and volumes.

The third generalization is concerned with scale, since traditional GIS contains few tools for moving between scales and handling data at multiple levels of geographic generalization. In the case of fields, generalization is often implemented as a form of filtering, but in the case of discrete objects much more complex methods are needed to emulate the work of a cartographer (Muller, Lagrange, and Weibel 1995). A multiscaled GIS would also work with hierarchies of objects, in which a single object at one scale was explicitly linked to multiple objects at larger scales. Such methods of complex object representation exist in some GIS but not in others.

The fourth generalization is concerned with the curved nature of the Earth's surface. Maps and images must be flat, and we have developed complex systems of map projections to allow us to represent the Earth's surface in hard-copy form. But in the digital computer there is no need to flatten the Earth, except when making paper maps. Although the screen of a computer monitor is almost flat, sophisticated methods of visualization are already used to create the impression of a three-dimensional solid in the mind of the user. This is particularly easy if the solid can be manipulated by the user. Yet our methods of analysis, modeling, and representation are dominated by the need to emulate the handling of flat maps and images, and we do not yet have a complete suite of methods for working with the surface of the planet. This is unfortunate, because it means that the products of GIS are frequently distorted (Willmott, Robeson, and Feddema 1991).

Finally, the fifth generalization of GIS data models is concerned with the handling of uncertainty. Like the previous four, the lack of explicit methods for representing and handling uncertainty in GIS is inherited from map tradition. Recent research has led to effective methods for modeling uncertainty, capturing it in databases, visualizing it in displays, propagating its effects during analysis, and reporting its consequences to the user (Hunter, Caetano, and Goodchild 1995). Hopefully, future generations of GIS will have the tools available for explicit and objective analysis of the uncertainty present in almost all geographic information.

These five generalizations are not equally important, and it will probably not be feasible to support all of them in GIS, at least in the near future. All five are already supported to some degree in computer software, and there has been steady progress over the past decade in integrating such software with GIS through common interfaces and other kinds of linkage. Three-dimensional applications, for example, can now be supported by linking GIS to specialized 3D packages, and data can be exchanged through a common two-dimensional interface. How many of the five can be supported in GIS itself remains to be seen, and important questions remain about the appropriate balance between feasibility and priority, which are the most important generalizations and what will it take to expand current products by extending their data models?

Data Access

Much effort in the research community is currently being directed to problems of data access. Many of these are institutional, concerned with issues of copyright, intellectual property, protection of privacy, and confidentiality (Onsrud and Rushton 1995). Others are more technical and concerned with the techniques available to describe data, transfer it from one system to another, find it across the worldwide network, and share it effectively.

The term "metadata" is often used to refer to the descriptive material needed if data is to be effectively shared. It includes information analogous to a catalog record for a book—subject, title, author—as well as information needed to make the data readable, such as its format. Metadata also includes information needed to handle the data successfully, such as file names. Finally, it includes information on the data's quality, to help the user assess its fitness for use.

Metadata is not merely a description of data but a means of communication about data from the custodian to the potential user. As such, it must reflect not only the properties of the data but also the level of expertise of the user, and it must be expressed in terms whose meaning is known to the user. The more successful a metadata description, the more it will be possible to share data beyond the limits of the discipline of the custodian or creator. Thus metadata is the key to realizing the value invested in data, by extending its use to the widest possible community.

In 1995, the Federal Geographic Data Committee published its Content Standard for Geospatial Metadata, an attempt to provide a standard framework for GIS metadata. It has been widely studied and adopted, and similar standards are being developed by other U.S. jurisdictions and in other countries. Metadata handling is being added to many GIS software products, with the result that it will soon be possible to access and display metadata, to use it to provide automatic control of processes where appropriate, and to process it automatically when a data set is changed by a GIS process.

Concern for spatial metadata is only one part of a broadly based effort to improve access to geographic information. The Alexandria Digital Library (http://alexandria.sdc.ucsb.edu) is a research project dedicated to the development of services on the Internet analogous to those provided by a map library but without the latter's physical access limitations. Alexandria uses the power of the digital environment to overcome many of the traditional impediments to the handling of maps and images in the library. For example, geographic location is used as the basis for an additional search and browse capability, in addition to the traditional subject, author, and title.

GIS in the Data Life-cycle

One legacy of the first era of computer technology discussed earlier has been the view that GIS's primary purpose is the analysis of existing data, in support of modeling, prediction, and decision-making. Many GIS projects have been built around existing sources of data in map and image form. But analysis is only one stage in the life cycle of geographic data, which extends from field data collection through compilation, interpretation, classification, and digitization to final archiving. As a technology for handling geographic data, GIS has the potential to play a role in all of these stages. The previous section's discussion of metadata, for example, can be seen as an extension of GIS technology into a broader concern for data access mechanisms.

The recent development of portable and hand-held computer hardware opens the potential for GIS support of field data collection to a much greater extent than has been possible in the past. With appropriate functionality, such systems can support acquisition of field knowledge by providing capabilities for capture of a broad range of types of information onto a preloaded base, such as an ortho-rectified digital image. GPS and digital cameras allow other important types of geographic information to be acquired and stored directly in the GIS database. Technologies such as this may help open the channel of communication between the field scientist and the analyst.

Modern computer technology allows far more than raw data to be communicated. Object-oriented concepts of encapsulation allow processes to be coupled with data, suggesting that the field scientist of the future might be able to specify the operations that make sense for data and thus help the eventual user understand the data's meaning.

Conclusion

There seems little doubt that the early efforts to develop GIS documented in the chapters of this book have paid off and will continue to do so as GIS expands both in capabilities and in applications. Whether or not GIS continues to be the appropriate umbrella term, geographic information technologies clearly have an increasing role to play, both in the professional workplace and in everyday life. The GIS research community will continue to find ways of getting around the impediments presented by current GIS and solving the issues that underlie its use.

Histories are normally written after the events they describe have receded sufficiently in time to dull their impact. In that sense, this history of GIS may be premature, because even the earliest developments in the field continue to have an influence on ways of thinking about GIS and on the technology we see today. Both GIS and GPS were being designed and researched 30 years ago, and yet both are only now beginning to impact everyday life. Although many might see GIS as an overnight sensation, it is the cumulative efforts described in these pages that have allowed this immensely complex and challenging computer application to reach the stage it is in today. Many problems remain to be solved as GIS struggles to become easier to use through efforts to build better user interfaces and to make the products of different vendors interoperable (http://www.regis.berkeley.edu:80/ogis/) by taking advantage of wider trends in the computer industry towards open standards. The results of these efforts will be seen in the next generation of GIS products and will be recorded the next time the history of GIS is written.

Bibliography

Coppock, J. T., and D. W. Rhind. 1991. "The History of GIS." In D. J. Maguire, M. F. Goodchild, and D. W. Rhind, eds., *Geographical Information Systems Principles and Applications*. Harlow, U.K.: Longman Scientific and Technical, 21–43.

Gingrich, N. 1995. "Citizen's Guide to the Twenty-first Century." In A. Toffler and H. Toffler, eds., *Creating a New Civilization: The Politics of the Third Wave*. Atlanta: Turner Publishing Inc., 13–18.

GIS World Inc. 1996. *GIS World Sourcebook 1996*. Fort Collins, CO: GIS World Inc.

Goodchild, M. F. 1992a. "Geographical Information Science." *International Journal of Geographical Information Systems,* 6 (1): 31–45.

Goodchild, M. F. 1992b. "Geographical Data Modeling." *Computers and Geosciences,* 18 (4): 401–408.

Gore, A. 1993. Remarks at the National Press Club.

Hunter, G. J., M. Caetano, and M. F. Goodchild. 1995. "A Methodology for Reporting Uncertainty in Spatial Database Products." *Journal of the Urban and Regional Information Systems Association*, 7 (2): 11–21.

Langran, G. 1992. *Time in Geographic Information Systems*. London: Taylor and Francis.

Muller, J. C., J. P. Lagrange, and R. Weibel, eds. 1995. *GIS and Generalization: Methodology and Practice*. London: Taylor and Francis.

Negroponte, N. 1995. *Being Digital*. New York: Knopf.

Onsrud, H. J., and G. Rushton, eds. 1995. *Sharing Geographic Information*. New Brunswick, N.J.: Center for Urban Policy Research.

Pickles, J. 1995. *Ground Truth: The Social Implications of Geographic Information Systems*. New York: Guilford Press.

Stern, N. B. 1981. *From ENIAC to UNIVAC; An Appraisal of the Eckert-Mauchly Computers*. Bedford, MA: Digital Press.

Turner, A. K. 1992. *Three-Dimensional Modeling with Geoscientific Information Systems*. Dordrecht, Germany: Kluwer.

Willmott, C. J., S. M. Robeson, and J. J. Feddema. 1991. "Influence of Spatially Variable Instrument Networks on Climatic Averages." *Geophysical Research Letters*, 18 (12): 2249–2251.

Contents

B

C

D

E

F

G

H

I

J

K

L

M

N

O

P

Q

R

S

T

U

V

W

X

Z

W

X

Z